World Regional Geography

World Regional Geography

VOLUME 2

Omer Lasme

Kruger Brentt
Publishers
2025

Kruger Brentt Publishers UK. LTD.
Company Number 9728962

Regd. Office: 68 St Margarets Road, Edgware, Middlesex HA8 9UU

ISBN: 978-1-78715-454-4 (Vol. 2)

For information on all our publications visit our website at http://krugerbrentt.com/

PREFACE

"World Regional Geography" offers a comprehensive exploration of the diverse landscapes, cultures, and environments that shape the world's regions. This book serves as an essential resource for students, educators, researchers, and anyone interested in understanding the complex interactions between human societies and their geographical contexts.

Geography is the study of the Earth's surface and the processes that shape it, including human activities, physical landscapes, and environmental dynamics. World regional geography provides a framework for understanding the spatial patterns, cultural diversity, and economic disparities that characterize different parts of the globe.

As editors of this volume, we have endeavored to compile a diverse array of perspectives, methodologies, and case studies from leading experts and practitioners in the field of world regional geography. Through a blend of theoretical foundations, empirical analysis, and real-world examples, this book covers a wide range of topics, including the physical geography of regions, cultural landscapes, economic activities, political systems, and environmental challenges.

Our overarching goal in presenting this book is to provide readers with a comprehensive understanding of the world's regions and the complex interactions between human societies and their environments. By offering insights into the geographical diversity, cultural richness, and interconnectedness of the world's regions, we aim to foster a deeper appreciation and understanding of global issues and challenges.

We extend our sincere gratitude to all the contributors who have generously shared their expertise, experiences, and insights in the field of world regional geography. It is our hope that this book will serve as a valuable reference and guide for students, educators, researchers, and anyone interested in exploring the dynamic and interconnected nature of the world we inhabit.

Omer Lasme

CONTENTS

PREFACE V

VOLUME–1

1. INTRODUCTION TO THE WORLD 1
   - *1.1: INTRODUCTION: AN OVERVIEW* 1
   - *1.2: The Interaction Between the Environment and Human Activity* 17
   - *1.3: Population and Culture* 30
   - *1.4: Globalization and Development* 42
   - *1.5 Concise Summary* 58
2. EUROPE 61
   - *2.1: INTRODUCTION: AN OVERVIEW* 61
   - *2.2. Patterns of Historical Development* 70
   - *2.3 Regions of Western Europe* 81
   - *2.4. Eastern Europe* 108
   - *2.5: Concise Summary* 124
3. RUSSIA 127
   - *3.1: Introduction: An Overview* 127
   - *3.2 The USSR and the Russian Federation* 133
   - *3.3: Regions of Russia* 144
   - *3.4: Concise Summary* 155
4. NORTH AMERICA 157
   - *4.1: Introduction: An Overview* 157
   - *4.2. United States- Early Development and Globalization* 162
   - *4.3. United States- Population and Religion* 176
   - *4.4. Canada* 185
   - *4.5. Regions of the United States and Canada* 198
   - *4.6. Concise Summary* 214

5. MIDDLE AMERICA 215

5.1. Introduction: An Overview 215

5.2 Mexico 225

5.3. Central America 239

5.4. The Caribbean 253

5.5. Tropical Cyclones (Hurricanes) 269

5.6. Concise Summary 275

6. SOUTH AMERICA 277

6.1. Introduction: An Overview 277

6.2. Urban North and Andean West 296

6.3. Brazil 311

6.4. The Southern Cone 327

6.5 Concise Summary 340

VOLUME–2

7. SUBSAHARAN AFRICA 343

7.1. Introduction: An Overview 343

7.2. Human Geography of Subsaharan Africa 356

7.3. West Africa 371

7.4. Central Africa 388

7.5. East Africa 405

7.6. Southern Africa 423

7.7. Concise Summary 441

8. NORTH AFRICA AND SOUTHWEST ASIA 443

8.1 Introducing the Realm 443

8.2. Muhammad and Islam 455

8.3. North Africa and the African Transition Zone 464

8.4. Israel and Its Neighbors 481

8.5. Arabs, Islam, and Oil 497

8.6: Iraq, Turkey, and Iran 513

8.7. Central Asia and Afghanistan 528

8.8. Concise Summary 547

9. SOUTH ASIA 551

9.1. Introduction: An Overview 551

9.2. The Peripheral States of South Asia 562

9.3. Pakistan and Bangladesh 574
9.4. India 591
9.5. Religions of India and South Asia 604
9.6. Concise Summary 615
10. EAST ASIA 617
10.1. Introduction: An Overview 617
10.2. Emerging China 630
10.3. China's Periphery 642
10.4. Japan and Korea (North and South) 654
10.5. Concise Summary 671
11. SOUTHEAST ASIA 673
11.1. Introduction: An Overview 673
11.2. The Mainland Countries 679
11.3. The Insular Region (Islands of Southeast Asia) 696
11.4. Concise Summary 715
12. AUSTRALIA AND NEW ZEALAND 717
12.1. Introduction: An Overview 717
12.2: Australia 721
12.3. New Zealand 732
12.4. Concise Summary 741
13. THE PACIFIC AND ANTARCTICA 743
13.1. The Pacific Islands 743
13.2. Antarctica 758
13.3: Concise Summary 770
INDEX 771

Chapter 7 SUBSAHARAN AFRICA

7.1. INTRODUCTION: AN OVERVIEW

There is significant variability in the delineation or identification of the regions of Sub-Saharan Africa on maps. The discussion is not on the regions of Sub-Saharan Africa, but on the specific countries to be included under each category. The regions exhibit both similarities and disparities. Cultural geography exhibits significant variation between countries and ethnic groups; nevertheless, there are also common cultural trends throughout all Sub-Saharan African regions. Colonialism has significantly influenced the development of nations. Families are extensive, and a swift transition from rural to urban areas is taking place throughout all regions. Each area contains significant metropolitan centers—typically port cities that serve as key hubs, bolstered by an extensive surrounding rural hinterland.

Globalization has integrated these previously isolated places with the global community. Progress in communication and transportation technology has established networks linking Africa to global markets. Sub-Saharan Africa possesses a youthful population that is actively pursuing possibilities and advantages. The political landscape is fluid: alterations in leadership via coups or military usurpations are prevalent, as is autocratic governance. Sub-Saharan Africa contains several of the world's most impoverished nations. Poverty is manifest in rural areas and urban slums of major cities. Acute civil conflicts are an integral aspect of the historical narrative of any region. Violence and wars persist in certain regions, but others demonstrate political stability and flourishing economy. The diversity in human geography is the most significant aspect of Sub-Saharan Africa. The diversity of ethnic groups, languages, and religious affiliations generates significant centripetal and centrifugal forces that engage within a vibrant cultural mosaic.

The majority of the population adheres to an agrarian lifestyle; yet, some individuals are acquiring the skills required to adapt to the swift wave of globalization that is introducing new technologies and ideas to the continent. The metropolitan core regions of the continent are the primary concentration of global trends in technology and communication. These urban core areas demonstrate the characteristic dynamics of the core-periphery spatial connection. Sub-Saharan Africa comprises numerous core regions and several periphery areas. The principal urban areas wield political influence due to the social elites who possess connections to the global economy and frequently control political endeavors.

Figure 7.1 The Continent of Africa

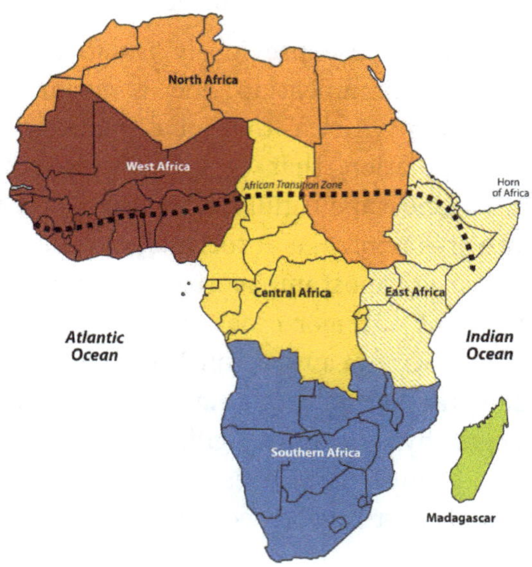

Figure 7.2: Regions of Africa To help identify places discussed in this book, Africa has been divided into regions along traditional boundaries. Updated from map courtesy of Andreas 06.

These central metropolitan regions frequently attract individuals from extensive families in rural outskirts seeking economic opportunities. Numerous individuals in Africa pursuing employment are prepared to relocate to urban areas or even other nations in search of work. Rural immigrants frequently belong to different ethnic groups than those in authority, establishing a foundation for discriminatory policies that penalize numerous minority groups unaffiliated with the government. These dynamics can incite protest activities aimed at toppling the dominant class. Diverse proposals have been suggested to equalize the socioeconomic landscape. A notable possibility is the establishment of a democratic government, when the majority of the populace participates in electing leaders and those in positions of authority.

Recurring patterns and dynamics of individuals and locations are mirrored across the Sub-Saharan region. The regions exhibit similar demographic characteristics, including big family sizes, agrarian lifestyles, and low income levels. Economic patterns characterized by agricultural production and mineral extraction, along with significant shifts in political leadership, are prevalent across the continent. Each region possesses a variety of ethnic groups and a multitude of distinct languages. South of the African Transition Zone, predominant belief systems are Christian and animist, whereas north of the zone, Islam is prevalent. Conflict and social discord may arise at the intersection of competing religious groups vying for political dominance. These themes will be reiterated throughout the chapter. This chapter will delineate the fundamental tendencies and patterns of the intricate cultural mosaic of Sub-Saharan Africa, accompanied by specific instances for contextual clarity.

Physical Geography of Subsaharan Africa

The African Transition Zone divides North Africa from the rest of Africa because of climatic and cultural dynamics. Cultural conflicts and desertification are common in the zone. Dry, arid type B climates, common in the Sahara Desert, are dominant north of the zone. Tropical type A climates prevail south of the zone. Global climate changes continue to shape the continent. The shifting sands of the Sahara are slowly moving southward toward the tropics. Desertification in the zone continues as natural conditions and human activity place pressure on the region through overgrazing and the lack of precipitation. Type B climates resurface again south of the tropics in the southern latitudes. The **Kalahari** and **Namib Deserts** are located in Southern Africa, mainly in the countries of Botswana and Namibia.For a continent as large as Africa, Subsaharan Africa does not have extended mountain ranges comparable to the ranges in North or South America, Europe, Asia, or Antarctica. There are, however, on the **Ethiopian Highlands** the **Ethiopian Plateau** that reach as high as 15,000 feet in elevation. East Africa has a number of well-known volcanic peaks that are high in elevation. The tallest point in Africa—**Mt. Kilimanjaro** in Tanzania near the border with Kenya—is 19,340 feet high. Nearby in Kenya, **Mt. Kenya** is 17,058 feet high. The **Rwenzori Mountains** on The Congo/Uganda border reach more than 16,000 feet in elevation and create a rain shadow effect for the region. Permanent glaciers exist on these ranges even though they are near the equator. On the western side of the continent, **Mt. Cameroon** in Central Africa is more than 13,000 feet in elevation. The South Africa's **Cape Ranges** are low-lying mountains no higher than about 6,000 feet. The continent of Africa consists of basins and plateaus without long mountain chains. The plateaus can range more than 1,000–2,500 feet in elevation. The only continuous feature is the eastern rift valleys that run along the tectonic plate boundaries from the Red Sea through to South Africa.

Figure 7.3: A Giraffe with Mt. Kilimanjaro in the Background, Located in Tanzania on the Border with Kenya Africa has many scenic areas and many national parks set aside as game preserves, which have become major tourist attractions for world travelers. Diana Robinson –Two elephants doing a mock battle– CC BY-NC-ND 2.0.

The main rivers of Africa include the **Nile**, **Niger**, **Congo**, and **Zambezi.** The Nile River competes with the Amazon for the status as the longest river in the world; the **White Nile** branch begins in **Lake Victoria** in East Africa, and the **Blue Nile** branch starts in **Lake Tana** in Ethiopia. The Niger flows through West Africa; its mouth is in Nigeria. The **Congo River** crosses the equator with a large tropical drainage basin that creates a flow of water second only to the Amazon in volume. The **Zambezi River** in the south is famous for the extensive **Victoria Falls** on the Zambia and Zimbabwe border. Victoria Falls is considered the largest waterfall in the world. Other significant rivers exist such as the **Orange River**, which makes up part of the border between South Africa and Namibia.

There are a number of large lakes in Subsaharan Africa. The largest is Lake Victoria, which borders several East African countries and is considered to be the second-largest lake in the world in surface area. Only Lake Superior in North America has a greater surface area. A number of large lakes are located in the rift valleys of the east. Three of the largest lakes along the western rift are **Lake Malawi, Lake Tanganyika**, and **Lake Albert.** To the northeast of these in Kenya is **Lake Turkana**, which reaches to the Ethiopian border. **Lake Chad** is located in the African Transition Zone on the border between Chad, Mali, and Nigeria. Lake Chad has been severely reduced in size in recent years.

The equator runs through the center of Subsaharan Africa, providing tropical type A climates. These regions usually have more rainfall, resulting in lighter, leached-out soils that may not be as productive as regions with richer volcanic soils, such as those found in the rift valleys. Root crops are common in Africa, as are millet and corn (maize). The savanna regions of the east and south have seasonal rains that affect the growing season. Soils in savanna areas are usually not as productive and cannot be depended on to fulfill the agricultural needs of growing populations. Savannas are usually grasslands or scrub forests with a seasonal precipitation pattern. Cattle and livestock grazing are common in savannas, and migrations are frequent to follow seasonal grazing conditions. In specific areas of Southern Africa, larger farming operations exist in type C climates. However, Subsaharan Africa is not blessed with the large regions of rich alluvial soils found in the Northern Hemisphere. The ever-growing agrarian population has always depended on the land for food and sustenance, but these conditions are not favorable for Africa's future. Populations are growing faster than any increase in agricultural production.

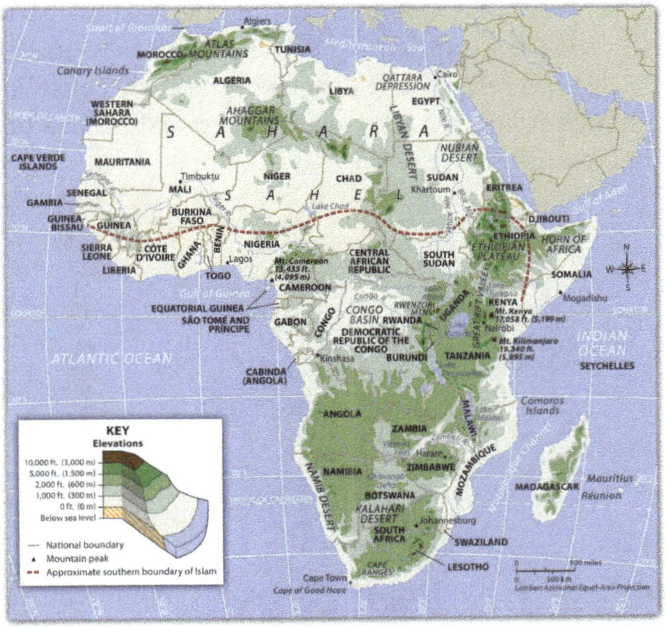

Figure 7.4: Physical Features of Subsaharan Africa

Increasing populations in Subsaharan Africa are taxing the natural environment. Where the carrying capacity has been exceeded, the natural capital is being depleted at an unsustainable rate.

Deforestation is occurring in areas where firewood is in high demand, and trees are cut down faster than they can grow back. Expanding human populations are also encroaching on the natural biodiversity for which the African continent is renowned. Large game animals such as rhinoceroses, elephants, and lions have been hunted or poached with devastating consequences.

The creation of game preserves and national parks has stemmed this tide, but poaching remains a serious problem even in these protected areas. Gorilla and chimpanzee populations have also been stressed by human population growth. These animals are being killed by humans for bush meat, and their habitats are being reduced by human activities.

Subsaharan Africa does not have long mountain chains such as those found in the other continents. The Nile, Congo, and Niger Rivers are the main waterways. The Namib and Kalahari are the main deserts south of the African Transition Zone. The Great Rift Valleys are the longest physical features in East Africa.

Kingdoms, Empires, and the Slave Trade

West Africa was home to a number of large empires that shaped the culture and people of present-day Africa. Gold, salt, copper, and ivory were important early trading commodities. Mining and acquiring these items brought enormous wealth to the region. Slavery became a component in the warfare between groups, as the winner usually took the captives to serve as servants or slaves. The **Ghana Empire** (790–1076 CE) was responsible for one of the early consolidations of power in West Africa. Though its power and wealth were vast, it was only the first of a number of empires to arise in the region.

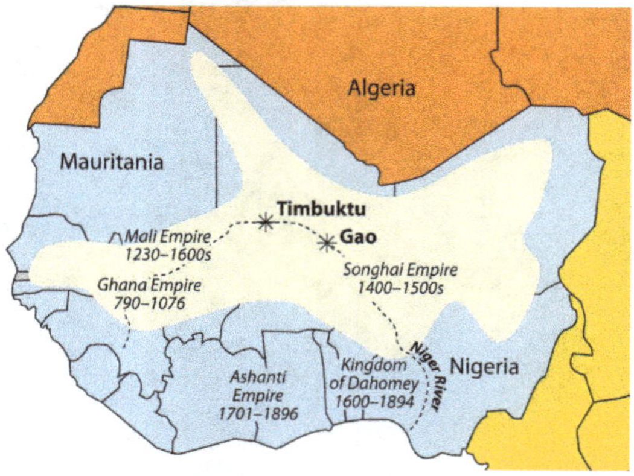

Figure 7.5: Empires of West Africa The Niger River and the main empires of West Africa are indicated on this map. The shaded region indicates the furthest expanse of the Songhai Empire at its peak in the sixteenth century. Updated from map courtesy of Andreas 06

After the Ghana Empire came the **Mali Empire** (1230–1600 CE), which was known for its wealth and size. The empire held major mining operations including gold, copper, and salt. Expanding along the Niger River and to the Atlantic coast, the Mali Empire extended over an area the size of Western Europe. Originally a federation of local groups, the empire incorporated many smaller provinces or kingdoms and had a major influence on the region's laws and cultural heritage.

One of the indigenous states that evolved out of the Mali Empire was that of the Songhai people, who were located to the east of the main Mali Empire. The Songhai had been based around the city of Gao for centuries, and in fifteenth century, they established the **Songhai Empire**, one of the largest empires in African history. The city of Timbuktu became prominent during this period as an urban center, and its educational institutions became well known. In the end, the Songhai Empire was defeated by Moroccan Berber forces that splintered the empire into many smaller kingdoms and ethnic clans.

South of the empires of the Niger River were coastal kingdoms, including the **Ashanti** and the **Dahomey**, which flourished during the colonial era. The local African groups had to contend with the colonial European traders and invaders, which had superior weapons and different technology. Besides the mining resources of gold copper, salt, and ivory, the slave trade was a driving force that attracted trade ships from various European countries.

The **Ashanti Empire** was located along the Atlantic coast in an inland region of what is now Ghana and Ivory Coast; this area served as a trading corridor where goods from the northern regions crossed to make their way to the coastal trading towns. The Ashanti Empire took advantage of the trade opportunities and became wealthy. By the time the Europeans arrived, Ashanti trade relationships were well established and capable of supplying the colonial ships with slaves, ivory, and gold. The coastal area of Ghana was known as the **Gold Coast** during colonial times. Similarly, the country of **Ivory Coast** (Côte d'Ivoire) got its name from its participation in the elephant ivory trade. The Ashanti were known for their involvement in the slave trade and today must live with that legacy.

This region of West Africa bordering the Atlantic was called the Slave Coast because of the high number of local people from this area who were sold into the European slave trade and shipped to the Americas. The once powerful **Kingdom of Dahomey**, east of the Ashanti Empire, was heavily involved in the slave trade. Slave trading with Europeans brought them rifles and other firearms that were useful in conflicts with their regional adversaries. A major slave state along the Slave Coast, the kingdom attracted the spite of its neighbors as a pariah state for selling its war captives into transatlantic slavery rather than killing them, as was the custom in warfare. The Kingdom of Dahomey became the country of Benin after the colonial era.

The slave trade brought wealth to many of the slave-trading kingdoms on the African coast. The demand for slaves pushed many of the African kingdoms to expand their slave-trading activities, capturing more individuals whom they could sell to European slave traders. Millions of Africans were captured by other Africans only to be sold into slavery, placed on ships, and sent to slave owners in the Americas. The total number sold into the Atlantic slave trade will never be known, but estimates range somewhere between ten and twelve million.

Figure 7.6: The Door of No Return Slaves kept for shipment were held in dungeons awaiting the arrival of the slave ships. This was the last doorway that slaves from Elmina, Ghana, saw on their way from the dungeon to the slave ship. Paul Arps – Dungeon entrance Elmina Castle – CC BY.

States and kingdoms existed throughout Subsaharan Africa. Early kingdoms flourished in the region that is now Ethiopia. East Africa's **Axum Empire** was an example of the powers of the early African savanna kingdoms located in the east. The struggles between Islam and Christianity divided the region and gave way to the African Transition Zone, which serves as the dividing line between the two religions. Ethiopia has retained its Christian heritage, while to the north and east Islam has prevailed. South of the equator, the ruins of **Great Zimbabwe** give testimony to the regional base of power that existed there in centuries past. The **Zulu Empire** thrived in the nineteenth century in the region that is now South Africa.

Colonialism in Subsaharan Africa

Wherever you go in the world, you cannot escape the impact of European colonialism. Subsaharan Africa was broadly affected by colonial activities, the legacy of which can be recognized to this day. Colonial powers of Europe ventured into Africa to claim colonies. In the discussion on Brazil in Chapter 6 "South America", Section 6.3 "Brazil", there is segment on slavery and the Trade Triangle of the North Atlantic.

Slavery impacted Africa in other ways. Many African groups were instrumental in capturing and holding slaves to trade with the European merchants. These groups still exist and have had to live with the fallout of their role in the supply side of the slave trade.

By 1900, European colonial powers claimed most of Africa. Only the Kingdom of Ethiopia and the area of Liberia, which was established as a home for freed slaves, remained independent. In 1884, Otto von Bismarck of Germany hosted the Berlin Conference, which to a great degree determined how Africa was colonized. This conference included fourteen colonial European countries and the United States, and its purpose was to divide Africa and agree on colonial boundary lines. Germany had few claims in Africa; Bismarck was hoping to benefit by playing the other countries against each other. At the time, more than 80 percent of Africa remained free of colonial control. On a large map of Africa, claims were argued over and boundary lines were drawn according to European agreements.

There was little regard for the concerns of local ethnic groups. Colonial boundaries divided close-knit communities into separate colonies. Ethnic boundaries were disregarded. Often-warring groups were placed together in the same colony. The Europeans, seeking profits from cheap labor and resources, did not consider the local people or culture and often employed brutal measures to subdue the local people. Most of the current borders in Africa are a result of the Berlin Conference, and many of the geopolitical issues that confront Africa today can be traced back to this event specifically and to colonialism in general.

Figure 7.7: Colonialism in Africa

European colonialism remained strong in Africa until the end of World War II, which left many European countries economically exhausted. In 1945, the United Nations (UN) was established. One of the primary UN objectives was to oversee the decolonization of European colonies. Still, colonialism in Subsaharan Africa lingered on. It was not until the 1990s that the last colony was finally freed. The transition from colony to independent nation itself caused conflict. Civil wars were fought over who would control the country after the Europeans were pushed out. The transition to full independence has exacted a heavy toll from African countries but has resulted in stronger political structures and greater democratic liberties in many cases. The first country to gain independence in Subsaharan Africa was Ghana in 1957.

During the era of independence after World War II, the Cold War between the United States and the Soviet Union enticed many of the African countries to support one or the other of the superpowers. The political pressure that divided African countries during the Cold War has persisted and is the basis of some of the current political problems. The European countries extracted raw materials and minerals from their African colonies, as well as slave labor. Varying degrees of attention were given to education, medical care, or infrastructure development. The dependency that a colony had had on a European country for economic income in some cases continued long after independence and may still continue today. On the other hand, major technology transfers from Europe to Africa infused greater efficiencies into Africa's economic activities. The imperial nature of colonialism and the divisive Cold War did bring about, in an unfortunate manner, the benefits of structured governments and greater democratic processes for many of the local areas of the realm.

Figure 7.8: European Colonies in Africa, 1884. Updated from map courtesy of Andreas 06

From Colonialism to Independence

Figure 7.9: Freedom Square in Ghana This monument honors Ghana's independence in 1957 from the British. Ghana was the first country in Subsaharan Africa to gain independence. James Cridland – Important monument – CC BY 2.0.

The transition from European colony to independent state has not necessarily been a civil transition for African countries; likewise, independent African countries have struggled to create stable governments or peaceful conditions—though stable governments and peaceful conditions have been established in some cases. In nearly all cases, removing the colonial powers from Africa was only half the battle toward independence. The other half was establishing a functional, effective government.

Though each country has taken a slightly different path, most former colonies have endured civil unrest, conflict, or warfare in their push for stable governments. Coups, fraudulent elections, military regimes, and corruption have plagued the leadership in a number of African countries. Civil unrest usually precedes a change in leadership: political power regularly changes hands by a military coup or an overthrow of the current leader.

The realm has had difficulty developing and maintaining effective political systems with democratic leadership. Political leaders have come and gone, many have been replaced before their terms were over, and a few have stayed long after their terms ended. When a leader is elected democratically, it is common to have widespread accusations of corruption, voter fraud, or ballot box scandals. Such government mismanagement and corruption have been common problems.

The vacuum left by retreating colonial powers at times has been filled by authoritarian dictators or by leaders who assumed control of the government and then proceeded to pillage and plunder the state for personal gain or for the enrichment of their clan or political cronies. After the colonial era, it was not uncommon for new leaders to be connected to the old colonial power and to adopt the language and business arrangements of the European colonizer. This gave them an advantage over their competitors and provided a means of income gained by the support of their former European colonial business partners, which often wanted to keep ties with their colonies to continue to exploit their resources for economic profits. Many of these leaders stayed in power because of military backing or authoritarian rule funded by the profits from selling minerals or resources to their former colonial masters.

A few countries are still struggling to bring about some type of order and unity, and in spite of all the negative issues, there have been democratic and relatively stable progressive governments in Africa that have emerged from the transition in recent years, which is a hopeful trend for the long term.

It is important to note that the objective of colonialism was to connect a colony with the mother country, not to connect African countries to each other. As a result, little cooperation occurs between African countries. Each individual country interacts more with its European colonial counterpart with regard to trade, economics, and cultural dynamics. European colonialism has isolated African countries from their neighboring countries and does not contribute to unity within or among African regions. Colonial powers often built new port cities to extract goods and resources from their colonies and transportation systems from the new ports to the interior to collect the resources and bring them to the port. However, the colonial powers did not build a network of transportation systems that connected the region as a whole. Colonial powers were intent on continuing dependence on the mother country so their colonies could be controlled and dominated.

The current wave of globalization based on corporate colonialism continues to encourage the same pattern of discouraging interaction between counties and instead encouraging trade with more economically powerful non-African countries. Most of the interaction between countries is a result of crisis or warfare, in which case large numbers of refugees cross the border into neighboring countries for personal survival or security. To begin to work together and promote trade, common transportation, security, and industry, fifteen West African states came together in 1975 to create the Economic Community of West African States (ECOWAS). Since then, additional political and economic agreements have been signed by various African countries. In 2001, the more expansive African Union was created to help African states compete in the international marketplace.

Key Takeaways

- Subsaharan Africa includes the African Transition Zone and the regions south of the Sahara. The African Transition Zone is a transitional zone between type A and type B climates and between the religions of Islam and Christianity.

- Subsaharan Africa is a realm of plateaus and basins with four main river systems. Mountain ranges, volcanic peaks, and large lakes are found in or along the rift valleys of eastern Africa. The rift valleys were created by tectonic activity.
- Before the colonial era, many African kingdoms dominated regions of Subsaharan Africa. West Africa had a number of great kingdoms. Kingdoms along the coast contributed to the slave trade.
- European countries colonized almost the entire realm of Subsaharan Africa. European powers made agreements during the 1884 Berlin Conference to divide the realm and create boundary lines between their African colonies. Many of the lines remain as the current borders today.
- Colonies were designed to provide labor and resources to the mother country. Port cities, transportation systems, and other infrastructure were implemented to benefit the imperial power without regard for development of the colony.
- African colonies received independence after World War II. The transition from colonies to independent countries has been plagued by civil war, political chaos, or economic devastation.

Discussion and Study Questions

1. What are the main boundaries that define Subsaharan Africa?
2. Where is the African Transition Zone? What are the main transitions that this zone represents?
3. What geological phenomenon has created the rift valleys of the east? What physical geographic features are found here?
4. What are some of the general environmental issues that can be found throughout the realm of Subsaharan Africa?
5. Name the African kingdoms that were involved in the supply of humans for the European slave trade.
6. How did imperial powers in Africa cultivate their colonies' dependence on the mother countries?
7. When did the colonies in Subsaharan Africa gain their independence?
8. What has been the usual pattern of transition from colonies to independent countries?

9. What country was the first in Subsaharan Africa to gain its independence? In what year did it gain independence?
10. How are the Subsaharan African countries continually isolated from their neighbors in the global economy?

Geography Exercise

Identify the following key places on a map:

- African Transition Zone
- Blue Nile
- Cape of Good Hope
- Cape Ranges
- Central Africa
- Congo River
- East Africa
- Ethiopian Plateau
- Kalahari Desert
- Lake Albert
- Lake Chad
- Lake Malawi
- Lake Tana
- Lake Tanganyika
- Lake Turkana
- Lake Victoria
- Madagascar
- Mt. Cameroon
- Mt. Kilimanjaro
- Namib Desert
- Niger River
- Rwenzori Mountains
- Southern Africa
- Timbuktu
- Victoria Falls
- West Africa
- White Nile
- Zambezi River

Activities

1. Determine which of the countries in Subsaharan Africa are now experiencing civil war or national unrest over political leadership.
2. Determine which countries of the realm have the most stable governments.

7.2. HUMAN GEOGRAPHY OF SUBSAHARAN AFRICA

Sub-Saharan Africa is a region characterized by diverse cultures, languages, and ecosystems, shaped by its rich history and complex socio-economic dynamics. The region is home to over 1 billion people, with more than 3,000 distinct ethnic groups and around 2,000 languages spoken. Urbanization is rapidly transforming landscapes, as cities like Lagos, Nairobi, and Johannesburg grow, presenting both opportunities and challenges in terms of infrastructure, employment, and social services. Rural areas, often reliant on agriculture, face issues such as climate change, land tenure conflicts, and migration, which influence food security and livelihoods. Additionally, Sub-Saharan Africa's colonial past continues to impact its political boundaries and economic development, creating a tapestry of resilience and adaptation in the face of ongoing challenges. Understanding this region requires an appreciation of its historical context, cultural richness, and the interconnections between its people and the environment.

Economic Geography

The demographic data for each country in each region of Subsaharan Africa indicate the region's human geography. Urban percentages, family size, income levels, and the other data that can indicate the lifestyle or development level are acutely helpful in understanding the trends in Subsaharan Africa. The index of economic development illustrates the dynamics and conditions that exist in the realm. The data indicate the consistency of economic and development trends across Africa. The data does not, however, indicate differences in cultural dynamics and uniqueness in the ways that local people live. The interesting part of studying Subsaharan Africa is the many ethnic and cultural groups in each country that bring to the surface a wide array of global diversity in our human community. Within each and every country are microcosms of human societies that hold particular customs that may be thousands of years old. Globalization and technological advancements challenge every cultural group to adapt and innovate to make a living yet provide continuity in their heritage. The remote cultural groups of Subsaharan Africa are most susceptible to the volatile nature of globalization, which threatens their current way of life.

Subsaharan Africa is a peripheral world region with neocolonial economic patterns. Peripheral regions usually supply raw materials, food, and cheap labor to the core industrial countries. Most of the population in Subsaharan Africa works in subsistence agriculture to make a living and feed their families. Families are large. In recent decades, there has been enormous rural-to-urban migration to the major cities, which are extremely overcrowded. Subsaharan Africa has more than 750 million people, and most earn the US equivalent of only $1–3 per day.

There is no single major core economic area in Subsaharan Africa. This realm has many core cities and the rest is periphery. Many of the core cities are improving their technology and infrastructure and entering into the globalized economy. Even so, as much as 70 percent of the people still work in agriculture, leaving little time to develop a large educated group of professionals to assist with social services and administrative responsibilities. The realm depends heavily on outside support for technical and financial assistance. Computers, medical equipment, and other high-tech goods are all imported. African states have formed trade agreements and have joined the African Union to assist each other in economic development and trade.

Figure 7.10: Informal and Formal Sectors. The photo on the left is of side-street vendors in Senegal, who might contribute to the informal market system that is prevalent in many Subsaharan African countries. Cash and bartering are the main methods of payment. The photo on the right is of a shopping mall in Zimbabwe, which is part of the formal market system that is regulated and taxed by the government. Radoslaw Botev; Gary Bembridge - Zimbabwe Harare Eastgate Shopping Mall - CC BY 2.0.

Subsaharan Africa has nearly forty urban areas of more than one million people. At the center of the central business districts (CBDs) are modern high-rise business offices well connected to the global economy. Outside the CBD are slums with no services and miserable, unsanitary conditions. The informal sector of the economy—that which is not regulated, controlled, or taxed—has become the primary system of doing business in most of the cities. The informal sector comprises trading, street markets, and any other business without financial records for cash transactions.

The lack of government regulation and control prevents taxes from being assessed or collected, which in turn diminishes support for public services or infrastructure. The formal sector of the economy—that which the government can regulate, control, and tax—is forced to foot the bill to operate the government and support public services such as education, security, and transportation. In spite of the misery and unhealthy conditions of the slums where millions of people already live, more migrants from the countryside continue to shift to the city in search of jobs and opportunities. African cities are growing rapidly, many without organized planning.

If the index of economic development were applied to Subsaharan Africa, a clear pattern would emerge. Rural areas would be in the lower stages of development, and only a few developing areas would be higher than stage 3. A large percentage of this realm would be in stage 1 or 2 of this model. Countries with higher standards of living, such as South Africa or Botswana, would be working through stage 3 or 4. This region has one of the fastest-growing populations of the world and economically lags behind countries in the Northern Hemisphere, which have transitioned into the higher stages of the index of economic development.

Incomes, Urbanization, and Family Size

The socioeconomic data illustrate well the conditions for people in Africa in comparison to the rest of the world. African countries are at the lowest end of the statistics for development prospects. The larger cities are showing promise for advancement into the higher stages of development. Family sizes in the rural countries are some of the largest in the world. The average fertility rate for much of Africa is about 5; in Mali and Niger, the rate is higher than 7. One-third to half of the populations of these countries are under the age of fifteen. Children make up most of the population in many areas—an indication that heavier burdens are placed on women, meaning that women are not easily able to get an education or work outside the home.

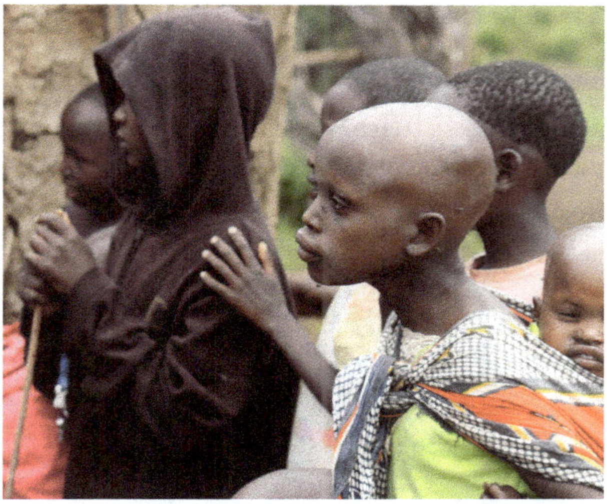

Figure 7.11: Young Women and Child of the Maasai in Kenya. Davida De La Harpe – maasai children – CC BY-ND 2.0.

The populations of West African countries are increasing rapidly and will double in about thirty years at the current rate. This trend places an extra burden on the economy and on the environment. It is fueling one of the fastest rural-to-urban shifts in the world. West Africa is now only about 32 percent urban, and Burkina Faso and Niger are less than 20 percent urban—clear indications that agriculture is the largest sector of their economies and that most of the people live in rural areas or small villages. Personal income levels in West African countries are among the lowest in the world; as far as standard of living is concerned, these are poor countries. Few economic opportunities exist for the millions of young people entering the employment market.

The United Nations (UN) human development index lists all but two of West African countries in the lowest category of development. Sierra Leone is the lowest of the world. Ghana and Senegal were the two countries listed in the medium range.

Senegal is the lowest country in the medium category, just barely rising about the lowest tier. A country listed in the lowest tier of the medium category translates into a region with low availability of opportunities and advantages for its people. As a peripheral world region, the economic base is structured around agriculture with supportive extractive activities. Agricultural activities are renewable, but agricultural profit margins are slim. These countries are in a subsistence mode with a rapidly expanding population and few industrial or postindustrial activities to gain income.

No Subsaharan African nation is in stage 5 of the index of economic development. Nevertheless, the fact that they are not consumer societies does not negate the rich cultural values and heritage of the realm. The energy of the people does not revolve around consumerism but is instead focused on the people themselves. High levels of social interaction and community involvement bring about different cultural standards than those of a consumer society, which focuses more on the individual and less on community. Unless there is social unrest or open warfare, which does exist in various places, the people work hard to bring about a civil society based on family and tradition.

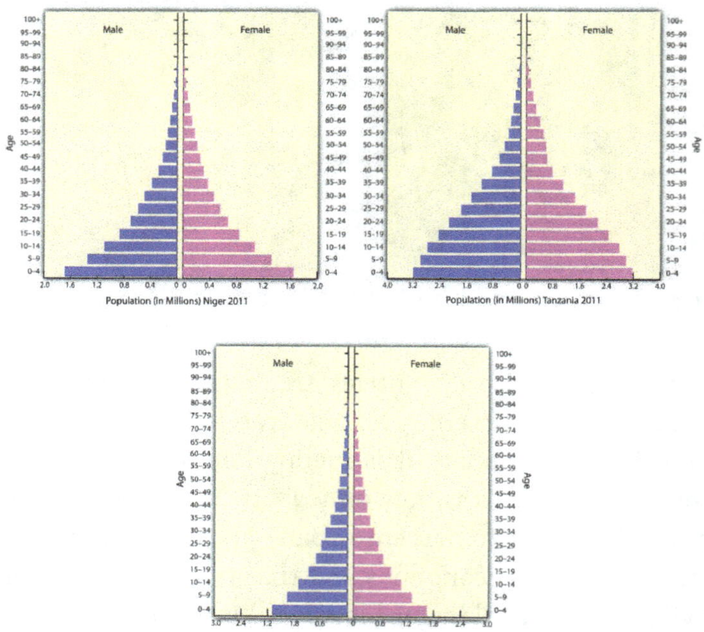

Figure 7.12: Population Pyramids of African Countries The population pyramids of Niger, Tanzania, and South Africa illustrate the population growth dynamics of Subsaharan Africa. Niger's population pyramid illustrates large family size and rapid population growth. Tanzania's pyramid is similar but shows signs of slowing population growth. South Africa, which is more urbanized and industrialized, shows signs of declining family sizes and fewer children. US Census Bureau International Programs – public domain.

Languages in Subsaharan Africa

Subsaharan Africa covers a large land area more than 2.3 times the size of the United States. Thousands of ethnic groups are scattered throughout the realm. There is immense diversity within the 750 million people in Subsaharan Africa, and within each country are cultural and ethnic groups with their own history, language, and religion. More than two thousand separate and distinct languages are spoken in all of Africa. Forty are spoken by more than a million people. Many local languages are not written down and have no historical record or dictionary. Local languages without a written history are usually the first to be lost as globalization affects the realm. Nigeria, with more than 130 million people, is the most populous country in Africa. It is about the size of Texas and Oklahoma combined, and the African Transition Zone cuts through the country's northern portion. More than five hundred separate languages are spoken in Nigeria alone. Three of the six dominant languages in Subsaharan Africa—spoken by at least ten million people or more—are spoken in Nigeria: Hausa, Yoruba, and Ibo. The three remaining major languages of Subsaharan Africa are Swahili, Lingala, and Zulu.

Colonial activity changed much of how the African countries operated economically, socially, and politically. Language is one aspect of culture that indicates a colonial relationship. Many African countries today speak a European language as the official language. Mauritania is the only country that has Arabic as its official language. Nigeria has English plus other local languages. The official languages of most of West Africa are either French or English, and Guinea-Bissau's official language is Portuguese.

This vestige of colonial power would seem inconsistent with the desire to be free of foreign domination. However, because often dozens to hundreds of local languages are spoken within the country, choosing the colonial language as the official language produces less of an advantage for one group wishing to dominate the political arena with its own local language and heritage.

For example, a language problem arises when a government needs to print material for the country. What language do they use? In Nigeria, there would be more than five hundred possible languages. What if the leadership used a language only spoken by a few people? The language of those in power would provide an advantage over those that could not understand it. What if there were more than five hundred separate languages in Texas and Oklahoma? How would they function? This is why many African countries have chosen a colonial language as their country's lingua franca, or national language. Ghana, which is the size of Minnesota, has more than eighty spoken languages. Ghana and Nigeria have both chosen English as their national language to provide a cohesive and inclusive method of addressing the language dilemma.

A portion of the two thousand languages spoken on the African continent will not survive. A large number of the languages are spoken by a small number of local groups that may or may not have a written text or alphabet. The influence of globalization causes the country's lingua franca to overshadow local languages, which are relegated to the older generations that may not be fluent in the languages of global business and commerce. Young people often learn the lingua franca and may or may not pass their local language on to the next generation. This is how languages become extinct. Similar dynamics can be applied to local religious beliefs. Outside influence can often erode local beliefs and cause an evolution of religious tenets that eventually transform indigenous beliefs into patterns similar to the larger, more mainstream religions.

Figure 7.13: Language Families of Africa There are six main language families with many variations of each. It is estimated that more than two thousand languages are spoken in all Africa. Map courtesy of Mark Dingemanse and Pmx – CC BY-SA 3.0.

Religion in Subsaharan Africa

Before the monotheistic religions from the Middle East were even in existence, the people of Africa followed traditional animist beliefs. The diffusion of Christianity and Islam to the African continent convinced many African people to abandon their animist beliefs. African religions spread with the slave trade and became a part of the African Diaspora. Examples can be found in the Santeria religion in Cuba,

Umbanda followers in Brazil, and Vodou (Voodoo) practitioners in Haiti. Many of these examples indicate a high rate of mixing between traditional religions and Christianity, something that is not as well accepted within Islam.

The current religious trends in Africa follow the pattern of the African Transition Zone. Most of the population north of the zone follows Islam, and most of the population living south of the zone follows Christianity. Large percentages of people in the region follow a wide array of traditional or animist beliefs. For example, as of 2010, more than 50 percent of the people of Togo still followed local religions not affiliated with Christianity or Islam. Only about 29 percent of the population claimed to be Christian, and even fewer claimed to follow Islam (Central Intelligence Agency).

Along the African Transition Zone, followers of one religion will clash with followers of the other. Countries such as Nigeria have a history of this type of social division, and Nigeria's government allows Islamic Sharia law to take precedence over civil law in the country's northern regions. For example, in 2002, the Miss World beauty pageant was to be held in Nigeria's capital. At the same time, Nigerian news reported a case of a young woman charged with adultery in Muslim-dominated northern Nigeria. The woman was to be stoned to death for her crime. Northern Nigeria is north of the transition zone and is staunchly Islamic. Southern Nigeria is mainly Christian or animist. The northern Muslims were protesting the "decadent" Miss World beauty pageant, and riots spilled over into the streets. Buildings were burned, cars were overturned, and more than one hundred people were killed. Meanwhile, people in the south were protesting the death sentence of the woman charged with adultery. Contestants for the Miss World contest began to pull out of the competition, some in protest and some for personal safety reasons. In the end, the woman sentenced to death was smuggled out of northern Nigeria to the safety of the south. The Miss World contest was moved to London.

Both Islam and Christianity have been on the rise in Africa. As the local beliefs are replaced with monotheistic religions, there is more integration with either the West (Europe and America) or the Middle East. Religious activity through Christian missionaries or the advancement of Islam sometimes coincides with economic support being brought in through the same channels, which is often welcome and enhances the global relationships that occur.

However, Africa is still full of traditional religions with rich spiritual histories. Spiritual forces are found in the environment. Deities of all kinds are worshipped throughout Africa. Christianity and Islam are latecomers to the region but have made deep inroads into the African culture. Both compete for the souls of the African people.

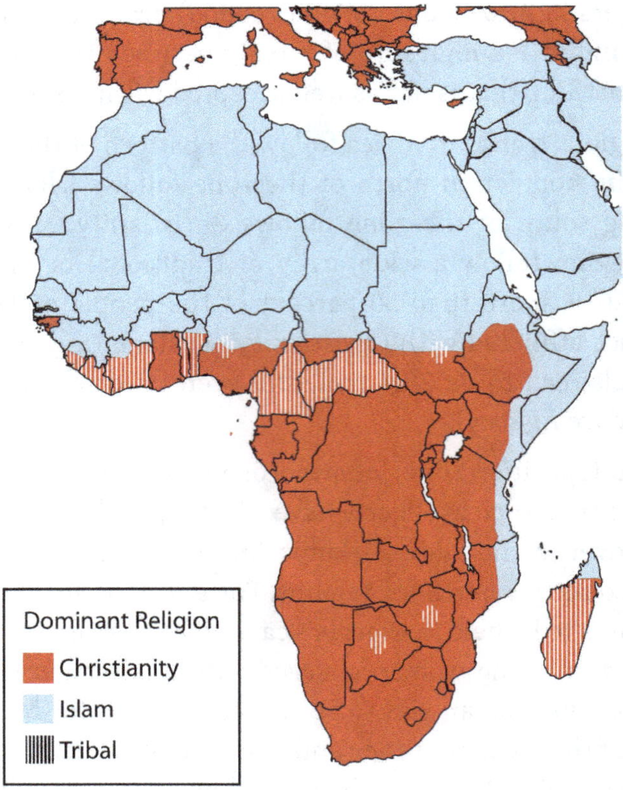

Figure 7.14: Dominant Religions in Africa Islam is prominent north of the African Transition Zone, and Christianity is more prominent south of the African Transition Zone. Updated from map courtesy of Andreas 06 – public domain.

Ethnic Divisions and Civil Wars

Subsaharan Africa is home to thousands of ethnic or traditional groups. Each has a separate identity and history, and often one group is in conflict with another. The slave trade and the establishment of colonial political boundaries or policies exacerbated historical ethnic hostilities. Major civil wars have been fought throughout the history of throughout Subsaharan Africa and continue at the present time. Central Africa has endured ongoing brutal conflict in the past decade, with no solution in sight. More than five million people have died as a result of the civil war in The Congo (Zaire). Fighting continues between various factions over political control or control over natural resources, such as diamonds or gold. Civil wars are devastating some African countries. Many other countries, such as Zimbabwe, Chad, and the Central African Republic, have also suffered economic disintegration as a result of severe political unrest.

Rwanda's Tutsi-Hutu conflict has been historic in its violence and in the senseless killing of innocent people. In 1994, the centuries-old conflict erupted into violence of unprecedented proportions. Hutu militias took revenge on the Tutsis for years of suppression and massacred anyone who did not support the Hutu cause. Tutsi rebels finally gained strength, fought back, and defeated the Hutu militias. More than a million people were killed, and more than a million defeated Hutus fled as refugees to neighboring countries, where many died of cholera and dysentery in refugee camps.

Figure 7.15: Tragedy in Rwanda The bones of victims killed in the Rwandan genocide are kept in this school in Rwanda. The total number of people killed in the war is unknown. Wikimedia Commons – public domain.

The civil war in Rwanda and the many refugees it created destabilized the entire Central African region. The shift in population and the increase in military arms along the Zairian border resulted in an extensive civil war in The Congo (Zaire) that has resulted in the deaths of more than five million people, many by disease or starvation. Over three million deaths are estimated to have been related directly to the war and another two million by the harsh conditions in the region (AmericanRenaissance.com). The civil wars in The Congo from 1996 to 2003 changed the cultural and political landscapes and destroyed valuable infrastructure. One of the driving forces in these wars is the control of valuable mineral resources found in the Great Rift Valley along the eastern boundary of The Congo. Diamonds, gold, copper, zinc, and other minerals are abundant in this region, and wealth that can be gained from the mining of these products attracts political forces to compete for their control.

Civil wars have wreaked havoc on the countries of Angola, Sierra Leone, Liberia, and Somalia. All have experienced some amount of serious conflicts in the past decade. Many of the civil wars are not reported by the news media worldwide, even though the number of people affected, injured, and killed is deplorable. In places such as Ivory Coast, Nigeria, and Zimbabwe, the streets have become a battleground.

Creating stability in parts of Africa has been challenging, as civil unrest and political corruption continue in many African countries. The core industrialized countries have been hesitant to step in or invest in the peace and stability of Africa. Governments of more than a few African countries have been unable to bring stability or to provide for their people. For example, Somalia has no central government; rather, it is ruled by warlords and village chiefs. Corruption, dictatorial rule, and military force have been major components of government rule in these cases.

HIV and AIDS in Subsaharan Africa

As if Subsaharan Africa did not have enough to work through to achieve stability, it must also cope with high numbers of people infected with HIV and AIDS. From South Africa to Kenya, there is a line of countries with some of the highest percentages of HIV-infected people in the world. The HIV virus has infected more than 24 percent of adults in Botswana, and some villages have lost an entire generation of adults to AIDS. The AIDS pandemic has become a major health crisis for Subsaharan Africa. According to the World Health Organization, as many as thirty million people in this realm are HIV-infected.

It's possible to live with HIV for years before dying of AIDS. HIV-infected individuals can pass the virus to others without knowing they have it, and millions of people die of AIDS in Subsaharan Africa without ever knowing they were infected. The lack of education, HIV testing, and medical services hinder progress toward stopping this deadly disease. Prevention is an ideal that has not materialized yet. Many people do not want to be tested out of fear of rejection by their families and friends if they are infected. AIDS will surely kill millions more in Africa before a solution is found.

Many other diseases are common in Subsaharan Africa. Some are spread by vectors in the environment. Mosquitoes spread illnesses such as malaria and yellow fever, which are common throughout the realm. Sleeping sickness is spread by the tsetse fly, which can also infect cattle and livestock. Hepatitis is widespread. Schistosomiasis, tuberculosis, and typhoid are also common. Unsanitary conditions and polluted water are breeding grounds for microbes that cause diseases. People living in less-than-sanitary conditions are more likely to come in contact with and become infected with these diseases and are also less likely to obtain medical attention at an early stage of the disease.

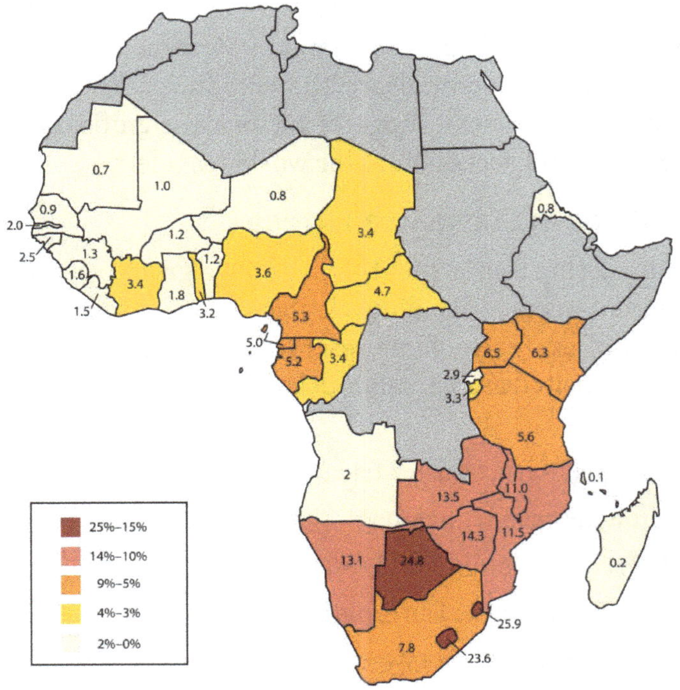

Figure 7.16: HIV and AIDS This map shows the estimated percentage of adults with HIV in Subsaharan Africa for each state based on best estimates by the World Health Organization and the UN. Map courtesy of Andreas 60 - public domain; data courtesy of UNAIDS 2010 Global Report - public domain.

Tourism Potential

Subsaharan Africa has great potential for the development of tourism. Tourism is considered a postindustrial activity with mixed income opportunities. If Africa can manage its resources and provide a safe environment for travelers, tourism will have a strong impact on Africa's economic growth and will play a significant role in its future.

Tourism is a growing sector of the global economy. Travel and tourism jobs are increasing worldwide. Africa as a whole attracts less than 5 percent of total world tourists and accounts for only a small percentage of international tourism income. Given the region's slim share of the tourism market and the expected dynamic growth of the sector worldwide, Africa can expect to substantially increase its share of global tourism activity.

Subsaharan Africa has a strong supply-side potential to attract tourists. Beach resorts alone create a large draw for tourists. The coastal waters of the Indian Ocean boast some of the finest beaches in the world, with plenty of opportunities for sailing, diving, or other water sports. South Africa is proud of its secluded beaches and beautiful coastline. Other well-known coastal tourist destinations include Zanzibar-Tanzania, Benguerra Island in Mozambique, and the Seychelles. Other areas with tourism potential are the wildlife parks and game preserves. Cultural locations with a rich heritage of historical significance are growing in their attractiveness to and accessibility for world travelers.

Tourist Destinations in Subsaharan Africa

Africa excels in attracting tourists to its wildlife and game reserves. Safari tourism highlights exotic creatures, including elephants, lions, rhinos, hippos, and big game. Africa is full of extensive game reserves and national parks. The following are some of the most visited animal safari destinations (* indicates a UNESCO World Heritage Site):

- Kruger National Park, South Africa
- Selous Game Reserve, Tanzania*
- Amboseli Reserve, Kenya
- Serengeti National Park, Tanzania*
- Kalahari Game Reserve, Botswana
- Berenty Nature Reserve, Madagascar
- Virunga (Gorilla) National Park, Congo (Zaire)*

Subsaharan Africa is replete with natural features or attractions that tourists gravitate toward, particularly tourists who are in search of outstanding scenic sites or desire an environmental adventure. There are dozens of awe-inspiring national parks throughout Africa. The following are some of the more interesting physical geography locations in Subsaharan Africa (* indicates a UNESCO World Heritage Site):

- Mt. Kilimanjaro, Tanzania*
- Mt. Kenya National Park, Kenya*
- Ngorongoro Crater, Tanzania
- Victoria Falls, Zambia/Zimbabwe
- Kamale Pinnacle, Cameroon Mountains
- Zuma Rock, Nigeria
- Table Mt. National Park, South Africa

Cultural and heritage tourism has grown immensely in recent years. Cultural assets and heritage landscapes have always provided for excellent tourist attractions. Subsaharan Africa has a great wealth of culture and heritage to attract tourists. The following are a few of the most well-known cultural destinations in the realm (* indicates a UNESCO World Heritage Site):

- Timbuktu (Tombouctou), Mali*
- St. Peter's Basilica, Ivory Coast
- Ashanti-National Cultural Center, Ghana*
- Great Zimbabwe Ruins, Zimbabwe*
- Robben Island, South Africa*
- Soweto City Tour, South Africa
- Rock City of Lalibela, Ethiopia*
- Ancient City of Abomey, Benin*

Figure 7.17: African Elephants Wildlife and big game are natural resources for tourism. Countries in Subsaharan Africa have created many protected areas for wildlife. Michael Jansen – Afrikanische Elefanten / African Bush Elephants – CC BY-ND 2.0.

Every African country, urban center, or rural village is its own unique tourism magnet. The tourism business, however, is broader than just the sites themselves. Considerations need to be made for transportation to and from the country and final destination. Hotels and guest accommodations, such as food services, restaurants, and the availability of other types of consumer goods, need to be considered. Services need to be available that link the various components of a trip, such as guide services in national parks or city bus tours. The attractiveness and competitiveness of each tourism destination will depend on the site's quality and accessibility.

A serious financial investment is needed to bring Subsaharan Africa up to par with the global marketplace. Africa has huge potential for growth in its tourism market. Unfortunately, the same problems with tourism that were discussed in the lesson on tourism in the Caribbean (Chapter 5 "Middle America", Section 5.4 "The Caribbean") can be applied to Africa. There are many positive and negative aspects to tourism, and a trade-off is usually needed. Heavy tourism traffic might have a negative impact on the environment, cultural stereotypes tend to be exploited, and the disparity between wealthy tourists and service workers earning a modest wage may lend itself to divisions and social friction.

Tourism demands higher levels of security and public health at all levels. Money spent on tourism development is money not spent on schools or clinics. On the other hand, without the tourism income, there are no jobs. Tourism brings to the surface both centrifugal and centripetal cultural forces. To be successful, Africa will need to balance out the economic need for tourism with its willingness to comply with the requirements of the tourism industry.

Key Takeaways

- Subsaharan Africa's many core cities and peripheral regions create a high rural-to-urban shift. Much of the realm's economic activity is conducted in the informal sector.
- Most countries in this realm are in the earlier stages of the index of economic development. The integration into global markets has brought urbanization and development to many of the major core cities.
- The informal sector dominates many of the economic activities in Subsaharan Africa. The emergence of stable governments and economic structures usually results in a stronger formal sector of the economy, which in turn provides improved national infrastructure.
- Subsaharan Africa is realm with a high diversity of languages and religions. In an attempt to provide an equal footing for all languages, many countries have chosen their colonial language as their official language to avoid giving speakers of one language any political advantage over another.
- Parts of Subsaharan Africa continue to experience social unrest and civil war. Armed conflicts in Rwanda and The Congo (Zaire) have either killed or affected millions of people.
- A high percentage of people in Subsaharan Africa are infected with HIV—up to one-fourth of the population of some countries. Health care, education, and cultural forces are important factors in combating AIDS and the tropical diseases common in the realm.
- Subsaharan Africa has enormous potential to exploit their physical features and cultural heritage to expand their tourism market. It is often difficult to balance the investments for adequate infrastructure and services required for tourism with other needs, such as education and health care.

Discussion and Study Questions

1. How does the core-peripheral spatial relationship apply to Subsaharan Africa?
2. What is the difference between the formal and informal sectors of the economy?
3. How do the countries of Subsaharan Africa fit into the index of economic development model?
4. What roles do women play in the realm's socioeconomic environment?
5. Approximately how many languages are spoken in Africa? How many are spoken in Nigeria alone?
6. Why would a former colony under European imperialism agree to use the language of its colonizer as its national language when dozens or hundreds of languages are already spoken in the country?
7. How are Christianity and Islam distributed across Subsaharan Africa?
8. How are the armed conflicts in Rwanda, The Congo (Zaire), and Somalia different from each other?
9. Why is the HIV/AIDS situation so difficult to combat or address?
10. What are some of the obstacles in increasing tourism development in Subsaharan Africa?

Activities

1. Determine the ten largest cities of Subsaharan Africa and locate them on a map.
2. Select a country in this realm that is indicated to have a higher percentage of people who do not follow Christianity or Islam. Research and then outline the variance in the indigenous religions that are adhered to by the people who live there.
3. On a map of Africa, locate the tourist attractions mentioned in this section.

References

AmericanRenaissance.com "Crisis Caused 5.4 Million Deaths in Congo, Report Says," http://www.amren.com/mtnews/archives/2008/01/crisis_caused_5.php; "DR Congo War Deaths 'Exaggerated,'" BBC, http://news.bbc.co.uk/2/hi/africa/8471147.stm.

Central Intelligence Agency "The World Factbook," https://www.cia.gov/library/publications/the-world-factbook/geos/to.html.

7.3. WEST AFRICA

The region of West Africa includes the southern portion of the bulge of the continent, which extends westward to the Atlantic Ocean. This region is bisected by the African Transition Zone, which borders the southern edge of the Sahara Desert. The main physical features include the Sahara Desert and the Niger River. The **Cameroon Highlands** are located on the eastern border between Nigeria and Cameroon. At 4,100 miles long, the Nile River is the longest, while the Congo River is Africa's second longest at 2,922 miles in length. The Niger River is Africa's third-longest river and extends more than 2,600 miles from the Guinea Highlands through Mali, Niger, and Nigeria before reaching the Atlantic Ocean in the Gulf of Guinea.

Some geographers include the country of Chad or portions of it within the region of West Africa. In this textbook, Chad is listed with Central Africa. The portions of Chad located north of the African Transition Zone share similar characteristics with North Africa. Off the coast of Mauritania are the Cape Verde Islands, which are united as an independent country associated with Africa. Cape Verde was once a Portuguese colony but received its independence in 1975. Western Sahara has been in conflict with Morocco over independence and is most often associated with the region of North Africa because of the influence of Islam and because of its connection to Morocco.

The African Transition Zone cuts across the region of West Africa, indicating a division between Islam and Christianity and between the Sahara Desert and the tropics. This diversity in religion and climate is usually exhibited with a north/south division. Islam is the dominant religion on the north side of the African Transition Zone; Christianity is more dominant to the south. The two religions often clash in the areas where they meet. Traditional beliefs and animist religions are also practiced in the African Transition Zone.

Figure 7.18: West Africa This map shows the region of West Africa as defined in this chapter. The African Transition Zone crosses the middle of this region. Updated from map courtesy of University of Texas Libraries.

The terms "state" and "country" are often used interchangeably by the world community outside of the United States. Both are meant to refer to a physical unit with a sovereign and independent government. In Sub-Saharan Africa the term "state" is commonly used to refer to a country in any one of the various regions.

Niger

The former French colony of Niger is landlocked, with the Sahara Desert making up its largest portion. Niger is a land of subsistence farmers, and most of the population lives in the southern regions. The country is less than 20 percent urban. Other economic activities include uranium mining, which is the country's main export. The world demand for uranium has not been strong in recent years. Oil exploration has begun and international oil corporations have garnered contracts for drilling.

The Sahara Desert is moving southward, and the agrarian culture at the base of Niger's society is often plagued by drought and famine. The Niger River flows through the southwestern region and provides fresh water, but the northern region is mainly the Sahara Desert, and large portions are covered with sand dunes.

The country has extreme demographics. Niger has the highest fertility rate (7.6) in the world, and half the population is under the age of fifteen, causing a population explosion that taxes the sparse natural resources and brings even more poverty to a country at the bottom end of the index of economic development. Infant mortality rates in Niger have been the highest in the world.

The mainly Sunni Muslim country has a rich cultural base but suffers from economic problems that appear to increase with the increase in population and desertification. Heavy national debt has hindered social services and has required a considerable amount of foreign aid from a number of sources. As a former colony, Niger, France has been a main contributor to providing economic aid along with the United Nations (UN).

The political conditions in Niger are typical of the region. For the first thirty years after independence from France, the country was ruled by a single political party and military rule. There have been several coups and various political leaders have been in power. A dispute remains with Libya over its northern border. Ethnic infighting with a minority Taureg group has emerged in recent years, bringing conflict and discord. The Taureg, found throughout northwest Africa in the interior Saharan region, have many ethnic clans and have been masters of camel caravans, often with a nomadic heritage. Their clothing is often made of cloth colored with an indigo dye that distinguishes them from other ethnic groups. The Taureg fought historic battles for an independent homeland against the French during the colonial era.

Mali

To the west of Niger is Mali, another landlocked Muslim country dominated by the Sahara Desert. Mali was home to various ancient empires. Wealth was historically gained from the mining of gold, salt, and copper. The Niger River flows through the entire southern region, providing a means of transportation and fishing as well as fresh water. The ancient city of **Timbuktu**, once a port on the Niger River, was a center of commerce and trade for the region and was used by the Taureg as early as the tenth century. Timbuktu has played an important role as an educational focal point for many of the peoples of the region. The University of Timbuktu is noted for its educational activities and comprises various colleges and madrasas (Islamic schools of learning). At one point in its history, it claimed to hold the largest collection of books in Africa. The capital city of Mali is not Timbuktu but Bamako, which is located in the southwest. Timbuktu remains a main tourist destination for the country and the region.

International Culture and the Power of the Soap Opera

Timbuktu (Tombouctou) is a key city for the country of Mali in the middle of the Sahel. The city has been stereotyped by people in Western culture as the farthest place from civilization on Earth. In reality, it is a thriving city that is well connected to the globalization activities of the rest of the planet. It was once one of the wealthiest cities in the world. Gold was plentiful, and the city was home to one of the largest universities in the ancient world. Today the city exhibits traits similar to North Africa but is in the African Transition Zone. Modern technology has found its way into the lives of the people of Timbuktu. International exchanges are providing interesting insights into the globalization process. Even in the rural areas, the people have found innovative ways to share in the international dynamics of global cultures.

Dr. Ibrahim N'diaye, originally from Mali and as of 2011 a history instructor at Morehead State University in Morehead, Kentucky, tells a story about his experience while visiting the rural Mali village of Koriyome as a member of the planning committee for the "Tombouctou 2000" millennium commemoration festival the Mali Ministry of Culture was sponsoring at the time. He noticed that in the evening at a certain time, the people of the village started up gas-powered electric generators and then gathered around television sets connected to a satellite feed throughout the village. The evening feature was a Brazilian soap opera called *Mari Mar*. It was broadcast in Portuguese, a language no one spoke in the village. The seminomadic villagers were totally immersed in the soap opera and had even altered the names of the key characters to match local names. For example, they referred to the main character Mari Mar as "Mariyama" (which is the same as Mari/Mary); Mari Mar's husband, Sergio, as "Saajo"; the cruel stepmother Angelika as "Jelika"; and so on.

Everyone in the village seemed to understand the plot and reveled in the drama of the characters and their exploits. The production became a centripetal force that brought the villagers together in a way that would not have been expected by anyone in Brazil or the United States. Who would have believed that soap operas would have become a method of diffusing global culture to the planet? This story also indicates that human nature is similar in societies throughout the world.

Figure 7.19: Bamako, Mali This scene shows the various modes of transportation that are common on the road outside Bamako, Mali. Robin Taylor – looking north from the old bridge – CC BY 2.0.

Mali shares similar political dynamics and demographics with Niger and the rest of the region; however, Mali now has a stable democratic government with little political conflict. There are issues with the Taureg in the north and refugees from Ivory Coast in the south, but the country is quite stable politically in spite of the low standard of living and high population growth.

Mauritania

Mauritania is also dominated by Islam and the Sahara Desert. Niger, Mali, and Mauritania are the largest of the West African states and together would cover the area of the United States east of the Mississippi River. The term "state" refers to countries with independent governments. Mauritania's access to the Atlantic coast provided a shipping connection to the rest of the world. For years the country fought over control of portions of Western Sahara but relinquished those areas in 1975.

Unfortunately, the political and economic problems that are common in the region are evident in Mauritania. In 1960, the country gained its independence from the French, and a number of political coups and changes in government occurred as a result. Interethnic conflicts exist between the African groups of the south and the Arab-Berber groups of the north.

Political unrest and economic hardships continue to add to the challenge of a growing population. Poverty, health care, and education have been major issues that require government support. Foreign aid has been critical to the operation of the country.

Large amounts of iron ore make up about 40 percent of the country's exports. Fishing off the coast has great potential but has been exploited by international fishing vessels; thus Mauritania is in danger of losing revenues and a declining resource base if protection is not secured. Even though the country is largely desert, most of the population works in subsistence agriculture.

Figure 7.20: Senegal and the Gambia The Gambia is totally surrounded by the country of Senegal. The Gambia was once a British colony and Senegal was once a French colony. CIA World Factbook – public domain.

Within certain communities in Mauritania, especially within the Arab Moor group, the concept of female beauty and prosperity was often associated with ample body size. Young girls were encouraged to eat high-fat foods in high quantities and at times were force-fed to put on weight. Being fat was an indication of wealth, and being thin was a sign of poverty and low esteem. This custom resulted in many women being considered overweight and obese by today's health standards. This trend is not as common today, though it still exists. Many Mauritanian women consider this old-fashioned, and natural eating habits are more the standard. Global views regarding women might be contributing to the country's current concept of beauty and health.

Senegal and the Gambia

The country of Senegal on the Atlantic coast totally surrounds the independent country of the Gambia. Senegal was a French colony, while the Gambia was a British colony. The Gambia is an enclave of Senegal and extends on both sides of the **Gambia River** for about 186 miles. It is the smallest country on the mainland of Africa and is about the size of the US state of Connecticut. The two countries of Senegal and the Gambia were united into the confederation of Senegambia from 1982 to 1989 when it was dissolved. They have kept separate political identities ever since.

The capital of Senegal, Dakar, is located on the Cap Vert Peninsula, which is most extreme western extension of the African continent. Dakar was a main colonial port for French West Africa during colonial times. The French influence remains: the common currency in what were former French colonies is the West African CFA franc (Central African franc), which has a fixed rate in relation to the euro.

Guinea and Guinea-Bissau

Freeing itself from the grip of Portuguese colonial rule was a hard-fought battle for Guinea-Bissau, which became independent in 1974. The political upheaval that followed up to the present has resulted in various coups, military rule, changes in leadership, and the assassination of their president in 2009. Without political stability, it is difficult to develop a growing economy. Graft, corruption, and civil war have devastated the country's infrastructure and hindered its economic development, making it among the poorest nations in the world. The country has to depend on outside aid to supply its basic needs.

The French colony of Guinea, which is larger than Guinea-Bissau, holds more promise due to its generous natural resource base, but Guinea has suffered the same types of political and economic disasters as its smaller neighbor. Though mineral resources are abundant, there is potential for increased agricultural production, where most of the people make their living. Guinea has a large amount of bauxite and other minerals, including diamonds and gold. The lack of infrastructure and political stability has discouraged investors from helping to develop these natural resources and converting them into national wealth.

Sierra Leone

Sierra Leone has been devastated in the past decades due to brutal civil war and political turmoil; as a result, the country was the lowest in the world on the human development index in 2010. There has been little political stability. Military factions have wreaked havoc on the country since it gained independence from Britain in the early 1960s. After independence, the country's contested elections and multiple coups resulted in a ban on all but one political party and established military rule. At the core of the conflicts was competition for the control of the diamond industry, which was also a primary factor in the civil war.

The abuses of power escalated to a breaking point in the late 1980s, which set the stage for a decade-long civil war in the 1990s. A group calling themselves the Revolutionary United Front (RUF) pushed to take control of the eastern diamond mining sector through terrorist tactics and brute force. Thousands of people died and thousands more were mutilated by having their arms, legs, or other body parts cut off by machetes. Whole villages were destroyed and the residents killed, tortured, or maimed.

The government fell under various military coups and eventually resorted to hiring mercenaries to help push back the RUF forces. By 1998 the whole country resembled a military camp with arms and ammunition being trafficked to all sides. There was a total breakdown in state structures and institutions. At various times, there were agreements between the sides in hopes that a stable government could emerge, but the agreements broke down and the country continued to scramble for civility. Nigerian forces were actively fighting in the conflict. In 2000, British troops were employed to help evacuate foreign nationals and establish order. Military forces from neighboring Guinea also entered the country to attack RUF bases.

Figure 7.21: Diamond Fever A miner in Kono District, Sierra Leone, searches his pan for diamonds. Diamonds have been used as currency to fund armed conflicts in this region, hence the term *conflict diamonds* or *blood diamonds*. The term *blood diamonds* was first used in the wars in Angola. United States Agency for International Development – public domain.

UN forces with US support eventually established a sense of control of the country. By 2003, major fighting was over and attempts were made to establish a civilian government. Approximately fifty thousand people were killed in this civil war and as many as two million people were displaced, many of them refugees. Sierra Leone set up a special court to address crimes against humanity, and the country has been working to establish a stable government and maintain a sense of order.

The diamond trade still dominates and political factions still vie for control. Much of the diamond mining is uncontrolled, allowing considerable smuggling operations to operate. After the civil war, revenues from diamond mining increased from less than $10 million in 2000 to an estimated $130 million in 2004. The civil war destroyed the infrastructure of the country, and most of the resources were looted by forces on one side or another. The economy has had a difficult time recovering. Most of the people make their living in subsistence agriculture. Medicine, food, and goods have been in a short supply. Many people have died due to the lack of these items during and after the civil war. Today the country struggles to recover and work toward stabilization.

Liberia

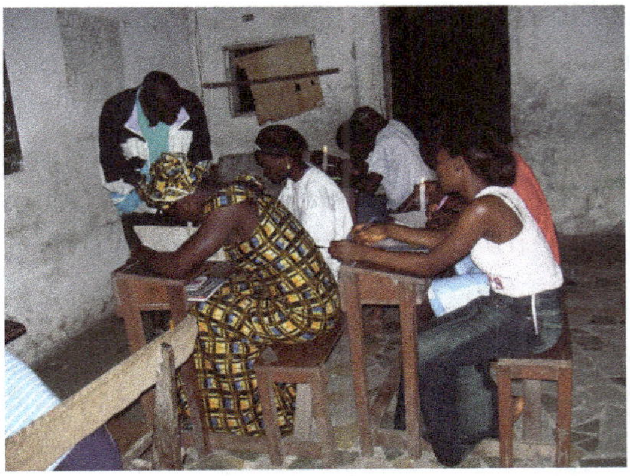

Figure 7.22: School Scene Students in Liberia study by candlelight to catch up on missed time during the civil war. United States Agency for International Development – public domain.

To fully understand where Liberia is today as a country and as a people, one has to understand its background and its geopolitical history. Liberia has a unique background in that it was not a European colony and was not included in the Berlin Conference of 1884 as a country up for grabs in the scramble for African territory. The conference of colonial European powers who divided up Africa did establish borders for the majority of African countries. Liberia, which means "Land of the Free," was a destination for freed slaves from the United States. In 1847, Liberia became an independent country patterned after the US government, even naming its capital Monrovia after President James Monroe, the fifth US president. Many of the freed slaves from the United States returned to Africa and relocated to Liberia. The slavery experience in the United States gave returning Africans a different culture and history from those indigenous to the African continent. Assimilation between the two groups was difficult. The division between the Americanized settler communities along the coast and the indigenous Africans of the interior continued to widen over the years.

Centrifugal forces within the country surfaced in 1980 when a military-backed coup overthrew the government. This event became a turning point that escalated into social division and political unrest. Elections were held in 1985, which resulted in accusations of election fraud. Successive coups by various factions continued to plague the political arena. By 1989, civil war had commenced. Military forces from the neighboring countries of Ghana, Guinea, Senegal, and Nigeria entered the conflict. Whoever was in charge of the government at any one time had plenty of opposition from the various factions. A warlord by the name of Charles Taylor became a major player and eventually took control of the government in 1997 following a bloody insurgency in the capital.

US marines were deployed to protect the US embassy and personnel. Nigerian military troops pushed into the conflict. By 2003, more than fifteen thousand UN peacekeeping soldiers were active in Liberia. The power of the government was reduced to a minimum, and the warlord president Taylor was allowed asylum in Nigeria and later brought before the UN court in The Hague for war crimes against humanity. In the end, more than two hundred thousand people were killed and the country was devastated. With UN troops bringing stability, the 2005 presidential election brought Ellen Johnson-Sirleaf, an economist with a Harvard degree. She became the first elected female head of state in Africa.

Poverty and the lack of goods and services have been a persistent problem for the people of Liberia. An agrarian society, few global corporations have made investments because of the long-standing civil war. As was true for Sierra Leone and other African countries such as Angola, the diamond trade helped fund the civil war. Diamonds from Sierra Leone had an estimated export value of more than $300 million annually in 2010. Diamonds from these countries are often called blood diamonds because of their use as currency to fund death and destruction. The concept of blood diamonds originated in the civil war in Angola years earlier. The UN banned the export of blood diamonds from Liberia during the war and finally lifted the ban in 2007. Looting and corruption by warlords and military forces pillaged the country's other resources, leaving most of the population with few opportunities and advantages for the future.

Ivory Coast (Côte d'Ivoire)

The former French colony of Ivory Coast is often referred to by its French name: Côte d'Ivoire. Before independence in 1960, Ivory Coast was one of France's most prosperous West African colonies. During the 1960s and 1970s, the economy grew, and its production of cocoa beans became the largest in the world. Its coffee production became the third highest in the world after Brazil and Colombia. Ivory Coast is the leading exporter of palm oil and pineapples in Africa. The potential for economic prosperity remains but has been diminished over the past two decades because of the economic downturn in commodity prices and political instability that resulted in a devastating civil war.

From 1960 to 1993, the country only had one political leader. Though this brought continuity to the government and the economy grew in the early years, authoritarian rule resulted in civil unrest and dissatisfaction with the conditions in the country. Coups in 1999 and 2001 triggered the beginning of political division. Rebel leaders from the northern part of the country challenged the government's legitimacy and pushed for land reform and a change in citizenship qualifications. Rebel groups took over the northern regions and vied for control over the prime cocoa-growing lands. Fighters under warlords and militias from Sierra Leone and Liberia encroached on the western sectors to gain a foothold. By 2003, French troops patrolled the western border region to provide security and stability. The country experimented with a unity government, but this broke down when the rebel groups failed to disarm. By 2007 the government and rebel leaders from the north worked out an agreement to reunite the country and dismantle the dividing zone between the north and south. French and UN troops remain to help implement the peace process.

Figure 7.23: A Woman in Abidjan, Ivory Coast Muslims, Christians, and animists are about evenly split in Ivory Coast. An estimated sixty-five languages are spoken here.

Family size has been quite large but is slowly decreasing with urbanization. Steve Evans –

The people of Ivory Coast are highly diverse. Religious affiliations are evenly split between Muslim, Christian, and animist. The economic success in agricultural production and recent oil exploration have attracted people from neighboring countries seeking opportunities and advantages. After the 2007 political agreements, approximately 20 percent of the population was made up of immigrant workers from neighboring countries, which are predominantly Muslim. The small percentage of non-Africans are from France, Lebanon, Vietnam, and Spain. There are also a high number of missionaries from North American Protestant churches. Attacks against non-native-born citizens have occurred; qualification for citizenship is a major concern, and natural-born citizens often support stronger restrictions on nonnatives becoming citizens or holding political offices.

Ghana

Named after the former Ghana Empire, the region of the Gold Coast was home to the Ashanti kingdom, which thrived on the slave trade. Many of the European colonial powers had sought to build fortified beachheads on the coast and profit from the lucrative trade in slaves, gold, and ivory. In the end it was the British that dominated the region and established the Gold Coast colony. The push for independence came in 1957. Ghana was the first European colony to gain independence in Subsaharan Africa.

Ghana remains a poor country, but it is endowed with natural resources. Even with half the population employed in agriculture, the country is experiencing positive economic growth. Gold and other mining operations contribute to the economy. Individual remittances and foreign aid remain a necessary component of the country's economic well being.

Kwame Nkrumah was the first prime minister and the first president. He advocated for a Pan-African concept that would pull the African countries together, oppose neocolonial activity, and increase trade and interaction within the African community. His ideas were welcome and are still celebrated, but his authoritarian rule brought about his downfall. In 1966, he was ousted in a military coup and exiled to Guinea. The political dynamics that followed were typical of the region: a pattern of corruption, coups, and authoritarian rule. It was not until 1981, when Jerry Rawlings came to power, that all political parties were banned and a new constitution was introduced. By 1992, the political parties were reinstated and the country began to experience political stability and democratic rule, which has endured to the present, making Ghana one of the most stable democracies in Subsaharan Africa.

Figure 7.24: Fishing Boats on the Coast of Ghana Ghana is a former British colony and was one of the first African colonies to gain independence (1957). Fishing is still a major economic activity along the coast. Stephen - kokrobite catch - CC BY-NC-ND 2.0.

Ghana is diverse in its people and in its physical geography. The physical terrain ranges from coastal plains and interior hills to the large **Volta basin** in the center. The **Volta River** was dammed up to produce hydroelectric power, resulting in the world's largest artificial lake, **Lake Volta**, which covers a large portion of eastern Ghana. The cultural diversity is exemplified by many ethnic groups, more than eighty languages, and at least double that in the number of dialects spoken within the country. English is the official language and is used in all public education. Funding for education has been available, and the people have access to quality educational services and an increasing agenda of social services. Ghana is showing promise in providing its people with a stable and thriving prospect for growth and opportunities.

Burkina Faso (Formerly Upper Volta)

Upstream on the Volta River is the former French colony of Upper Volta. In 1960, the former French colony attained full independence, and its name was changed in 1984 to Burkina Faso, meaning "the country of honorable people." Burkina Faso is a landlocked nation without a port city. The country lacks natural resources and has few industries. Subsistence agriculture is the main economic activity for about 90 percent of the population. Droughts and desertification continue because of overgrazing of the land and the natural southward shift of the Sahara Desert. Rainfall varies from forty inches in the south to only ten inches in the north. This is one of the poorest nations on Earth.

Political unrest and a series of military coups after independence have not provided stability. The current president has been in power since 1987 and continues to win in presidential elections. New limits may restrict his term in office past 2015. Many people seek employment in the neighboring countries of Ivory Coast and Ghana. This extremely poor nation occupies the second-to-last place on the UN's human development index. At less than 25 percent, Burkina Faso has the lowest literacy rate in the world. The poor economy offers few resources to increase educational services.

Togo and Benin

Both Togo and Benin were French colonies at the time of independence. The Berlin Conference in 1884 accepted Germany's control over the coastal region that became Togoland in 1905. After World War I, Germany lost this colony and the territory was administered between the British and the French. The French colony of Dahomey eventually became independent in 1960 and changed its name to the People's Republic of Benin in 1975.

Togo and Benin are both elongated countries with major port cities on the Gulf of Guinea along the Atlantic coast. Benin is about the same size in area as Liberia or the US state of Kentucky. Benin is twice the size in physical area as Togo. The south is a few degrees cooler than the north, where the scrub forests and grasslands of the savanna-type landscape can be more arid, as it is closer to the African Transition Zone. A nature reserve and a national park in Benin's northern portion attract tourists to see big game animals such as elephants, lions, and hippos in their natural environment. Both countries have a poor economy with most of the people working in agriculture. The political dynamics are similar to other West African states.

Nigeria

Africa's most populous country is Nigeria. The exact population has been difficult to determine, but 2010 estimates report the population to be just short of 150 million. This is a country of more than 250 different ethnic groups with twice that many separate languages or dialects. English is the official language, along with **Hausa**, **Yoruba**, and **Igbo (Ibo)**, all of which are spoken by ten million people or more. The distribution of the major ethnic groups is illustrated by the different regions of the country. Hausa groups are found mainly in the northern region, **Kanuri** groups in the northeast, Yoruba in the southwest, and Igbo (Ibo) in the southeast. Ethnic division has also caused serious confrontations and violence. In 1967, a separatist movement in the eastern Igbo region created the independent nation of Biafra. This sparked an all-out civil war that lasted more than two years and resulted in more than a million deaths. Political divisions did not end even after the two sides reconciled and the country was united under the same government in 1970. Political instability in Nigeria has resulted in an almost endless number of military coups and government leaders being removed from office.

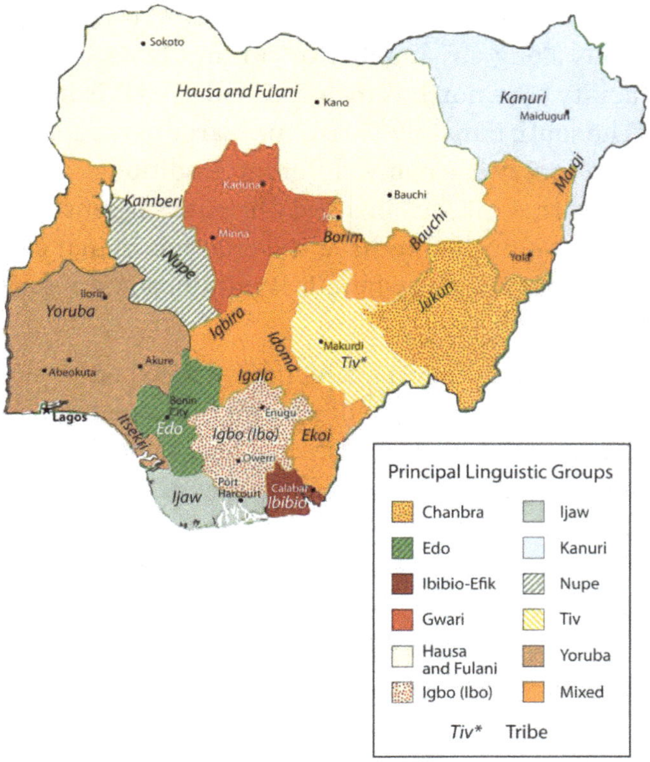

Figure 7.25: Linguistic Groups in Nigeria There are many linguistic groups in Nigeria but only four main ethnic groups: Yoruba, Hausa, Kanuri, and Igbo (Ibo), each in the four corners of the country. Map courtesy of University of Texas Libraries.

Figure 7.26: Making Music in Nigeria Yoruba drummers perform at a local event (*left*). A Hausa harpist (*right*). The Hausa are located in the northwest part of the country, which is mainly Muslim. The Yoruba are located in the south and are mainly Christian. Melvin "Buddy" Baker – Nigerian Drummers – CC BY 2.0.; Melvin "Buddy" Baker – The Harpist – CC BY 2.0.

The size and diversity of this country create a host of centrifugal forces that can bring about divisions along any number of cultural lines. Religious issues add to the political instability. The north is mainly Muslim, as it is located in the African Transition Zone. The south is mainly Christian. A large percentage of the population follows animist religions with many different traditional beliefs. Clashes have erupted in the streets that pit Muslims and Christians against each other. Several northern provinces have pushed to have the Sharia criminal code made into the area's civil law. There are regions where it is not uncommon for people to have a mix Christian and animist beliefs.

Family size has been large in Nigeria, which has caused an exploding population. Statistically, Nigeria has more people of African heritage than any other country in the world. The population density is equivalent to having half the population of the United States pushed into an area the size of Texas and Oklahoma. Most of the population makes their living on subsistence agriculture, but millions are employed in the growing urban service sector. Nigeria's main economic engine is the oil industry, which accounts for up to 80 percent of government revenues and is the number one export product. Nigeria is a member of the Organization of Petroleum Exporting Countries (OPEC) and is one of the world's top ten oil-exporting countries.

The lucrative oil business has attracted many seeking to share in the profits. The government's distribution of oil revenues has brought about even more infighting, corruption, and mismanagement within the political leadership. Rebels and militant groups along the **Niger Delta**, where most of the oil activity is located, have attacked the oil industry's infrastructure and taken workers hostage. They want a larger share of the revenues to remain in Nigeria and to go to their people rather than the government or CEOs of international corporations.

Poverty, a low standard of living, the lack of opportunities and advantages, a poor educational system, or the lack of social services does not diminish the human spirit. Confronting all these issues and more, the people of West Africa and Nigeria are vibrant, energetic, and hard working and value the institutions of family and religion. Just because they have not transitioned to a consumer society does not mean they cannot find fulfillment and happiness in their lives. A survey conducted by social scientists in 2001 and reported by BBC News in 2003, indicated that Nigeria had the highest percentage of happy people of any country in the world at that time. The status of each country may change from year to year, but the interesting part is that the survey confirmed that money or income level does not always equate to happiness. Countries such as Nigeria, with a low level of per-capita income, can still highly value their heritage and the traditions that revolve around their family and community and transcend the global push for economic gain and the possession of consumer items. Much can be said for the vibrant cultural attributes of the people of West Africa and its developing countries.

Key Takeaways

- Most of West Africa lies in the African Transition Zone, with portions north and south of the transitional region. Dry, desert conditions exist to the north and tropical conditions exist to the south.
- The main economic activity in the region is subsistence agriculture. Minerals, diamonds, or oil are also extracted in varying amounts in West Africa.
- West Africa has a large number of independent states that share similar economic qualities of poverty with rapidly expanding populations. Foreign aid and international assistance have been extensive.
- The transition from colony to independent nation has not been without serious political conflicts. Bloody civil wars, military coups, and political unrest have plagued the region.
- In spite of the political difficulties, poverty and the lack of resources, the people of West Africa hold vibrant cultural values that revolve around family and religion. The diversity of languages, religious beliefs, and rich cultural traditions provide an African heritage that is celebrated and valued by its people.

Discussion and Study Questions

1. What has been the general political pattern after independence in West Africa?
2. Which country has the highest number of ethnic groups? What are the four main ethnic groups there?
3. Why are Senegal and the Gambia two separate countries? What river flows through this area?
4. How do most of the people in West Africa make a living?
5. What are blood diamonds? What two countries in West Africa have had civil wars based on them?
6. How did Liberia become a country? Why was Liberia never colonized by Europeans?
7. With more than eighty indigenous languages, why is English the official language of Ghana?
8. What was the Pan-African concept, and how did it affect West Africa?
9. How have natural resources contributed to the wealth of the region?
10. How will these countries address their high population growth in the future?

Geography Exercise
Identify the following key places on a map:

- Cameroon Highlands
- Gambia River
- Lake Volta
- Niger Delta
- Timbuktu
- Volta River

7.4. CENTRAL AFRICA

Central Africa is a region marked by its rich biodiversity, cultural diversity, and complex historical narratives. It encompasses countries like the Democratic Republic of the Congo, Cameroon, and Gabon, which are home to vast rainforests and vital river systems, such as the Congo River. These natural resources play a crucial role in the livelihoods of local communities, yet they also attract exploitation, leading to environmental degradation and conflict over land use. The region is characterized by a mosaic of ethnic groups, each with unique traditions and languages, contributing to a vibrant cultural landscape. However, Central Africa also grapples with challenges like political instability, economic disparities, and health crises, including the impact of diseases such as Ebola and malaria. Efforts toward regional cooperation and sustainable development are essential for addressing these issues and harnessing the region's potential for growth and prosperity.

Central African Republics

Central Africa covers a large physical area that can range from desert conditions to the north in Chad to tropical rain forests and mountains in the equatorial region of The Congo. The entire region is roughly the same size as the United States west of the Mississippi River. The countries included in the region vary with different organizations or geographic perspectives; in this textbook, the countries include Cameroon, Congo, Chad, the Central African Republic, Gabon, Equatorial Guinea, the Congo (Zaire), Rwanda, and Burundi. Located off the west coast of Central Africa is the small twin island country of São Tomé and Príncipe. Burundi and Rwanda are often included in the region of East Africa, but their connection to the Congo makes them more relevant to the Central African region for the purposes of this textbook.

The equator runs through the middle of Central Africa. Type A climates are dominant in the region, complete with tropical rain forests and jungle environments. In the north is the African Transition Zone, which runs through Chad. In Chad, the arid region of the Sahara Desert transitions into the humid tropics. Central Africa usually only includes portions of Subsaharan Africa south of the African Transition Zone. Southern Chad exhibits qualities similar to Central Africa, and the northern areas exhibit qualities similar to North Africa or the northern regions of Niger, Mali, or Mauritania.

The physical geography varies with each country in Central Africa. The most prominent physical landscape is the tropical rain forests of the equatorial region. Highlands can be found in both the western and eastern regions of Central Africa. The **Cameroon Highlands** are a product of a geologic rift in tectonic plates that created São Tomé and Príncipe, the island portion of Equatorial Guinea, and the mountainous portions of the mainland on the border between Nigeria and Cameroon. There are a number of volcanoes in Cameroon. The only currently active volcano and the highest in elevation is Mt. Cameroon, at more than thirteen thousand feet. Mt. Cameroon emitted a cloud of carbon dioxide in 1986, killing more than 1,700 people. The volcano last erupted in 2000. In the crater of one of the volcanic peaks is Lake Nyos (Mount Cameroon).

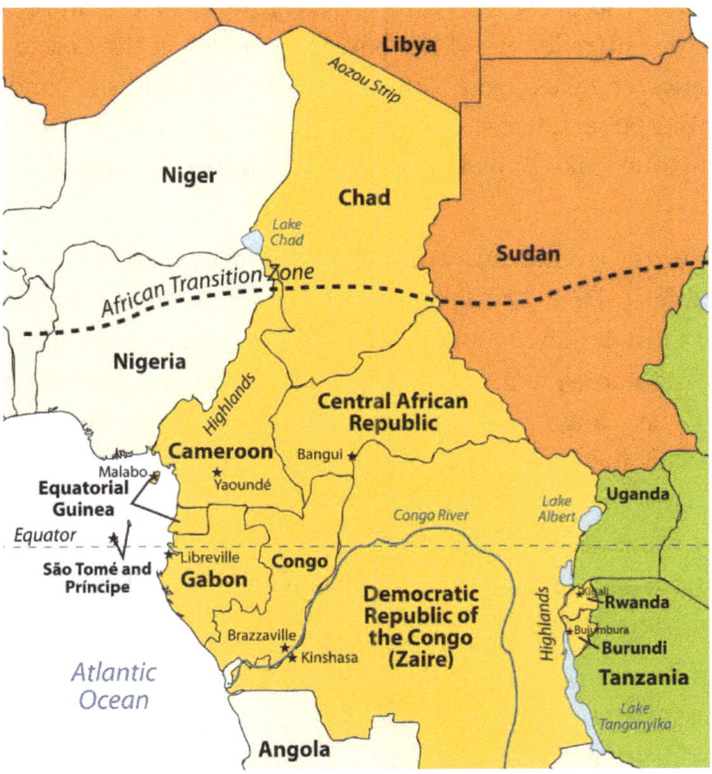

Figure 7.27: Central Africa Chad is often included in other regions, but in this text it is included with Central Africa. The Congo River is the main river of this region. Updated from map courtesy of University of Texas Libraries.

Lake Chad in the north is a large, shallow body of water that lies on the border of Chad, Niger, Nigeria, and Cameroon. The size of Lake Chad varies widely because it expands and contracts with the seasonal variations in precipitation. Lake Chad provides water for local livestock and fishing for millions of people. Its location on the border of four countries has caused political infighting over local water rights, which are valuable commodities in such an arid climate. The average depth of Lake Chad ranges from five to thirty five feet. Climatic conditions and diversion of the lake for human purposes have caused the water to recede. If these conditions continue, the lake might virtually disappear by the end of the twenty-first century, which would have disastrous effects on both the human population around the lake and the biodiversity. Waterfowl, crocodiles, fish, and a host of other creatures depend on Lake Chad for their survival, and its loss would create an environmental catastrophe.

On the eastern border of the Congo is a portion of the **Great Rift Valley**, which extends from Southern Africa in Mozambique to Lebanon in the Middle East. **Lake Tanganyika** and **Lake Albert** are two of the larger lakes in Central Africa along the western section of the Great Rift Valley. Lake Tanganyika is more than 418 miles long and runs the entire length of the boundary between the Congo and Tanzania. These are deepwater lakes. Lake Tanganyika is the world's second-deepest lake, with a depth of 4,800 feet. Because of its depth, it is also the world's second-largest lake by volume after Lake Baikal in Russia, which has the record for both volume and depth.

Fish and fresh water from Lake Tanganyika support millions of people who make their living directly from the lake's resources or who live in the surrounding area. The Great Rift Valley is bordered by high mountains such as the **Rwenzori Range** and its highlands, which include active volcanoes. A number of volcanoes are located in the Congo, not far from the border with Rwanda. In 2002, Mt. Nyiragongo, a volcano with an elevation of 11,358 feet in the eastern highlands of the Congo, experienced a series of eruptions that killed a few dozen people, destroyed thousands of buildings, and caused the evacuation of as many as four hundred thousand people. Various volcanoes in this range have active lava lakes in their craters at their peaks.

At the heart of Central Africa are the massive Congo River and all its tributaries. It is the deepest river system in the world and has some stretches that run more than seven hundred feet deep, providing habitats for a wide range of organisms and fish species. The **Congo River basin** is second only to the size of the Amazon basin in South America. Home to Africa's largest tropical rain forest, this region is host to a massive variety of plant and animal species, which create an extensive environmental resource base. Human activity has been encroaching on this valuable environmental region filled with extensive biodiversity. Logging, slash-and-burn agriculture, and civil war have devastated large areas of the Congo basin, resulting in the loss of habitat for many tropical species.

Figure 7.28: Living on the Edge This mountainside home is located in the Rwenzori Mountain range in the eastern region of the Congo. Elevations can reach as high as 16,761 feet. Some peaks are snow-capped with permanent glaciers. Dylan Walters – DSC_0615 – CC BY 2.0.

Deforestation activities have been on the rise in Central Africa and the Congo basin. Endangered primates such as chimpanzees and gorillas, along with monkeys and game animals, have struggled to adjust to the shrinkage of their habitat and the decline of their population. The endangered mountain gorillas in the northern region of the highlands have gained attention through the efforts of professional scientists such as **Dian Fossey** who have attempted to understand and preserve their ecosystems. Mountain gorillas have been dwindling in numbers because of poaching, civil war, and hunting. It will take concerted preservation activities for gorillas to survive in their native habitat.

Conflicts and Unrest in Central Africa

The two landlocked countries of **Chad** and the **Central African Republic** have endured unstable conditions in their transitions to independent, stable democratic states. Chad had been in dispute with Libya over the **Aozou Strip** bordering their two countries, an area deemed rich in minerals and uranium, but in 1994, Chad was awarded sovereignty over the Aozou Strip by the United Nations (UN) International Court of Justice. Chad is a temporary home of more than 250,000 refugees from the ethnic cleansing campaign in the Darfur region of the Sudan. Thousands more from the Central African Republic have sought refuge in Chad. Meanwhile, Chad's government has been plagued with corruption and mismanagement, a state of affairs that has hampered its efforts to offer humanitarian aid to its refugees. Similarly, the Central African Republic is troubled by a history of unstable and short-lived democratic governments. Military coups and transitional governments are frequent. Civil unrest erupts into chaos. Rebel groups control large parts of the countryside. It is difficult for the people to access reliable public services such as health care, education, and transportation systems when the government is not functioning adequately.

Figure 7.29: A Tailor in Chad Chad is a poor country with existing political conflicts and many refugees from Sudan. The man in this photo is using an old manual sewing machine to make clothes. Mark Knobil – Tayor shop – CC BY 2.0.

Cameroon and **Gabon** have more stable governments than Chad or the Central African Republic. Still, they are not without political issues. Cameroon was a German colony as a result of the Berlin Conference of 1884 where the colonial European countries divided up Africa. It remained so until after World War I, when it was divided between the British and the French. Finally, in 1961, the two sides were merged into one country under one government. The two hundred or so different ethnic groups place pressure on the government to address social concerns. Even though the government has become stable more recently, social pressures between traditional groups and groups with a European colonial background have erupted into social unrest. The European Anglo factions have gone so far as to threaten to separate the once-British portion and secede from Cameroon.

Formerly a French colony, **Gabon** has transitioned to independence successfully. The country's small population of about 1.5 million people, along with adequate natural resources, has facilitated Gabon's development into a country with a relatively stable democratic government and a higher standard of living. Gabon is geared toward attracting more foreign investments and continues to progress forward in the index of economic development. As of 2010, Gabon was edging toward stage 3, the highest level in Central Africa for a country as a whole.

Equatorial Guinea and the island nation of **São Tomé and Príncipe** are small countries on the west coast of Central Africa. Serious violence erupted in the former Spanish colony of Equatorial Guinea after it gained independence in 1968. The first elected president governed through authoritarian rule, and after a few years in office, he unleashed a reign of terror that resulted in the death of more than one-third of the population and totally neglected the country's public service sector and infrastructure. He was overthrown and executed by his successor, who then implemented authoritarian measures to make sure he would remain in power and continue to control all the revenues from the extensive oil activity in the offshore waters around his country. The billions in oil revenues have remained in the hands of the president and his family-controlled cabinet. Equatorial Guinea is Subsaharan Africa's third-largest oil exporter; however, most of the country's citizens have benefited little from the vast wealth of the lucrative oil industry.

Figure 7.30: Cultural Artifacts A mask from Gabon, in the Ogowe River region, made by the Tsogo people. The mask, made in the late nineteenth to early twentieth century, is made out of wood with natural pigments. Masks are a common part of cultural rituals or festival events. Cliff – Mask, Tsogo peopls, Ogowe River region – CC BY 2.0.

The former Portuguese colony of São Tomé and Príncipe consists of two small islands off the coast of Gabon. The islands received their independence in 1975, but that's when their problems in establishing a stable government began. Political turmoil delayed the arrival of democratic reforms for a decade. Elections for leadership were not held until 1991, and after the elections, leadership changed repeatedly because of political infighting. In 1995 and 2003, two coups were attempted without success. The democratic process has been complicated by offshore oil discovery, which has brought an influx of outside workers and media attention to the country. Multinational oil companies have begun to invest heavily in the development of oil production in the region.

Rwanda has been severely affected by the difficulties typically associated with the transition between colony and independent nation. Ethnic divisions manipulated by the colonial masters erupted to challenge the country's stability and future. The division between the Tutsis and the Hutus has deep historic roots. In 1994, the centuries-old conflict between the two ethnic groups erupted into violence of unprecedented proportions and resulted in the senseless killing of hundreds of thousands of innocent people. The Hutus amassed large militias and took revenge on the Tutsis for years of oppression. Hutu militias rounded up and killed all Tutsis, moderates, and anyone not supporting the Hutu cause. The killing of hundreds of thousands of people continued from city to city throughout the countryside. Within a few months, the genocide is estimated to have caused the death of as many as a million people.

The atrocities of the Rwandan genocide extended across the population. Men, women, and children were encouraged to kill their neighbors by hacking them to death with machetes. If they refused to comply, they themselves were threatened with death. Victims were herded into schools and churches, where they were massacred and the buildings would be burned to the ground. As people fled the region, the number of Tutsi refugees entering neighboring countries ballooned to more than a million.

Figure 7.31: Refugee Camp An estimated 1.2 million Rwandan refugees fled to the Congo (Zaire) after a civil war erupted in Rwanda. Many are still in the Kibumba refugee camp, shown here covered a smoky haze. CDC – public domain.

Tutsi rebels finally gained strength, defeated the Hutu militias, and ended the slaughter. Fearing retribution for the more than one million Tutsis who had been massacred, more than a million defeated Hutus fled as refugees across the borders into Uganda, Burundi, the Congo, Uganda, and Tanzania. Refugee camps housing thousands of people each were hastily constructed without proper sanitation or supplies of drinking water. Diseases such as cholera and dysentery swept through the camps and killed thousands of refugees. The country of Rwanda was left divided and devastated.

The aftermath of the genocide and conflict continues to incite civil unrest and political division in Rwanda. The search for common ground between the different ethnic groups is divisive and conflict ridden. Now Tutsi leaders dominate the Rwandan government.

The entire Central African region was devastated by the massive shifts of refugees and the brutal killing of so many people. The competition for the control of resources and the increase in military arms along the Zairian border became a major component of the civil war that plagued the Congo (Zaire) during the same period. Up to five million people died in the Congo because of warfare, disease, and starvation. The wars in the Congo involved military actions by the Tutsis from the Rwandan and Ugandan governments and Hutu militias.

The small country of **Burundi** also has been caught up in the conflicts in Rwanda and the Congo. Germany claimed the region of Burundi as a colony in the European scramble for Africa's resources. After World War I, they handed Burundi over to Belgium. Burundi and Rwanda were both part of Belgium's African colonial empire and were together called Ruanda-Urundi. The two countries were later separated. Burundi achieved independence in 1962. Unfortunately, the ethnic conflict between the Hutus and the Tutsis that tore apart Rwanda in the genocidal war in 1994 had also plagued Burundi. In 1965, the military was controlled by Tutsi leadership. When the Hutus revolted, they were repressed. The entire government came under Tutsi control. Hutu attacks in 1972 resulted in a systematic retaliation by the Tutsis that killed approximately 200,000 Hutus and forced another 150,000 to flee the country. Civil unrest between these two groups continues to plague the political and social structures of Burundi.

The names of the **Democratic Republic of the Congo** (a former Belgian colony) and the **Republic of the Congo** (a former French colony) are confusing. At one point after they had become independent, both countries chose the name *Republic of Congo*. To keep them straight, they were commonly referred to by their respective capital cities—that is, **Congo-Leopoldville** (the larger eastern country, also known as Zaire) and **Congo-Brazzaville** (the smaller western country). The larger former Belgian colony has since become simply **The Congo** or is referred to unofficially as the **Belgian Congo**, and the smaller former French colony has become **Congo.** In 1966, **Joseph Mobutu**, the political leader of the Congo, officially changed the nation's name to **Democratic Republic of the Congo.** In 1971, the name was changed to the **Republic of Zaire.** In 1997, after a bitter civil war and the overthrow of Mobutu, the new president, **Laurent Kabila**, changed Zaire back to the Democratic Republic of the Congo, often referred to as the Congo. To keep the two names separate, many refer to the Congo as **Congo-Zaire.**

The confusing name situation exemplifies the difficulties and changes in government leadership that have transpired since colonial times in the two countries. The former French colony to the west side of the **Congo River** has survived with fewer conflicts than its eastern neighbor but has not escaped civil war. From 1997 to 1999, Congo had a harsh civil war, and the end result was the overthrow of a democratically elected president and the installation of a former president. Tens of thousands are reported to have been killed. Related conflicts erupted in various regions that extended into 2003 before they were finally resolved.

The former Belgian colony of The Congo is nearly as large as the United States east of the Mississippi River and is a challenge to govern. The population—distributed between about 250 different ethnic groups and about as many languages—was estimated at about seventy million in 2010. Authoritarian rule from political leaders such as Mobutu from 1965 to 1997 polarized the country's many factions and played world leaders against each other. The Congo acted as a *swing state* in the Cold War between the United States and the Soviet Union. As mentioned, Mobutu changed the name of the country and pillaged the country's finances for personal gain. He stashed away for himself billions of dollars in public funds in foreign bank accounts that should have been spent on the people of the country.

The two wars in The Congo has resulted in the highest number of deaths since World War II. The **First Congo War** (1996 to 1997) occurred when President Mobutu was overthrown by militant forces led by rebel leader Kabila, who was a long-standing political opponent of Mobutu and was backed by Ugandan and Rwandan militant groups. Mobutu was eventually forced from office and fled the country. Kabila declared himself president and changed the official title of the country from Zaire back to Democratic Republic of the Congo. This transition in political power caused a shift in the militant rebel groups, which created the conditions for the **Second Congo War** (1998 to 2003), which was even more brutal than the first. The assassination of President Kabila in 2001 permitted his adopted son **Joseph Kabila** to take power and run the country up to the present.

Figure 7.32: Soldiers in The Congo In this photo, General Kisempia, former chief of staff for Joseph Kabila, inspects his troops. About five million people died in warfare in the Congo 1996–2003. Wikimedia Commons – public domain.

The **Second Congo War** raged through the Congo, bringing destruction to the country and the deaths of millions of people. Besides political control of the country, a main objective of the combatants was to dominate the resource-rich eastern sector of the country, where valuable reserves of zinc, diamonds, copper, and gold can be found in the eastern region. Before the war ended, the surrounding countries of Angola, Zimbabwe, Namibia, Uganda, and Rwanda had troops fighting on one side or the other. The civil war formally ended in 2003 when rebel groups and the government worked out a shared political arrangement. Many rebel groups remained engaged in the eastern region of the Congo long after the agreement was reached. The total number of deaths resulting from the civil wars in the Congo was estimated in 2008 at about 5.4 million—the majority by the war and the remaining by disease or starvation (AmericanRenaissance.com).

Armed military conflicts in the Congo did not end with the Second Congo War. Conflicts continue in the eastern region. These armed confrontations are often referred to as the **Kivu conflict** because they are taking place in the provinces of North and South Kivu in the Congo's eastern region along the borders with Rwanda and Burundi. Rebels are fighting against forces from the Congo and Rwanda. As recently as 2009, fighting continued deep in the interior of Central Africa near the Rwandan border between various militias. Such conflicts are not widely reported by news outlets in core areas such as the United States.

People in the Congo are still dying because of the devastation of the war. Those not killed in actual warfare are dying of hunger and disease where food, medicine, and health care are not available. The death rate in 2008 was estimated to be as high as forty-five thousand people per month in the eastern and central regions—many of them children. Survivors of the bloody civil war report horrendous terrorist campaigns conducted by various militias that used mass rape, mutilation, and torture as a means of control and social pressure. The UN had more than eighteen thousand troops in the Congo as of 2007 to help curb the civil unrest and militant activities.

Understanding that most wars are fought over the control of resources is a major step in comprehending the conflicts in Central Africa. The conflicts in the Congo are likely to continue because of vast caches of mineral wealth in the country that have yet to be extracted. Economic pressure to control the extraction activities of saleable raw materials is often the driving force behind rebel groups in places such as the Congo. Local factions usually sell the raw materials for prices well below market value.

The sale of precious minerals such as cobalt, coltan, gold, and diamonds has helped fund the wars in the Congo. The country produces more cobalt ore than any other country in the world. Cobalt is a valued metal used in aircraft engines, medical implants, and high-performance batteries. The major mining operations for cobalt ore are located in the southeastern Katanga state, where there are also large reserves of copper.

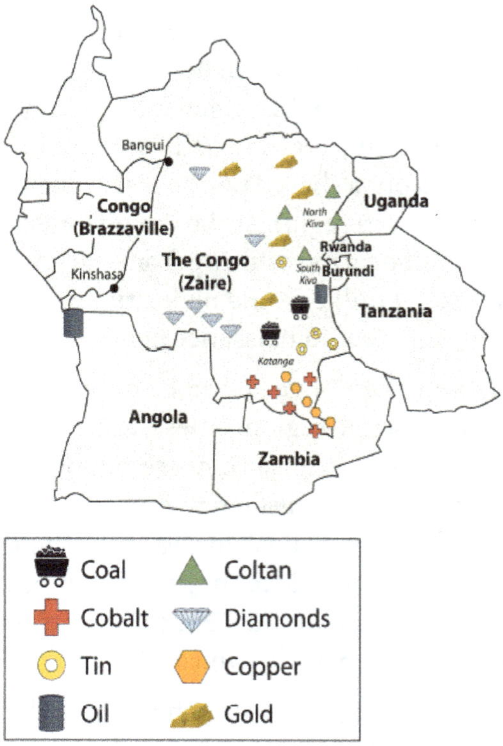

Figure 7.33: Minerals in The Congo This map shows the locations of major minerals in the Congo. Other minerals can also be found in the region. Updated from map courtesy of Andreas 06.

Coltan is a mineral that tantalum comes from, which is highly prized for its use in capacitors for electronic circuits. Tantalum is found in most modern electronic devices in high demand worldwide, from video game systems to cell phones. Most of the coltan in the world comes from mines in the Congo in the eastern Kiva regions. The eastern region is also home to a high percentage of the world's industrial diamond reserves. Other minerals and ores are also found in abundance in the region. The sale of precious gems and rare minerals can bring huge profits, but the wealth seldom reaches the hands of those that labor in the mines in the extraction process.

Learning about the geopolitics of Central Africa is critical to developing an understanding of how colonial activity gave shape to Central African countries and why multinational corporations are now highly involved in creating the demand and markets for the resources found there. The core economic regions of the world require the raw materials and resources that are extracted from peripheral places such as the Congo to fuel their economic activities and to bring profits to their shareholders. The core economic players in the global markets are also some of the largest arms manufacturers that sell weapons to the local factions involved in the battle for control of the valuable resources. Globalization connects the core to the periphery. In resource-rich places such as the Congo, this relationship is only going become more interactive.

The Other Side of Life in Central Africa

Figure 7.34: Football in Burundi A Sunday-afternoon football match takes place in Bujumbura, Burundi. The government has instituted a law that forbids any Burundians from participating in any activity that does not contribute to the communal good before 10 a.m. on a Saturday morning. Street cleaning, family events, and government tasks are examples of acceptable activities. Notice the livestock grazing on the edge of the field. Wikimedia Commons – CC BY 2.0.

Sports and entertainment seem to be a universal human activity. Football (American soccer) is the dominant sport in the Central African states. Football can be found in even the poorest areas. The low-cost sport is accessible to most citizens of each country. Other traditional sports include wrestling, archery, and track-and-field events. In Chad, for example, a type of freestyle wrestling in which the wrestlers wear animal hides and cover themselves with dust to prepare for the match is widely practiced. Canoe racing can be seen in Cameroon. Board games such as mancala, a count-and-capture African stone game, are also common. Western-style sports such as basketball, baseball, and cricket are played in many of the urban areas. Olympic sports are practiced in countries as skill, ability, and funding provides.

A great variety of cultural traditions spring from the thousands of ethnic groups, distinct languages, and different religious beliefs found in Africa. Cuisine and drink vary as widely as the people's traditions and are important elements of any celebration or festivity. Every country has its own unique way of promoting and celebrating its cultural diversity. Even poor countries such as Chad possess a rich cultural heritage. Chad has opened its own Chad National Museum and the Chad Cultural Center. Rwanda, with its civil wars and ethnic divisions, still honors its holidays with ceremonies and celebrations. The people of Central Africa find time and energy to celebrate life.

Figure 7.35: African Beer Tusker is a popular African beer sold throughout the world. Photo by R. Berglee – CC BY-NC-SA.

Food and drink differ with location, but most African countries brew some type of beer or alcoholic beverage. Ginger and honey beers are commonly brewed. Local resources, customs, and tastes determine the types of beer brewed in a particular region. In Cameroon, palm wine or millet beers are the traditional mealtime drinks. In Burundi, when close friends gather, they may drink beer from a large container, each with his or her own straw, as a symbol of unity. Popular in the Subsaharan side of Chad is a red millet beer called *billi-billi.* Beer brewed from white millet in Chad is called *coshate.* The Muslim populations in most countries do not traditionally consume alcohol, but alcoholic beverages are more common south of the African Transition Zone, where there is a mainly Christian population.

Many rural and urban women in the Central African Republic transform food crops into beverages for sale in the informal economy. Hard liquor and beer are common products that can be brewed or distilled from grains, sorghum, or other local crops. For religious reasons, some people do not drink alcoholic beverages. Juices and soft drinks are available as well as coffee and tea, which are the more traditional beverages of the region. Coffee is grown in the tropics, a climate prevalent in many of the countries in Subsaharan Africa. Coffee or tea is a major export product of almost all the Central African countries. In many areas, tea was introduced through colonial diffusion and has become not only a preferred beverage but an export product. Coffee has been the number one export product of Rwanda and Burundi for many years. Neighboring Uganda is the third-largest coffee producer in Africa after Ethiopia and Ivory Coast; all three are among the top dozen coffee producers in the world.

Figure 7.36: African Coffee Growing coffee is a major economic activity in Africa. Rex Roof – African And Costa Rican Coffee – CC BY 2.0.

The cuisines of Central Africa can be as diverse as each local population, but they have retained much of their local tastes and flavors based on available ingredients. The slave trade introduced items that are now quite common, such as chili peppers, peanuts, and cassavas. Nevertheless, food preparation is based mainly on traditional methods. Plantains and cassavas are mainstays often served with spicy sauces and grilled meat. Tomatoes, onions, and spinach are common ingredients. Root crops of yams, sweet potatoes, and maniocs are widely served. Fruits and vegetables are popular. Specialty stews are choice entrees that may contain ginger, okra, chicken, and various spices. Bambara is a favorite sugary porridge with peanuts and rice that is common in Central Africa. Traditional meats of chicken, goat, or beef are universal, but wild game such as antelope, monkey, crocodile, or whatever is available is widespread. Fish is a main protein source served with other spicy dishes or dried to be eaten later.

The Role of Women

Women's issues remain a concern in Central Africa regarding social equality, economic development, and population growth. The treatment of women reflects the attitudes and traditions of a society. Attitudes and traditions vary with location and with the cultural heritage of an ethnic group indigenous to a given area. Most of the Central African countries in are male dominated, a trend that can be witnessed in local activities or sports, in the construction of the government, or in the sex ratio of the business sector. With an average family size of five or more, few opportunities exist for women to break out of their traditional roles within their culture and receive a secondary education or enter a professional field of employment.

The Central African region has different types of marriages that vary from country to country. Some traditional societies still arrange their children's marriages, but this practice is becoming less common as tradition confronts more contemporary views. Marriages between two people with different ethnic backgrounds are becoming more widespread in the urban areas. Cities are places where people with diverse backgrounds come together to live and work, which can lead to an increase in mixed marriages. Cosmopolitan areas are also more likely to have connections with outside cultures through media and communication technology and might be more tolerant of diverse traditions.

One of the marriage traditions that exists in regions of Central Africa is polygamy, which is the practice of a man or a woman having more than one spouse. In Africa, as in most places in the world, it is far more likely for a man to have more than one spouse than for a woman to have more than one spouse. Polygamy is practiced in areas where traditional beliefs allow it. Approximately 30 to 40 percent of women in the Central African Republic are in polygamous unions. The law sanctions polygamy as long as spouses are aware of the arrangement and agree to it before they marry. In Cameroon, both monogamous and polygamous marriages are practiced. The average family is large and extended. Monogamous relationships are the standard in Christian areas such as Burundi and in more contemporary urban societies.

Women caring for large families do not always have the opportunity to earn an education or obtain the training required for employment opportunities or professional positions. Circumstances shaped by culture, tradition, or economic necessity are not necessarily supportive of women advancing in the private or public sectors. Even if the legal code of the country does not discriminate against women, traditional practices and the social status of men often keep women from making socioeconomic strides. Many societies are male dominated, and violence against women is still common and even accepted in Central Africa.

Figure 7.37 Congo Woman Weaving Di bo di – woman_weaves_basket – CC BY 2.0.

One of the worst cases of violence and discrimination against women exists in the eastern regions of the Congo, where abuse against women, including sexual violence and rape, is listed by UN humanitarian officials as the worst in the world. The civil wars in the Congo have created a difficult climate for women, often with tragic results. Violence against women has become an accepted practice in some areas. During the Congo wars, women were often raped and taken as slaves to serve the soldiers. If and when the women were released from captivity, it was not uncommon for them to commit suicide. Improvements in health care, education, and child care are needed to improve living circumstances for women and may in turn positively affect population growth in the region.

Key Takeaways

- Central Africa covers a large physical area about the size of the United States west of the Mississippi River. Most of the region is in the tropics and has a type A climate.
- Central Africa is a relatively poor region economically. National wealth has been garnered traditionally from the export of food crops such as coffee or tea but more recently from mineral and oil extraction.
- The civil war in Rwanda in 1994 was the result of conflict between the Tutsi and Hutu ethnic groups. The historic division was manipulated by colonial powers and erupted into brutal violence that led to the death of as many as a million innocent people; nearly as many refugees took shelter in neighboring countries.
- The civil wars in the Congo are based on political control as well as control over valuable resources in the eastern states. The two main Congo wars resulted in an estimated 5.4 million deaths from either direct military conflict or the devastation that resulted through starvation and disease.
- Central Africa resonates with cultural and ethnic diversity. Hundreds of ethnic groups live and work together with a high degree of interaction. Traditional cuisine and beverages such as coffee, tea, or beer can be found throughout the region.
- Women's issues range from cosmopolitan in the urban areas, where mixed marriages are common, to harsh conditions in which violence and abuse is commonplace. Large family sizes and male-dominated traditions make it difficult for women to transition to positions of equality.

Discussion and Study Questions

1. What are the main physical geographic features in Central Africa?
2. What is unique about Lake Chad and Lake Tanganyika? How are they different?
3. What caused the Rwandan civil war? Which two groups were in conflict?
4. What caused the First and Second Congo Wars? Why does the Kivu conflict exist?
5. What is the relationship between the poor, remote peripheral regions of the Congo and the core economic areas of the world with high standards of living and urban lifestyles?
6. What are some of the food and beverage choices for people in Central Africa?
7. Are there any differences between marriage traditions in the Central African Republic and where you live?

8. What are some of the main women's issues in Central Africa?
9. What is the relationship between family size and women's issues?
10. Which country in Central Africa has the highest standard of living and a stable government?

Geography Exercise
Identify the following key places on a map:

- Aozou Strip
- Cameroon Highlands
- Congo River
- Great Rift Valley
- Lake Albert
- Lake Chad
- Lake Tanganyika
- Rwenzori Range

Activities

1. Find out what is occurring at the present time in the countries of Rwanda and the Congo in regard to armed conflicts and civil war.
2. Search to see if any news is available where you live about current affairs in Central Africa.

References

AmericanRenaissance.com "Crisis Caused 5.4 Million Deaths in Congo, Report Says," http://www.amren.com/mtnews/archives/2008/01/crisis_caused_5.php; "DR Congo War Deaths 'Exaggerated,'" BBC, http://news.bbc.co.uk/2/hi/africa/8471147.stm.

"Mount Cameroon," http://www.cameroonconsul.com/mount.html; "Mt Cameroon Volcano—John Seach," Volcano Live, http://www.volcanolive.com/mtcameroon.html.

7.5. EAST AFRICA

East Africa is a vibrant region known for its stunning landscapes, rich cultural heritage, and diverse ecosystems. Home to countries like Kenya, Tanzania, Uganda, and Ethiopia, it features iconic landmarks such as Mount Kilimanjaro and the Serengeti National Park, drawing tourists and researchers alike. The region boasts a tapestry of ethnic groups, languages, and traditions, reflecting a history shaped by trade, colonialism, and migration. Economically, East Africa is emerging as a hub for innovation and entrepreneurship, particularly in technology and agriculture, though it still faces challenges like poverty, political instability, and climate change. The interplay of urbanization and rural livelihoods creates a dynamic environment where traditional practices coexist with modern development, making East Africa a fascinating study of resilience and adaptation in the face of ongoing changes.

Physical Geography

East Africa is a region that begins in Tanzania in the south and extends north through the great grasslands and scrub forest of the savannas of Kenya and Uganda and then across the highlands of Ethiopia, including the Great Rift Valley. The region also comprises the countries of Somalia, Djibouti, and Eritrea, which are located in the African Transition Zone between North Africa and Subsahran Africa. Rwanda and Burundi are physically in East Africa but are covered in the lesson about Central Africa because of their border activities with the Congo. The world's second-largest lake by surface area is **Lake Victoria**, which borders Uganda, Tanzania, and Kenya. (Lake Superior, on the border between the United States and Canada, is considered the lake with the largest surface area.) Lake Victoria provides fish and fresh water for millions of people in the surrounding region. The **White Nile** starts at Lake Victoria and flows north to the city of Khartoum in Sudan, where it converges with the **Blue Nile** to become the **Nile River**. The source of the Blue Nile is **Lake Tana** in the highlands of Ethiopia.

Figure 7.5.387.5.38: East Africa. Updated from map courtesy of University of Texas Libraries.

The highest mountain in Africa, **Mt. Kilimanjaro** (19,340 feet), is located in Tanzania near the border with Kenya. The second highest peak, **Mt. Kenya** (17,058 feet), located just north of the country's capital of Nairobi, near the equator, is the source of Kenya's name. Both mountains are inactive volcanoes and have permanent snow at their peaks. They provide fresh water, which flows down their mountainsides, to the surrounding areas. Mountain ranges in the Western Highlands of the Congo have a greater effect on climate than these two massive peaks. For example, the Rwenzori Mountains on the Congo–Uganda border have permanent snow and glaciers and reach elevations of more than sixteen thousand feet. These ranges create a rain shadow effect that cuts off moisture for the region from the westerly equatorial winds.

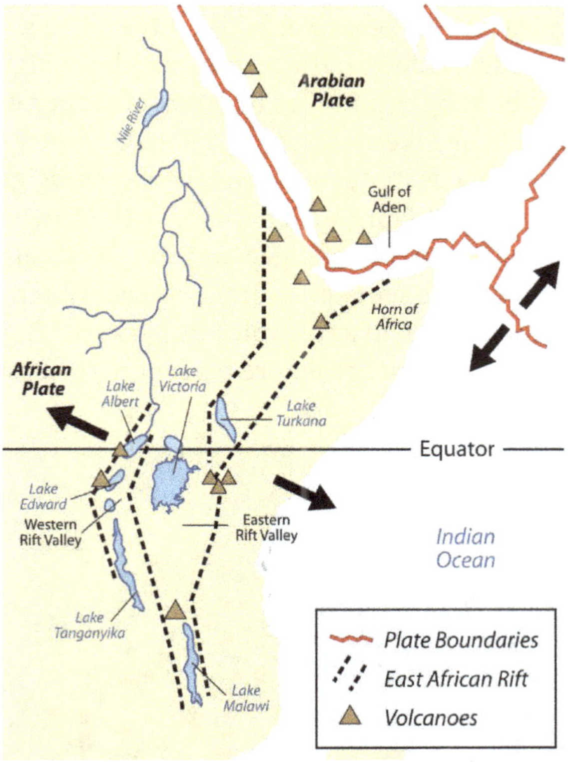

Figure 7.39: Great Rift Valley of East Africa The African portion of the rift extends from Mozambique to the Red Sea. The valley is created by the movement of tectonic plates. USGS – public domain.

This same effect is created by the highlands of Ethiopia, which reach as high as fifteen thousand feet in elevation and restrict precipitation in areas to the east. The lower level of rainfall transforms much of the region from a tropical rain forest into a savanna-type landscape with few forests, more open grasslands, and sporadic trees. Dry desert-like conditions can be found in a number of places along the **Great Rift Valley**.

The Great Rift Valley

The Great Rift Valley provides evidence of a split in the African Plate, dividing it into two smaller tectonic plates: the Somalian Plate and the Nubian Plate. The Great Rift Valley in East Africa is divided into the **Western Rift** and the **Eastern Rift**. The Western Rift runs along the border with the Congo. A series of deepwater lakes run along its valley. On the western edge of the Western Rift are the highlands, which have a series of high-elevation mountain ranges, including the **Rwenzori Mountains**, the highest in the series. The **Virunga Mountains** on the Congo–Uganda border are home to endangered mountain gorillas. The Western Rift includes a series of deepwater lakes, such as Lake Tanganyika, Lake Edward, and Lake Albert. Lake Victoria is located between the Western Rift and the Eastern Rift.

The Eastern Rift does not have deepwater lakes; rather, it is a wide valley or basin with shallow lakes that do not have outlets. The lakes have higher levels of sodium carbonate and mineral buildup because of a high rate of evaporation. The differences in water composition of the lakes along the Eastern Rift vary from freshwater to extremely alkaline. Alkaline water creates an ideal breeding ground for algae and other species of fish, such as tilapia, which thrive in this environment. Millions of birds feed off the abundant supply of algae and fish. Birds attract other wildlife, which in turn creates a unique set of environmental ecosystems. The eastern edge of the Eastern Rift is home to the inactive volcanic peaks of Mt. Kilimanjaro and Mt. Kenya. A number of other volcanic peaks are present in the Eastern Rift, such as **Ol Doinyo Lengai**, an active volcano.

Figure 7.40: The Great Rift Valley Hiking into the Great Rift Valley region in Tanzania. Guillaume Baviere – 2007-08-21 – CC BY 2.0.

The erosion patterns of the highlands have caused a buildup of sediments on the rift valley floor, creating a favorable environment for the preservation of biological remains, including both human and animal remains. Important fossils and bones of several hominid species have been found in the Great Rift Valley. One of the most famous finds came in 1974 when the nearly complete skeleton of an australopithecine nicknamed "**Lucy**." Lucy was discovered by anthropologist Donald Johanson. Noted anthropologists Richard and Mary Leakey have also done significant work in this region. Since the 1970s, remains of hominids from about ten million years ago were discovered in the northern region of the Great Rift Valley. Discoveries at the thirty-mile-long **Olduvai Gorge** indicate that early hominid species might have lived in the region for millions of years.

Serengeti and Game Reserves

The Great Rift Valley and the surrounding savannas in Kenya and Tanzania are home to some of the largest game reserves in Africa, with a broad variety of big game animals. One of these large regions is the vast **Serengeti Plain**, located in northern Tanzania and southern Kenya. The governments of Tanzania and Kenya maintain national parks, national game reserves, and wildlife sanctuaries in their countries, most notably in the Serengeti Plain. Legal protection for as much as 80 percent of the Serengeti has been provided. The protections restrict hunting and commercial agriculture and provide protection status for the wildlife. The word *Serengeti* means "Endless Plains."

The Serengeti Plain is host to an extraordinary diversity of large mammals and fauna. The largest migration of land animals in the world occurs in the Serengeti. Every fall and spring, as many as two million wildebeests, antelope, and other grazing animals migrate from the northern hills to the southern plains in search of grass and food. The Ngorongoro Conservation Area, which is a UNESCO World Heritage Site, and the Ngorongoro Crater are located on the Tanzanian side of the border. The enormous crater is the basin of an extinct volcano that has been transformed into a protected national park for the animals that graze on the grassy plains. This is a dry region because the Ngorongoro Highlands create a rain shadow for the area.

Figure 7.41: The Serengeti Plains in the Ngorongoro Crater, Tanzania. Wikimedia Commons – CC BY 2.0.

Dozens of other protected areas throughout Eastern Africa have been established in an effort to protect and sustain the valuable ecosystems for the large animals that have found their habitat encroached upon elsewhere by the ever-expanding human population. Kenya has more than fifty-five nationally protected areas that serve as parks, reserves, or sanctuaries for wildlife. The Amboseli National Reserve and Mt. Kenya National Park are two of the more well-known protected areas. The "big five" game animals—elephants, rhinoceroses, lions, leopards, and buffalo—and all the other unique animals found in the same ecosystems, translate into economic income from tourists from around the world that wish to experience this type of environment. The national park systems in Uganda and Ethiopia have made provisions to provide more sanctuaries for wildlife in areas where the human population is growing and the political situation has not always been stable.

Tanzania

Germany gained control of the region of what is now Tanzania in their scramble for African territory in the late 1800s. Germany relinquished the colony to Britain after World War I. Off the coast of East Africa is the island of **Zanzibar**, which has been an island trading post for centuries and drew in shipping trade from the Middle East, India, and other parts of Africa. The spice trade attracted European ships throughout colonial times. During British occupation, the mainland region was called Tanganyika, named after the large lake on the eastern edge. In 1960, the colony gained independence from Great Britain, and four years later Zanzibar and Tanganyika came together to form the country of Tanzania. Zanzibar remains an important travel destination; major tourism infrastructure has been developed there. The coastal city of **Dar es Salaam** is the primate city of the country and acts as the capital. In 1996, **Dodoma** was declared the official capital. Dodoma is a type of *forward capital*, because the declaration of Dodoma as the official capital was intended to move the political power inland, toward the country's center. The parliament meets in Dodoma, but major government offices remain in Dar es Salaam, making Dar es Salaam the de facto capital of the country.

Tanzania is an agricultural country; as much as 80 percent of the people make their living off the land. The rural nature of the population signifies that the country is at the lower end of the index of economic development with larger families and lower incomes. An emphasis on tourism is a growing trend in Tanzania. The government has stepped up efforts to expand the tourism sector of the economy. Oil and natural gas exploration has also been emphasized in hopes of raising the level of national wealth.

Figure 7.42: Elephant Tusks in Mombassa, Kenya Wildlife and game reserves are a large part of the tourism economy in places such as Kenya and Tanzania. Wikimedia Commons - public domain.

There are more than one hundred ethnic groups in Tanzania. Swahili, an indigenous language, is the lingua franca, and English is used in the higher legal courts and in the universities for higher education. Swahili is used as a second language throughout much of East Africa and serves as a major cultural connection between the many ethnic groups. Tanzania is unique in this aspect; an indigenous language—that is, Swahili—was chosen as the lingua franca rather than the colonial language. Most people learn at least two or three languages, depending on their circumstances. The religious balance of Tanzania's population of more than forty million people is almost evenly split three ways between Christianity, Islam, and traditional religions.

Kenya

During colonial times, the British considered the land area now called Kenya to be a Crown protectorate area. The coastal city of **Mombassa** has been an international shipping port for centuries and is now the busiest port in the region. Persian, Arab, Indian, and even Chinese ships made port in Mombassa during its earliest years to take part in the lucrative trade of slaves, ivory, and spices. Portugal sought early control of the trade center but eventually lost out to Britain. Arab and Middle Eastern shippers brought Islam to the region; Europeans brought Christianity. Hinduism and Sikhism from India found their way into the country with workers brought over by the British to help build a railroad to Uganda. Kenya gained independence in 1963 and has worked throughout the latter part of the twentieth century to establish a stable democratic government.

Nairobi, Kenya's capital, has become a central core urban area that serves the greater East African region as an economic hub for development and globalization.

The largest city in the region, Nairobi is an ever-expanding city that draws in people from rural areas seeking opportunities and advantages. It also has become a destination for international corporations planning to expand business ventures into Africa. Kenya has experienced economic growth and decline as market prices and agricultural production have fluctuated. The Kenyan government has been working with the International Monetary Fund (IMF) and the World Bank to support its economic reform initiatives and reduce waste and corruption in its fiscal processes. The countries of Uganda, Tanzania, and Kenya developed the East African Community (EAC) as a trading bloc to support mutual development and economic partnerships.

Kenya has no one culture that identifies it. There are more than forty different ethnic groups in Kenya, each with its own unique cultural history and traditions. Of the many ethnic groups in Kenya, the **Maasai** have gained international attention and are often given wide exposure in tourism information. The Maasai are a small minority of Kenya's population but are known for wearing vivid attire and unique jewelry. Their historical lands have been the border region between Kenya and Tanzania. Cattle, a sign of wealth, have been at the center of Maasai traditions and culture and provide for their subsistence and livelihood. Tourism brings to the surface the diversity of cultures that coexist within Kenya's environmental attractions, and the country is working to enhance its international draw in the tourism marketplace.

Uganda

Uganda is a small, landlocked country on the northern shores of Lake Victoria. The Western Rift borders it on the west, forming both high mountains and deep lakes. Lakes Albert, Edward, and George are three of the larger bodies of water. The Nile River flows out of Lake Victoria through Lake Kyoga and Lake Albert on its way north, providing an abundant fresh water supply and a transportation route. The Rwenzori Mountains and the Virunga Mountains shadow the country from the west. Mountain gorillas, an eastern gorilla subspecies, inhabit this region. They are extremely endangered: only about seven hundred mountain gorillas live in Africa. One of the two main populations of gorillas lives in the national parks of the Virunga Mountains in Rwanda, Uganda, and the Congo. The second population is found only in a national park in Uganda. The lowland eastern gorilla population is also threatened; their population continues to decline. Poaching, habitat loss, disease, and warfare have devastated their populations in the past century.

Uganda is a poor country and has experienced serious political and ethnic conflicts in recent decades. In 1971, the brutal dictator **Idi Amin** sought to rid the country of his opponents and many foreigners. He killed many of Uganda's own people and destroyed the economy in the process. He was ousted in 1979 and lived in exile in

Saudi Arabia until his death. Uganda was in conflict with Sudan in the 1990s, during the bloody civil war in neighboring Rwanda. Uganda sided with the Tutsi groups in the region and has had to deal with ethnic division within its own borders. Uganda has many troops engaged in the conflict along the unstable border region of The Congo.

Figure 7.43: Silverback Male Western Gorilla It has been reported that a number of eastern mountain gorillas in Virunga National Park were slaughtered and eaten by rebel troops in January of 2007. Wikimedia Commons – public domain.

Uganda is an agrarian state with natural mineral wealth. The potential for added national wealth through mineral extraction exists, but there is no way to fund the extraction operations. Agriculture is the principal employment of Uganda's labor force, and most workers earn fewer than two dollars per day US equivalent. Coffee has been and continues to be a main export crop. Uganda is about the same size in terms of land area as the US state of Wyoming, but whereas Uganda has an estimated population of more than thirty-two million people, Wyoming has fewer than half a million. Population growth without economic growth places a heavy strain on Uganda's natural resources.

Ethiopia

With more than eighty-five million inhabitants, Ethiopia has the second-largest population in Africa; Nigeria has the largest population. Ethiopia was never colonized by the Europeans in the scramble for Africa, but during World War II, it suffered a brief occupation by Italy (1936–41). From 1930 to 1974, **Emperor Haile Selassie** ruled the country until he was deposed in a military coup. Many Rastafarians in Jamaica consider Selassie to be the second coming of Christ and a messiah for the African people. According to Rastafarian traditions, Ethiopia was the biblical Zion. After Selassie was deposed, the government shifted to a one-party Communist state. Successive years in Ethiopia were filled with massive uprisings, bloody coups, and devastating droughts, which brought about massive refugee problems and civil unrest. Famine in the 1980s caused the deaths of more than a million people. The Communist element in Ethiopia diminished when the Soviet Union collapsed in 1991. The country's first multiparty elections were held in 1995.

The region of Eritrea had been a part of a federation with Ethiopia. In 1993, Eritrea declared independence, sparking a boundary war with Ethiopia that eventually concluded in a peace treaty and independence in 2000. The final boundary is still disputed. The breaking off of Eritrea left Ethiopia a landlocked country with no port city.

The capital and largest city in Ethiopia is **Addis Ababa**, which is the center for various international organizations serving East Africa and Africa in general, such as the African Union and the United Nations (UN) Economic Commission for Africa. This city is the hub of activity for the country and for international aid for the region.

The Great Rift Valley bisects Ethiopia. Highlands dominate the northwest, and minor highlands exist southwest of the rift. The **Ethiopian Plateau** encompasses the Northwest Highlands and is home to Lake Tana, the source of the Blue Nile. Elevations on the Ethiopian Plateau average more than 5,000 feet, and the highest peak, Ras Dashan, reaches up to 14,928 feet. The climate includes sporadic rain in early spring. The typical rainy season extends from June to September, but the rest of the year is usually dry. The high elevations of the highlands cause a rain shadow effect in the deep valleys or basins on the dry side of the region. Eastern Ethiopia is arid, with desert-like conditions. The impact of overpopulation on the natural environment has been deforestation and higher rates of soil erosion; thus continued loss of animal species is inevitable. Fortunately, Ethiopia has established natural parks and game reserves to protect wildlife and big game.

Ethiopia has been inhabited by divergent kingdoms and civilizations, giving rise to a rich heritage and many cultural traditions—so much so that Ethiopia is home to eight UNESCO World Heritage Sites. More than 60 percent of the population claims Christianity as its belief system; about 30 percent of the population is Muslim. Many traditional religions prevail in rural areas. In contrast to other African countries, Christianity came to Ethiopia directly from the Middle East rather than from European colonizers or missionaries from Western countries. In Ethiopia, Christianity was structured into the **Ethiopian Orthodox Church**, a church that has endured through centuries. Religious tradition claims that the Ethiopian Orthodox Church is in possession of the lost Ark of the Covenant, which once rested in the holy of holies in the great temple of Solomon in Jerusalem. However, no direct evidence supports this claim. Ethiopia does not use the Gregorian calendar, which is the calendar used in the most of the Western world; rather, Ethiopia uses a calendar based upon the calendar of the Coptic Christian Church, which is about eight years behind the Gregorian calendar because of differences in how the year is calculated and in differences in the dating of the life of Jesus.

Figure 7.5.447.5.44: Old Man in Harer, Ogaden Region of Eastern Ethiopia. Gustavo Jeronimo – Anciano – CC BY 2.0.

The large and growing population of Ethiopia is made up of an array of ethnic groups. The three largest ethnic groups are **Oromo** (35 percent), **Amhara** (27 percent), and **Tigray** (6 percent). A number of minority ethnic groups make up the remaining 32 percent. The dominance of the Oromo, Amhara, and Tigray groups provides advantages when it comes to determining which language to use for primary education or in community politics in a region. At least eighty-four separate local languages are spoken in Ethiopia. The lingua franca for higher education and for common use is English. The tradition in many elementary schools is to use Amharic as the primary language for instruction, but this is breaking down as other languages increase in usage because of population increases.

Eritrea

Eritrea was an Italian colony before joining with Ethiopia in a federation in 1952. Since declaring independence in 1993, Eritrea has had a difficult time balancing positive economic growth with its border dispute with Ethiopia. The border war with Ethiopia drained this small, poor country of economic resources and destroyed valuable infrastructure. The agrarian culture and economic activities common in Africa also exist in Eritrea. Farming and raising livestock are the main activities of as much as 80 percent of its citizens. The government has controlled almost every aspect of business and industry within its command economy structure. In 2008, in an attempt to attract business and connect with global markets, the port city of Massawa opened a free-trade zone and has been attracting business connections. Mineral extraction is being explored by multinational corporations, and various foreign governments have been working to establish stronger ties.

Eritrea is located north of the African Transition Zone. Most countries north of the African Transition Zone have an Islamic majority, but that is not necessarily the case in Eritrea. Statistical data are difficult to confirm, but Christianity is believed to be as prominent as Islam in Eritrea. The main Christian denomination is the Eritrean Orthodox Church, with smaller percentages of Roman Catholics and Protestants. Most Muslims are Sunni. The government highly regulates religious activities and requires all churches to register with the state and provide personal information regarding its members. Members of religious groups not registered with the state may be subject to arrest or imprisonment for violation of this requirement. Examples of groups not approved by the government of Eritrea as of 2010 include Baha'is and Jehovah's Witnesses.

Djibouti

The **Bab el-Mandeb Strait**, between the Red Sea and the Arabian Sea, separates Africa from Asia on the southwestern tip of the Arabian Peninsula. The strait is a narrow strategic choke point for international shipping tranferring cargo from Europe to Asia through the Red Sea and the Suez Canal. On the African side of the strait is the small former French colony of the Republic of Djibouti. Natural resources within the country are sparse, and the republic depends on its relationship with countries such as the United States or its former colonizer to augment its financial status, keep the region stable, and protect valuable shipping in the nearby waterways. The United States has an important military base in Djibouti, which is the only major US military base in Subsaharan Africa. The French also have a major military base located in Djibouti. The significance of the Djibouti's location means it is a critical place for monitoring both the war on terrorism and incidences of piracy off the Horn of Africa. A main objective of having European and US military bases in Djibouti is to insure safe passage of oil tankers through the strait providing European countries and the United States with petroleum from the Persian Gulf states.

Like many other African countries transitioning from colonies to independent countries, Djibouti has experienced political infighting that has been detrimental to the country's economic situation. The government struggles with foreign debt and a lack of economic development opportunities. Most of the population lives in the capital city of Djibouti, where the unemployment rate is extremely high. Its vital location is the country's main asset, and foreign aid has been a major part of the country's economic equation.

Somalia

The country presently called Somalia resisted the forces of European colonizers scrambling for African land in the 1800s. The various kingdoms and their powerful leaders kept the colonial forces out well beyond World War I. Somalia's close vicinity to the Arabian Peninsula and the prevalence of Arab trade provided a direct connection through which Islam was quickly diffused from Arabia to Africa. Political alliances were fused between the Somalian kingdoms and the Ottoman Empire. The leaders of the Horn of Africa used Islam as a centripetal force in uniting the people against outside forces.

It was not until 1920, as a direct result of the use of airplanes in warfare, that the northern region of **Somaliland** buckled under to the British colonial forces. The eastern and southern regions were soon dominated by Italy. Britain finally withdrew from **British Somaliland** in 1960. The country then joined with the Italian portions of the region to form a new nation, **Somalia.** An authoritarian socialist regime established power in 1969 and lasted until 1991. The socialist regime in Somalia initiated a territorial war in 1978 in an attempt to gain back territory in the eastern parts of Ethiopia's Ogaden region, which was once part of the various Somalian kingdoms. The war only intensified the divisions in the region. After 1991, the country descended into political chaos, economic turmoil, and factional fighting.

Northern clans who opposed the central government broke away in a move toward independence, and the old British Somaliland became the **Republic of Somaliland** in 1991. Somaliland operates independently from the rest of Somalia and prints its own currency. The northeast region of the country referred to as **Puntland** also broke away from the central government. Somaliland sought total independence, while Puntland was in favor of belonging to a national union but wanted to maintain autonomy. Both autonomous regions have their own governments and are relatively stable, but they have had conflicts regarding the territory between them. Neither is recognized by outside entities as independent countries. Puntland's port of **Bosaso** has developed into a rapidly growing economic city and has expanded from fifty thousand to five hundred thousand people since

2000. Plans are in the works for a new airport and an economic free-trade zone, which should attract additional business and an influx of additional immigrants from other parts of Somalia and the region.

Southern Somalia has been broken down into regions ruled by warlords who have pillaged the country and control vital transportation links. A total government breakdown occurred in 1991, leading to a meltdown in all areas of society. Food distribution was hampered through pillaging and a lack of fuel and structured transportation. Electrical power was lost and clan warfare became the rule of law. A famine subsequently caused an estimated three hundred thousand deaths. The UN stepped in with food and peacekeeping troops, and in 1992, the United States dispatched marines in Operation Restore Hope, which stabilized the distribution of food but was unsuccessful in stabilizing the political situation and establishing a legitimate government. In 1993, nineteen US soldiers were killed in a battle in the capital, Mogadishu, after of which the US withdrew its troops. The UN withdrew its peacekeeping troops in 1995.

Figure 7.45: Somalia, with Its Autonomous Regions of Somaliland and Puntland. CIA – public domain.

Figure 7.46: Super 64 Military Helicopter Piloted by Michael Durant over Mogadishu, 1993 This helicopter crashed and was the basis of the Hollywood movie *Blackhawk Down*, depicting warfare in Somalia. US Army - public domain.

Despite a lack of effective national governance, the informal economy in Somalia continues to thrive. Trading through personal transactions and the private marketplace continues to provide for the needs of the people. The main sources of revenue are agricultural goods and livestock, as well as money sent by people from outside the country to their families in Somalia. Banking and businesses have struggled to adapt to the continual conflicts, yet various hotels and service agencies continue to operate under the protection of private security militias. Telecommunication firms continue to offer service with low calling rates for urban areas. The educational system is heavily supported by the private sector because of the collapse of the central government. Secondary education at the university level is struggling in the south but is more established in Puntland and Somaliland, where they are funded by central authorities. Income from piracy on the high seas has brought in millions to the private warlords that manage the operations. Stabilizing the country and showing economic progress will remain a difficult task for whoever leads Somalia.

Rural-to-Urban Shift

The average family size in East Africa is about 5.5, which is typical of the entire continent of Africa and translates into exploding population growth. In many areas of Africa in general and East Africa in particular, most of the population (as much as 80 percent) makes its living off the land in agricultural pursuits. Large families in rural areas create the conditions for the highest levels of rural-to-urban shift of any continent in the world. The large cities—with expanding business operations complete with communication and transportations systems that link up with global activities—are an attractive draw for people seeking greater employment opportunities. In East Africa, each of the three largest cities—Nairobi, Dar es Salaam, and Addis Ababa—is more populous than Chicago, the third-largest US city. In West Africa, the city of Lagos, Nigeria, is more populous than New York City and Chicago combined. These cities are all riding the worldwide wave of globalization and are core centers of economic activity for the business sector and corporate enterprises.

Figure 7.47: Urban Slums of Nairobi People walking to and from work along the railroad tracks in a poor neighborhood in Nairobi, Kenya. ND Strupler – Nairobi 2008 – CC BY 2.0.

The other regions of Africa all have central cities that act as economic core areas and attract the multitudes from rural areas looking for employment and opportunities. International connections are indicative of local economic development, which is causing the urban areas to grow at exponential rates. Cities heavily affected by a high level of rural-to-urban shift often cannot build infrastructure fast enough to keep up with demand. Self-constructed slums and squatter settlements, which lack basic public services such as electricity, sewage disposal, running water, or transportation systems, circle the cities. All the large cities of Africa are expanding at unsustainable rates. Traffic congestion, trash buildup, higher crime rates, health problems, and air pollution are some of the common results.

As of 2010, the US population was about 80 percent urban. This is the opposite of places such as East Africa, where about 80 percent of the population remains rural. In the next few decades, Africa could witness the growth of megacities that might continue to expand and grow for another century. For example, if the current rates continue, Ethiopia's population of eighty-five million people in 2010 will double to more than 160 million by 2040. Urban areas will continue to be target destinations for employment opportunities, whether they exist or not, and rural-to-urban shift will drive the populations of cities such as Addis Ababa to double, triple, or quadruple in size by 2020.

Figure 7.48: Modern City of Nairobi Since this photo was taken in 1980, the city center of Nairobi has expanded and grown ever larger to become a major core business hub for East Africa. The central business district at the center of this core locale has an enormous pull factor for individuals from rural areas seeking employment. Brian Snelson – Nairobi 1980 – CC BY 2.0.

Key Takeaways

- The physical geography of East Africa is dominated by the Great Rift Valley, which extends through the middle of the region from north to south. Associated with the rift valleys are vast savannas such as the Serengeti Plain, large lakes, high mountains, and the highlands of Ethiopia.
- East Africa has extensive habitat for large herds of big game animals, mountains for great apes, and enough prey for the big cats such as lions and cheetahs. Countries are establishing game preserves, national parks, and conservation areas to protect these animals and their natural habitats and to provide economic opportunities to gain wealth through tourism.
- Most of the populations of countries in East Africa earn their livelihood through agricultural activities because the region has a high percentage of people living in rural areas. Large families fuel an ever-increasing rural-to-urban shift in the population, causing the cities to grow at an unprecedented rate.
- Areas along the African Transition Zone have experienced a higher degree of political conflict and civil unrest. Somalia has the highest level of political division and fragmentation based on religion and ethnic backgrounds, which contributed to the secession of the two autonomous regions of Somaliland and Puntland. The central government is neither stable nor strong enough to govern the fragmented country.
- Diversity in the region is remarkably high; there are hundreds of different indigenous ethnic groups, each with its own language and culture. Religious

activity is split between Christianity, Islam, and traditional beliefs. Islam is more favored in the African Transition Zone and to the north, while Christianity and animism are more prominent to the south of the zone.

Discussion and Study Questions

1. Where do the White Nile and the Blue Nile begin? Where do they converge?
2. What is the Great Rift Valley? How is the Western Rift different from the Eastern Rift?
3. What has been found at Olduvai Gorge? Why is this significant?
4. How have Kenya and Tanzania attempted to preserve and protect the environment?
5. Where is the Serengeti Plain? How does it bring national wealth to its home countries?
6. What are the main religions in Ethiopia, Eritrea, and Somalia? How did belief systems diffuse to Ethiopia, Eritrea, and Somalia? What regions did they diffuse from?
7. What is Tanzania's lingua franca? How is this different from other African countries?
8. What are the main political regions of Somalia? How does Somalia's central government function?
9. What is Djibouti's main asset? Why would the United States or France want to control Djibouti?
10. What two main factors cause a high level of rural-to-urban shift?

Geography Exercise
Identify the following key places on a map:

- Bab el-Mandeb Strait
- Blue Nile
- Bosaso
- Dar es Salaam
- Eastern Rift Valley
- Ethiopian Plateau
- Great Rift Valley
- Lake Albert
- Lake Edward
- Lake Tana
- Lake Turkana
- Lake Victoria
- Mombassa
- Mt. Kenya
- Mt. Kilimanjaro
- Nile River
- Ol Doinyo Lengai
- Olduvai Gorge
- Puntland
- Rwenzori Mountains

- Serengeti Plain
- Somaliland
- Virunga Mountains
- Western Rift Valley
- White Nile
- Zanzibar

Activities

1. Find out what the news media are reporting regarding current events and political stability in Somalia.
2. Research the level of endangerment of mountain gorillas in East Africa.

7.6. SOUTHERN AFRICA

Southern Africa is a region defined by its rich natural resources, diverse cultures, and complex socio-political history. Comprising countries like South Africa, Namibia, Botswana, and Zimbabwe, it features a stunning array of landscapes, from the Kalahari Desert to the lush coastal areas of Mozambique. The region is marked by a rich tapestry of ethnic groups and languages, each contributing to a vibrant cultural mosaic. Economically, Southern Africa is a leader in mining and agriculture, but it faces significant challenges, including economic inequality, high unemployment rates, and the impacts of climate change. The legacy of apartheid in South Africa continues to influence social dynamics and governance, while regional organizations like the Southern African Development Community (SADC) work towards greater cooperation and development. Despite these challenges, the resilience and innovation of its people shine through, as they navigate the complexities of modern life and strive for a more equitable future.

Figure 7.6.497.6.49: Southern Africa and Madagascar. Updated from map courtesy of University of Texas Libraries.

The region of the African continent south of the Congo and Tanzania is named Southern Africa. The physical location is the large part of Africa to the south of the extensive Congo River basin. Southern Africa is home to a number of river systems; the **Zambezi River** is the most prominent. The Zambezi flows from the northwest corner of Zambia and western Angola all the way to the Indian Ocean on the coast of Mozambique. Along the way, the Zambezi River flows over the mighty **Victoria Falls** on the border between Zambia and Zimbabwe. Victoria Falls is the largest waterfall in the world based on selected criteria and is a major tourist attraction for the region.

Southern Africa includes both type B and type C climates. The Tropic of Capricorn runs straight through the middle of the region, indicating that the southern portion is outside the tropics. The **Kalahari Desert**, which lies mainly in Botswana, is an extensive desert region with an arid mixture of grasslands and sand. When there is adequate rainfall, the grasslands provide excellent grazing for wildlife. Precipitation varies from three to ten inches per year. The Kalahari is home to game reserves and national parks. Large areas of dry salt pans stretch over ancient lake beds. The salt pans fill with water after heavy rainfall but are dry the remainder of the year. The **Namib Desert**, found along the west coast of Namibia, receives little rainfall. Moderate type C climates are found south of the Kalahari Desert in South Africa, where conditions are suitable for a variety of agricultural activities, including fruit orchards and an expanding wine industry.

Madagascar is located to the east of the continent, in the Indian Ocean. Madagascar is the world's fourth-largest island and is similar in area to France. Surrounding Madagascar are the independent island states of the Seychelles, Comoros, and Mauritius. Madagascar is included in this lesson on Southern Africa but does not share its cultural geography or biodiversity. Madagascar broke away from the mainland more than 160 million years ago and developed its own environmental conditions and cultural heritage. The early human inhabitants of Madagascar can trace their ancestry to the regions of Malaysia and Indonesia in Southeast Asia. People from the African mainland also joined the population. The whole island later came under the colonial domination of France but won its independence in 1960.

Madagascar's unique physical environment is home to many plants and animals found nowhere else in the world. At least thirty-three varieties of lemurs and many tropical bird species and other organisms are found only on Madagascar. It is an area of high biodiversity and is home to about 5 percent of all the animals and plants in the world. Tropical rain forests can be found on the eastern edge on the windward side of the island. The western side of the island experiences a rain shadow effect because of the height of the central highlands, which reach as high as 9,435 feet. The western side of the island has a smaller population and receives less precipitation.

Figure 7.50 Black-and-White Ruffed Lemur from Madagascar at the Dudley Zoo in England

Since 1990, the eastern tropical rain forest has experienced a sharp decline because of extensive logging, slash-and-burn agriculture, mining operations, and drought. Population growth has placed a heavier demand on the environment, which in turn puts stress on the habitats of many of the unusual organisms that are unique to the island. Typical of many African nations, agriculture is Madagascar's main economic activity. About 80 percent of the twenty million people who live on the island earn their living off the land.

Deforestation is occurring on all parts of the island and is more severe in areas where human habitation leads to a high demand for firewood used in cooking. In other parts of Africa, important environmental areas have been protected or transformed into national parks and wildlife preserves. Though protected areas do exist on Madagascar, efforts to protect the environment and the wildlife have been hampered by the lack of available funding and the population's high demand for natural resources.

The countries on the Southern African mainland share many of the demographic qualities of the rest of Africa: large family size, agrarian economies, multiple ethnic groups, rural populations, political instability, and a high rate of rural-to-urban shift. Southern Africa is set apart from other Subsaharan African regions because of its mineral resources, including copper, diamonds, gold, zinc, chromium, platinum, manganese, iron ore, and coal. Countries in Southern Africa are quite large in physical area, except three smaller landlocked states: Lesotho, Swaziland, and Malawi. The larger countries—South Africa, Botswana, Mozambique, Zimbabwe, Zambia, Namibia, and Angola—all have extensive mineral deposits.

The vast mineral resources make this one of the wealthiest regions of Africa with the greatest potential for economic growth. A physical band of mineral resources in Southern Africa stretches from the rich oil fields off the northwest coast of Angola, east through the diamond-mining region, and into the northern **Copper Belt** of Zambia. A region of rich mineral deposits continues from the unique geological formation called the **Great Dyke** in central Zimbabwe through the **Bushveld basin** into South Africa. From here, it extends southwest through the central gold- and diamond-mining regions of South Africa toward the southern coast. Mining activity exists on both sides of the belt of mineral resources. Diamond mining is found in parts of Botswana and along the coast of Namibia. Coal can be found in central Mozambique. The counties that are able to conduct the necessary extractive processes are creating national wealth and increasing the standard of living for their people.

Angola

The largest country in Southern Africa is Angola. Located on the west coast of the continent, the country includes the small exclave of **Cabinda** to the north, which borders the Congo River and is separated from the main body of Angola. Cabinda is a major oil producer and remains a foothold for rebel groups seeking greater control over oil resources. Offshore oil activity has increased in the northern region of Angola and has attracted international oil companies from many countries. In 2007, Angola was the largest oil exporter to China, which has become one of Angola's largest financial supporters. Oil revenues have helped Angola rebuild after a bitter civil war (1975–2002) devastated the country and its infrastructure. Nevertheless, the country remains one of the world's poorest; life expectancies were a mere thirty-eight years in 2010. Angola is also burdened with thousands of refugees seeking safety from the civil wars in the Congo.

Angola was a Portuguese colony and fought hard for independence, which it received in 1975. After independence, strong factions clashed to obtain political power. At the same time, the Cold War was at a zenith. The United States and the Soviet Union both used their influence to support political leadership in Angola and other parts of Africa to reflect their respective ideologies. Even Cuba had a large number of military troops in Angola in support of a socialist agenda. As the factions within Angola vied for power, the country was deeply divided. Thus began a twenty-seven-year civil war that finally ended in 2002. The country has been working to recover from this turmoil ever since. Foreign aid and charitable organizations have helped feed the people, and oil revenues are beginning to support recovery. The government of Angola has suffered from serious corruption. Much of the wealth is centralized in the hands of an elite few and does not filter down to the general population.

Natural resources were a major factor in the long, drawn-out civil war. Angola was one of the original countries in which the term *blood diamonds* was used. Resistant groups would mine the diamonds, sell them on the world market, and use that income to fund their military pursuits. Revenue from the clandestine diamond trade was fueled the death and destruction of the newly independent country. Diamonds are still being mined in Angola and bringing in considerable national income. The industry is not always controlled or managed to lessen the number of diamonds reaching the market from unsubstantiated sources. Natural resources have helped the country look to a future in which a more stable government can work to rebuild the country's infrastructure and improve the standard of living for its people.

Namibia

Namibia is a dry, arid land with few people. Its population density of about 6.5 people per square mile in 2009 was the second-lowest in the world after Mongolia. The physical geography of this large country is dominated by arid plateaus and desert regions with changes in elevation between them. The **Namib Desert**, characterized by enormous sand dunes, runs along the entire west coast. It meets up with the **Great Escarpment**, which is a sharp rise in elevation of as much as a mile high toward the interior. A central plateau dominates the north-south interior, and the Kalahari Desert is to the east. Natural resources are abundant: uranium and diamonds are the major export products, and lead, zinc, tin, silver, copper, and tungsten are also mined. The coastal waters support a substantial fishing industry. Namibia's constitution contains language that addresses environmental conservation and the protection of wildlife habitat. The country has created nature conservancies, which have advanced the activities of ecotourism, a source of national wealth promoted by the government.

Figure 7.51: Klein Namutomi Waterhole, Namibia Wildlife are becoming an excellent asset for tourism. Wikimedia Commons – CC BY-SA 3.0.

The Berlin Conference of 1884, when the colonial European powers met to divide up Africa, resulted in the German colonization of the region of Namibia. The German influence remains evident in the dominant religion: most of the Christians in Namibia, which now make up about 80 percent of the population, are Lutherans.

By 1920, the region known as South West Africa had been transferred to South Africa by the League of Nations. South West Africa had to endure the political policies of apartheid while under the jurisdiction of South Africa. By 1966, local uprisings and a push for sovereignty resulted in a socialist independence movement that created the South West Africa People's Organization (SWAPO). This organization was entrenched in a guerilla war with South Africa, which led to the return of Namibia to United Nations (UN) supervision. The name *Namibia* was adopted during this era. Socialist countries such as Cuba provided supported SWAPO's movement for autonomy.

The 1980s brought about a transition in the governing process in the Namibian War of Independence. In 1985, South Africa installed an interim government in Namibia. A UN peace plan was reached that included Namibia in 1988, and Namibia finally gained its independence in 1990 following South Africa's withdrawal from Angola in their civil war. SWAPO has become the dominant Namibian political affiliation. As a recently independent country, Namibia has had to work through a natural transition process to create a stable government to sustain itself as an independent country.

Zambia

The landlocked country of Zambia was known as Northern Rhodesia during colonial times. From 1891 to 1923, it was administered by Britain's South Africa Company. Then it became a colony of Great Britain. The name was changed to Zambia when the country was granted independence in 1964. Lusaka is its primate city and capital.

Mining in the Copper Belt to the north provides both economic activity and employment opportunities for the people, but long distances from major seaports and the fluctuation in world commodity prices do not provide a stable economic situation for the mining industry. Though most of the population makes their living by subsistence agriculture, the country has a large urbanized population that heavily depends on the mining industry for employment and economic support.

The country is expected to double in population in about thirty-five years if current rates continue. The average family size is more than five members. The larger towns and cities along the major transportation routes are the main population centers. Unemployment remains high in these urban areas; there are few opportunities for economic development. More than seventy ethnic groups can be found in Zambia, including a small but growing Chinese population. English is the official language, but many other languages are spoken. Most of the people are Christians with a wide variety of denominations. Animist and traditional beliefs are also common.

Malawi

Malawi is a landlocked nation that became independent in 1964. The British controlled the region in 1891 and named their colony Nyasaland after **Lake Nyasa**, a large lake in Malawi; the lake is also called **Lake Malawi.** The lake serves as the eastern border of the country. Another name for Lake Malawi is Calendar Lake, named so because of its physical dimensions, which are 52 miles wide and 365 miles long. Lake Malawi is a deepwater lake with depths reaching 2,300 feet. The great depths of the lake are caused by the Great Rift Valley, which created the depression the lake rests in. The lake provides for the livelihoods for millions of people who live along its shores and depend on it resources of fish and fresh water.

Figure 7.52: Tea Plantation in Malawi. David Davies – Mulanje tea estate – CC BY-SA 2.0.

Malawi is not a technologically developed country. Only about 15 percent of the population lives in urban areas. About 90 percent of the country's exports are agricultural products of tea, sugar, coffee, and tobacco. Malawi's government requires substantial assistance from the international community, such as the World Bank and the International Monetary Fund (IMF). Complicating the situation are instances of government corruption that have led to the withholding of international aid. Another major problem for Malawi is the significant number of HIV-infected people. Also, family size is commonly large, and rapid population growth places an ever-increasing demand on natural resources.

Mozambique

Before colonial Europeans arrived in the region of Mozambique along Africa's east coast, the **Bantu** people of a number of ethnic subgroups were the dominant people, and they still make up more than 99 percent of the population. Portugal had claimed the region as a colony as early as the sixteenth century. Portugal strongly resisted releasing its claims to Mozambique but did so after much opposition; the country became independent in 1975. The transition from colony to independent nation was a struggle for Mozambique. The rebel groups that had been instrumental in fighting a guerilla war for independence against Portugal remained active after 1975 and fighting continued after independence. The country fell into a violent civil war from 1977 to 1992. Central elements in this war were involvement or influence by Zimbabwe and South Africa on one side and Marxist support from the Soviet Union on the other. More than a million people died in this brutal war, which also devastated the country, destroyed valuable infrastructure, and created more than a million refugees who sought sanctuary in neighboring countries. At the same time, most people of Portuguese descent left because of safety concerns. A peace agreement was finally reached in 1992. Since then, the country has had to struggle to create a stable government and provide a safe environment for its people.

Foreign aid has remained a necessity to provide economic stability for Mozambique. The agrarian society includes a high percentage of the population that lives below the poverty line. Selective mining operations and the introduction of garment manufacturing augments the main agricultural activities. Since 2000, the government has worked to implement economic reforms and to curb excess spending. Both have had a positive effect on economic growth. It is not unusual for foreign debt to plague developing countries, and Mozambique has suffered from a high level of national debt that has threatened to bankrupt the country. Debt relief for Mozambique came through the IMF's debt forgiveness program; both debt forgiveness and loan rescheduling agreements were implemented. The positive growth pattern indicates that even desperately poor countries can work to improve the standard of living for their people if there are adequate resources and opportunities for employment. Still, rapid population growth potentially cancels out economic gains and threatens to deplete valuable resources, thereby creating an even more difficult path for future stability.

Zimbabwe

There is no better way to understand Zimbabwe than to become familiar with the history and heritage of the people who live there. The **Great Zimbabwe Kingdom** flourished from about 1250 to 1450, when it was eclipsed by succeeding kingdoms. Ruins from the extensive stone architecture of that era remain and are a major tourist attraction. These kingdoms were major trading centers for the region but later clashed with the colonial powers that desired to dominate regional trade for themselves. The Bantu civilization of Southern Africa established a number of kingdoms that existed in Zimbabwe up to the colonial era.

Zimbabwe experienced similar colonial activity to that which befell Zambia, its neighbor to the north. Britain arrived in the late 1800s, and by 1923 they called their newly controlled colony **Southern Rhodesia**, after Cecil Rhodes, who headed the British South Africa Company, a mercantile company that broke new ground in the region. The British were able to quell any resistance to their activity and consolidated their holdings. After Zambia become independent, Southern Rhodesia was renamed Rhodesia. In 1965, the white-dominated leadership of the Rhodesian government unilaterally declared its independence, but Britain did not recognize this action. The UN issued sanctions against the white leadership; the response was an internal guerilla uprising to fight for free elections that would include black Africans. Independence was finally granted in 1980, and the country's name was officially changed from Rhodesia to Zimbabwe.

In 1980, **Robert Mugabe** came to power as the first president and extended his authoritarian rule over the next three decades. During this time, the country experienced more than its fair share of civil unrest and political turmoil. Mugabe has been accused of corruption, election rigging, and pillaging of public funds for personal gain. Under his leadership, there have been accusations of government mismanagement, human rights abuses, and hyperinflation of the country's currency. In 2008, inflation led to serious devaluation of the currency. The Reserve Bank of Zimbabwe printed banknotes in the denomination of one hundred trillion dollars that were only worth a few US dollars in the international exchange. The people continue to work through these difficult economic conditions.

Figure 7.53: Towers of the Great Zimbabwe Kingdom This photo shows the ruins the Great Zimbabwe Kingdom, an ancient empire that gave Zimbabwe its name after the colonial era, during which it was called Rhodesia. Kevin Walsh - greatzim - CC BY 2.0.

Mugabe initiated a land reform policy that would take land owned by people of white European ancestry and redistribute it to people of black African ancestry. During the colonial era, white Europeans, who only made up a small percentage of the population, moved in and took control of most of the agricultural lands. Land reform was a progressive policy and was meant to provide a greater level of equality within the country. However, the disorganized methods used to carry out the plan resulted in violence and the confiscation of farmland with little regard for the rule of law.

Thousands of white farmers and their families left the country. Some were killed when their farms were taken over by force. Mugabe was accused of corruption in making sure his cronies benefited from the land reform without being concerned about the general population. The disruption in the agricultural sector resulted in a drastic reduction in agricultural output. The country shifted from a once sustainable, prosperous agricultural sector with extensive exports to an agricultural system that was in disarray. The result was food shortages and the loss of enormous agricultural export profits.

Shaping a stable post-Mugabe government will be a challenge for Zimbabwe. Solving the problems resulting from the transition from exploitive colonial rule to personal authoritarian rule will be an arduous undertaking. The country has serious economic problems that have lowered the amount of opportunities and advantages for its people. Life expectancies have dropped from about sixty years in 1990 to fewer than forty years two decades later. Health care, education, and social services have not reached the level needed to sustain a healthy society by world standards. The HIV/AIDS epidemic has hit Zimbabwe hard, affecting a higher percentage of its population than most of the other African countries. Political turmoil, civil unrest, and violence have reduced the level of law and order within the country. These issues are indicators that within the globalization process, countries such as Zimbabwe will be losing out on the opportunity to attract foreign investments, manufacturing expansions, or employment opportunities being developed elsewhere to meet the demands of the marketplace.

Botswana

The country of Botswana is relatively flat, and the Kalahari Desert covers as much as 70 percent of its land area. By the time it established independence from Britain in 1966, the lack of agricultural lands had reduced the country's economic level to the lowest in the world. At that time, the country was called Bechuanaland.

Botswana has transitioned more smoothly than other African countries into a stable political system with a fast-growing economy. It has an emerging service sector, extensive diamond mining, and expanding industrial ambitions. Personal incomes are rising, and the standard of living is reaching upward to the second highest in the region after South Africa.

Botswana is fortunate to have had uninterrupted civilian political leadership for the decades following independence. This stable government has implemented progressive social policies and attracted significant capital investments to create one of the most dynamic economies in Africa. Diamond mining has been the principle extraction activity dominating the country's export income. The stability of the country has enhanced the tourism market and created a growing ecotourism industry. The vast Kalahari Desert and the well-protected game reserves provide attractive tourist destinations.

Figure 7.54: San Bushmen in Deception Valley, Botswana. Karen Corby – Kgalagadi – CC BY-ND 2.0.

Far from the fast-growing urban development of the big cities in Botswana, there exists a dispute between the government of Botswana and the indigenous San people who live in the middle of the **Central Kalahari Game Reserve**. The government has established programs to move the San from the reserve to other locations where they would become less nomadic and settle into a more agrarian lifestyle. According to the government position, this is to keep the natural habitat and wildlife from being affected by human activity, thereby promoting tourism. The courts, however, ruled against the forced move of the San out of the Central Kalahari Game Reserve. Critics of the government program claim the real reason for the move is to clear the land of human habitation so the area can be developed for mining purposes. Estimates indicate that fewer than one hundred thousand San presently live in Southern Africa; about fifty-five thousand San live in Botswana.

Botswana's government has advanced one of Africa's most progressive and comprehensive programs for dealing with HIV/AIDS. Medical care, education, and social services are being strengthened. Botswana ranks at the top end in terms of the percentage of people who are infected with HIV/AIDS. This epidemic cuts across all levels of society and culture. The social stigma attached to being infected with HIV discourages individuals from being tested for the disease or seeking medical care. Addressing this epidemic is a challenge throughout Southern Africa.

These trees are so large that people have actually lived inside them.

Figure 7.55 Baobab Tree in Southern Africa

South Africa

Anchoring Subsaharan Africa to the south is the dominant country of South Africa. Its large land area and vast mineral resources support a population of about fifty million people. The **Cape of Good Hope** on the southern tip of the continent is a transition point from the Atlantic Ocean to the Indian Ocean. Its strategic location was important for the control of shipping during the early colonial era before the Suez Canal provided a shortcut between Europe and Asia, bypassing most of Africa. The European colonial era first brought Dutch explorers to the Cape of Good Hope, where they established the city of Cape Town as a stopover and resupply outpost on their way to rest of Asia.

South Africa is home to many indigenous ethnic groups and is demonstrative of the diverse pattern of human geography. The country has a history of both ethnic diversity and ethnic division. Two of the largest African groups are the **Xhosa** and the **Zulu**. The European component of the ethnic mosaic was enhanced by colonial and neocolonial activities. After the arrival of the first Dutch ships, other Europeans followed and competed with the African groups for land and control. The discovery of first diamonds and then gold prompted Britain's involvement in South Africa. The **Boer Wars** (1880–81 and 1899–1902) were fought between the Dutch-based Boers and Britain for control of South Africa's mineral resources. South Africa became a British colony dominated by a white power structure. The Boers (later known as Afrikaners) spoke **Afrikaans** and were prominent in the South African political system.

Figure 7.56: "Petty Apartheid" Beach Sign Written in Afrikaans and English. Wikimedia Commons – public domain.

Segregation first developed as an informal separation of the racial groups but evolved into the legally institutionalized policy of apartheid, which separated people into black, white, and coloured (mixed race) racial categories. A fourth category was developed for people from Indian or Asian backgrounds. Apartheid eventually found its way into every aspect of South African culture. In the larger scale of society, access and separation was based on race. Each racial group had its own beaches, buses, hospitals, schools, universities, and so on. The legal system divided the population according to race, with the white minority receiving every advantage. There were extensive and detailed rules for every aspect of daily activity, including which public restroom or drinking fountain could be used, which color an individual's telephone could be, and which park bench a person could sit on. The government also sanctioned separate homelands for people from different ethnic groups. People were physically removed from their homes and transported to their respective new homelands based on their racial or ethnic background. The policy of apartheid not only divided the country at that time but set up racial barriers that will take generations to overcome.

The controversial policy of apartheid in South Africa achieved international attention. Many countries condemned it and implemented economic sanctions and trade restrictions against South Africa. Opposition grew within the country and erupted into violence and social unrest. As a result, the white-dominated government of South Africa began to dismantle the apartheid system in the 1990s. The ban on political opposition parties, such as the African National Congress, was lifted, and after twenty-seven years in prison, Nelson Mandela was released from prison, where he had held for his resistance activities. The apartheid legislation was repealed, and a new era began. Mandela was the first African to be elected president of South Africa in the new multiracial elections of 1994. His presidency, which ended in 1999, set the stage for a multiracial society. The country is still working through the ramifications of all those years of racial separation. As can be expected, the transition has not been without difficulty.

South Africa has large, modern cities such as Cape Town, Johannesburg, and Durban, each about the size of the US city of Chicago or larger. The cities of Pretoria, East Rand, and Port Elizabeth are also major metropolitan areas and have more than one million people each. These urban centers all contribute to and support the extensive mining and agricultural activities that provide national wealth. As the country that exports more diamonds than any other in the world, South Africa has gained much national income from the extraction of mineral resources, which are being tapped by some of the largest mining operations on Earth.

Large parts of South Africa have a moderate climate and good soils, which combine to produce enormous quantities of agricultural products, both for domestic consumption and for export profits. Mining and agriculture have provided extensive opportunities for employment—opportunities that draw in migrants from neighboring countries that have experienced either political unrest or poor economic conditions. These immigrants add to the cultural dynamics of an already ethnically diverse country.

South Africa's manufacturing sector is not well developed. The country depends on Europe, East Asia, and the United States, the three main core economic areas of the world, to provide postindustrial goods. There are few well-paying manufacturing jobs to provide for a growing or stable South African middle class.

The structure and dynamics of the current economic activities have brought about a two-tiered socioeconomic system. Most of the population may work in the mines, in agricultural activities, or in the service sector, but they are not directly benefiting from the profits of their labor, other than earning a wage. These people find themselves in the poorer working-class majority of the population. The landowners, mining corporation executives, and social elite that control the service sector or are employed in activities such as banking or the commodity markets are receiving higher incomes and have created a wealthier upper class. Apartheid supported this class division. The current free and open legal system has yet to bring about a change in the socioeconomic structure of the population. Millions of poor ethnic minorities find themselves in living conditions similar to their economically depressed neighbors in other parts of Africa, while the wealthier upper class has established a good standard of living similar to that of the core economic areas of the world.

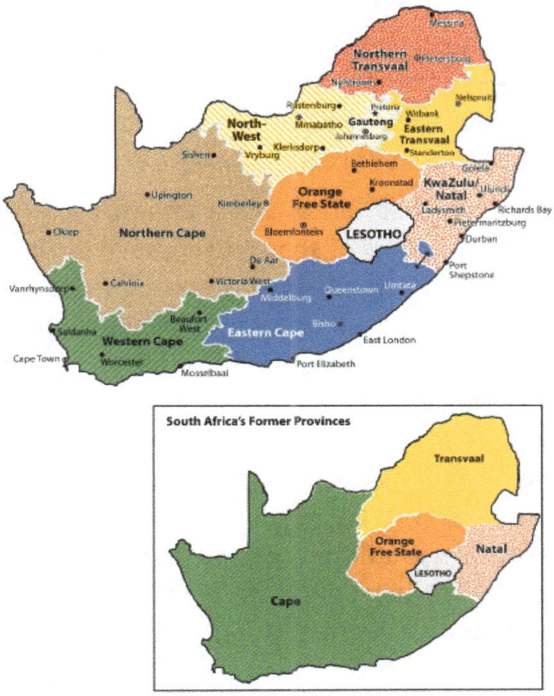

Figure 7.57: Provinces of South Africa after Reorganization in 1994

South Africa redesigned its internal provincial boundaries in 1994. The two large former provinces of Cape in the south and Transvaal in the north were divided up into seven smaller units. The province of Orange Free State was changed to Free State and the province of Natal was changed to Kwazulu-Natal; both kept their main boundaries intact.

South Africa redrew its internal provincial boundaries in 1994. The large former Cape Province in the south, which had been a former British center of power, was divided into three smaller provinces. The large Transvaal Province in the north, which had been a center for Afrikaners, was divided into four smaller provinces. Orange Free State changed its name to Free State and Natal Province changed its name to Kwazulu-Natal. The changes were made based on the size and political management of the country that had come with the transfer of power when Mandela became president. The European colonial pattern of settlement was adjusted to represent a more South African pattern of political administration.

The 2010 World Cup Soccer Games

In 2010, South Africa surpassed Morocco and Egypt in the political arena and were chosen to host the World Cup soccer games. The African teams in included Algeria, Cameroon, Ivory Coast, Ghana, Nigeria, and South Africa. The World Cup is held once every four years. The cost to host the games was enormous and also a testimony to the people and government of South Africa. The country had to build five new stadiums and upgrade five existing stadiums for the games; the largest stadium was Soccer City in Johannesburg, with a capacity of about 84,490. Overall construction costs were estimated to be more than one billion dollars. Transportation infrastructure and road networks were also improved.

The preparations and costs that South Africa undertook to hold the games were not without controversy. Sections of shack dwellers in Durban and in other places were relocated to be out of the sight of international visitors attending the games. This move was made to improve the country's image for tourism. There were also reports of strikes by the construction workers building the stadiums, many of them temporary workers making about three hundred dollars per month, who argued that their wages were too low. South Africa's unemployment rate at the time was about 25 percent, and the country continues to struggle with unemployment. The hope was that the infrastructure improvements would benefit the country for the long term and that tourism would increase if the individuals who came for the games encouraged others to visit. South Africa was banking on this event to boost its economic conditions and strengthen its standing in the global marketplace.

The impact of sports on a society—both positive and negative—can be enormous. Hosting the World Cup necessitated the strengthening of South Africa's security operations to meet FIFA (Fédération Internationale de Football Association) standards to handle the high influx of participants and visitors. It has been estimated that South Africa invested about $4.3 billion to host the games. These are billions that are not going toward providing water, medical clinics, and educational support to the poorest neighborhoods and communities. The financial investments indicate the power and attraction that the World Cup holds on a global basis. No African team was in the final match (Spain defeated the Netherlands), but soccer is a widely supported sport throughout Africa. The 2014 World Cup will be hosted by Brazil.

Figure 7.58 Cape Town Stadium, Taken from Signal Hill, South Africa; Seating Capacity of 64,100 Wikimedia Commons – CC BY-SA 3.0.

Swaziland and Lesotho

The two small landlocked enclaves of **Lesotho** and **Swaziland** are independent countries surrounded by South Africa. Mozambique borders Swaziland to the north. Under British occupation, the area of Lesotho was called **Basutoland.** Upon achieving independence in 1966, the area was renamed the Kingdom of Lesotho. The country is rather mountainous, and most of the land area is more than a mile in elevation. The **Sotho** people have a Bantu background and speak a Bantu language. More than 80 percent are Christian. Lesotho is the size of the US state of Maryland and has a population of more than two million people. The main economic activity has been based on agriculture, but a large percentage of income comes from remittances from Lesotho citizens who work in South Africa. Light manufacturing is beginning to emerge to enhance the economy. Diamonds and other minerals have been found here.

Lesotho is a constitutional monarchy: the government is led by a prime minister, but the king retains his ceremonial position. Reforming the country from a kingdom to a democratic-style government has not been without obstacles. The country has been subjected to political unrest, resulting in many years of military rule and violent internal conflicts. Lesotho is trying to stabilize its political situation and economy.

The **Swazi Kingdom** functioned under relative autonomy during colonial times, and Britain granted independence to Swaziland in 1968. The **Swazi** are a Bantu people who migrated into Southern Africa centuries ago with the Xhosas and Zulus. Swaziland has low mountains to the north with the grasslands and scrub forests of the savannas in the east and rain forests in the west. It is ruled by an absolute monarch, one of the last in the world. Political unrest has led to challenges to his rule. Reformers are pushing for modification of the government and for some element of democratic reform, but reforms are slow in materializing. The small country of Swaziland is about the size of the US state of New Jersey. There are more than one million people, and one-third of the population is infected with HIV/AIDS.

Key Takeaways

- Southern Africa is located along the Tropic of Capricorn and is host to various climate zones, including tropical, desert, and moderate type C climates. Extensive agricultural activity can be found south of the Tropic of Capricorn, where moderate climates prevail.
- Madagascar is a separated from the mainland of Africa and provides a habitat for animals not found on the mainland. The mainland region is large with many different physical landscapes.

- The mainland region of Southern Africa has extensive mineral deposits that provide resources for its countries to gain national wealth. Approaches to translating mineral deposits into wealth have varied among the countries.
- Many ethnic groups can be found in Southern Africa, ranging from the San in the Kalahari to the Zulu groups to the south. Colonialism made its footprint on the region by creating borders and providing colonial names. The transition from colony to independent nation has caused political and economic difficulties.
- South Africa had an apartheid system in place stemming from the colonial era. The apartheid system was dismantled after the 1994 election of President Nelson Mandela. South Africa went so far as to change its provincial system from four to nine provinces to more appropriately politically administer the country.

Discussion and Study Questions

1. What is threatening Madagascar's biodiversity?
2. What do the Great Dyke and the Bushveld basin offer for economic development?
3. What did Namibia do to address environmental conservation and protect its wildlife?
4. Name the country that provided support for SWAPO. Why do you think they did so?
5. Why is Lake Malawi called the Calendar Lake? What process created Lake Malawi?
6. How did the Cold War influence Southern African countries?
7. What economic condition plagues most developing countries that seek aid from the IMF and the World Bank?
8. Why has Botswana reached a higher standard of living for its people than other countries in Southern Africa?
9. How has the policy of apartheid affected South Africa? When did apartheid officially end?
10. How do the people of Swaziland or Lesotho fit into the two-tiered system of social stratification in South Africa?

Geography Exercise

Identify the following key places on a map:

- Bushveld basin
- Cabinda

- Cape of Good Hope
- Copper Belt
- Great Dyke
- Great Escarpment
- Kalahari Desert
- Lake Malawi
- Namib Desert
- Victoria Falls
- Zambezi River

7.7. CONCISE SUMMARY

Subsaharan Africa possesses numerous fast developing core cities and extensive rural peripheral areas undergoing significant rural-to-urban migration. The majority of nations are former European colonies that are cultivating their economy as developing countries. The several tribal tribes residing in the area, each with distinct languages and customs, frequently engage in conflicts with one another or with the government. The informal sector thrives in this environment as individuals pursue opportunities amidst civic upheaval and turmoil. The fragile formal sector in several nations sometimes fails to generate sufficient tax or fee revenue to sustain the public sector and deliver services to the populace. The region possesses significant potential for economic advancement through resource extraction and tourism. West Africa comprises numerous countries, each home to hundreds of distinct ethnic groups with unique languages. The considerable diversity hinders social unity and political cohesion. Situated in the African Transition Zone, there frequently exists tension between the Muslim north and the Christian south. Recent civil wars have ravaged numerous nations. Nigeria is the most populous nation in Africa, characterized by a rapidly expanding and diversified demographic. Despite its diversity and challenges, Sub-Saharan Africa maintains a vibrant celebration of family and community.

Central Africa encompasses a substantial area of tropical Africa, including the extensive Congo River basin. Adverse economic conditions have been partially attributed to numerous civil wars and conflicts in the region, where the dominant party or military authority has amassed the majority of wealth, leaving the populace in dire need of infrastructure enhancements and public services. The shift from colonial status to independence has frequently led to ephemeral and unstable governments, often supplanted by leaders who exploited their authority to divert substantial public funds into their foreign bank accounts. Fraud and corruption have frequently resulted in diminished possibilities and insufficient funding for public services for the majority of the populace.

The ethnic civil war in Rwanda resulted in around one million fatalities. The civil war in the Democratic Republic of the Congo, fought over land and resource control, resulted in nearly five million fatalities, the highest toll of any conflict since World War II. Numerous ethnic groups inhabit a region being explored for valuable mineral resources. Significant reserves of gold, diamonds, and rare minerals like coltan are found in the Democratic Republic of the Congo and its vicinity. The wealth they may create has incited several military confrontations among rival factions competing for control of these resources. Women in Central Africa have experienced significant regressions in their rights and position due to these conflicts. East Africa is characterized by the Great Rift Valley, with extensive lakes, towering mountains, and expansive savanna plains. The Serengeti and other wildlife reserves host vast populations of large animals that appeal to tourism. The majority of individuals are employed in agriculture. Various ethnic and religious communities are present, accompanied by a significant rural-to-urban migration of the populace. Somalia is fragmented, with multiple significant factions governing different regions of the country. Southern Africa, situated on the Tropic of Capricorn, has temperate weather and possesses abundant mineral resources. The nations possess advantageous means to acquire income from agriculture and minerals. A significant diversity of tribal groupings has been affected by colonial activities. South Africa maintained an apartheid system for several years, which ultimately disintegrated in recent decades. Zimbabwe exemplifies a nation ravaged by prolonged totalitarian governance by a single individual, devoid of democratic changes. Madagascar, a big island situated off the eastern coast, possesses distinct flora and fauna compared to the mainland due to isolation caused by tectonic activity. Madagascar was formerly attached to the African continent but separated due to tectonic plate movements in the region. The biodiversity of Madagascar is threatened by significant deforestation, which is obliterating the habitats of rare animal species.

8 Chapter NORTH AFRICA AND SOUTHWEST ASIA

8.1 INTRODUCING THE REALM

The nations of the realm possess three primary characteristics that affect all other human endeavors. The primary shared characteristic pertains to the region's climate. Although other climate types exist in this region, the arid type B climate predominates and encompasses the majority of the physical landscape. Additional climate classifications encompass the H highland climate, characterized by frigid temperatures at elevated altitudes and temperate conditions at lower elevations, found in the mountains of the Maghreb, Iran, and Central Asia, as well as the more temperate type C climate prevalent in coastal areas adjacent to the sea. The C climate of the coastal Mediterranean region fosters human development and hosts numerous significant port cities. A significant portion of each region consists of uninhabited desert. North Africa contains the world's largest desert, the Sahara, which adjoins the Libyan Desert and the Nubian Desert. Approximately one-third of the Arabian Peninsula constitutes the Empty Quarter of the Rub' al Khali (Arabian Desert). Kazakhstan, Uzbekistan, and Turkmenistan possess extensive desert areas with minimal or no population. This facet of the domain underscores the significance of water as a vital natural resource. The majority of individuals in the region are more reliant on the accessibility of water than on the supply of oil.

Figure 8.1 The Regions of North Africa, Southwest Asia, and Turkestan

The second characteristic is Islam: the majority of the population in the realm adheres to the Muslim faith. The practice of Islam in daily life manifests in diverse ways across the various sects of the faith.

The disparities among the divisions have led to friction or outright warfare. Islam functions as more than merely a religion. It functions as a potent cultural influence that has historically unified or divided populations. The contentious nature of the religion has frequently led to significant political conflicts among factions with varying Islamic views. Simultaneously, Islam serves as a unifying factor that connects Muslims with analogous views through shared bonds. Islam offers organization and stability in everyday life. Faith can offer solace and a framework for living. The sacred towns of Mecca and Medina are situated in Saudi Arabia. Additional sacred cities for various sects of Islam encompass Jerusalem and the two towns revered by Shia Muslims: Karbala and Najaf in Iraq. Islam predominates in the region, however other religions hold considerable importance in diverse areas. Israel is a Jewish state, although Christianity is prevalent in regions extending from Lebanon to Egypt. Additionally, adherents of the Baha'i faith, Zoroastrianism, and organizations such as the Druze exist, among others.

The third commonality among all three parts of the realm is the presence of substantial natural resources. North Africa, Southwest Asia, and Turkestan possess substantial quantities of oil, natural gas, and essential minerals. It is logical that not all countries own identical reserves, with some having minimal or none whatsoever. Nonetheless, regarding the acquisition of national riches, the export of oil has predominantly influenced economic activity in relation to the global community. This domain is a marginal realm. The resource that the realm can provide to the principal economic zones of the world is the energy necessary to sustain their economies and uphold their elevated standard of living. Substantial economic gains from the sale of these resources have historically been concentrated in the possession of the elite ruling leader or their clan, often failing to benefit the majority of the community. The management of and income derived from natural resources have emerged as the principal priorities of nations, inciting conflicts and military operations in regions like Iraq and Afghanistan.

Cultural Hearths

Availability and control of fresh water have typically resulted in the ability of humans to grow food crops and expand their cultural activities. Hunter-gatherer groups did not settle down in one area but were more nomadic because of their seasonal search for food. As humans developed the ability to grow crops and provide enough food in one place, they no longer needed to move. The earliest human settlements sprang up in what is the present-day Middle East. Early human settlements provide some indication of early urbanization patterns based on the availability or surplus of food. The shift to permanent settlements included the domestication of livestock and the production of grain crops. Fruits and vegetables were grown and harvested domestically. The activities of this era created humanity's earliest version of the rural-to-urban shift associated with the Industrial Revolution or present development. It is theorized that the ability to grow excess food provided the time and resources for urbanization and the establishment of organized communities, which often progressed into political entities or regional empires.

It has been estimated that some of the earliest cities in the world—Jericho, for example—were first inhabited around 10,000 BCE in the Middle East. In the same region, two cultural hearths provide significant historical value to the concept of human development: Mesopotamia and the Nile Valley in Egypt. Both areas were settings for the growth of human civilization and are still being examined and studied today. In Mesopotamia, a remarkable human civilization emerged along the banks of the **Tigris** and **Euphrates** Rivers in what is present-day Iraq, Syria, and southern Turkey. The climate, soils, and availability of fresh water provided the ingredients for the growth of a human civilization that is held in high esteem because of its significant contributions to our human history.

Figure 8.2 Head of Gudea, Sumerian Ruler from Mesopotamia, Circa 2121 BCE Wikimedia Commons - public domain.

Mesopotamia and the Fertile Crescent

Mesopotamia, meaning "land between rivers," is located between the Tigris and Euphrates Rivers. Neolithic pottery found there has been dated to before 7000 BCE. Humans in this area urbanized as early as 5000 BCE. People were settling in the Mesopotamia region, building magnificent cities, and developing their sense of human culture. Mesopotamia gave rise to a historical cradle of civilization that included the Assyrian, Babylonian, Sumerian, and Akkadian Empires, all established during the Bronze Age (about 3000 BCE or later). Famous cities such as Ur, Babylon, and Nineveh were located in the Mesopotamia region. The control of water and the ability to grow excess food contributed to their success. They developed extensive irrigation systems. Large grain storage units were necessary to provide the civic structure and to develop a military to protect and serve the city or empire. The human activity in this area extended around the region all the way to the Mediterranean Sea, which is where the term Fertile Crescent comes from.

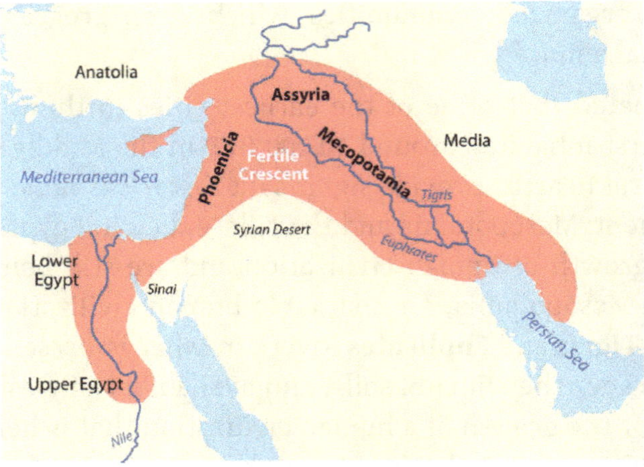

Figure 8.3: The Two Main Cultural Hearths in the Realm: Mesopotamia and the Fertile Crescent in Asia and Upper and Lower Egypt in Africa

Various ancient groups were well established on the eastern side of the Fertile Crescent along the Mediterranean coast. The cities of Tyre and Sidon were ports and access points for trade and commerce for groups like the Phoenicians who traded throughout the Mediterranean. Ancient cities such as Damascus and Jericho became established in the same region and were good examples of early human urbanization during the Bronze Age. These cities are two of the oldest continually inhabited cities in the world.

Nile River Civilization

Human civilization also emerged along the Nile River valley of what is now Egypt. The pyramids and the Sphinx in the **Giza Plateau** just outside Cairo stand testimony to the human endeavors that took place here. Spring flooding of the **Nile River** brought nutrients and water to the land along the Nile Valley. The land could produce excess food, which subsequently led to the ability to support a structured, urbanized civilization. The Nile River is the lifeblood of the region. In the fifth century BCE, the ancient Greek historian Herodotus suggested that Egypt was "the gift of the Nile." The dating for the beginning of the civilization along the Nile River is often in question, but Egyptologists estimate the first dynasty ruled both Upper and Lower Egypt around 3100 BCE. Upper Egypt is in the south and Lower Egypt is in the north because the Nile River flows north. The terms "Upper" and "Lower" refer to elevation. Geologists, using the erosion patterns of the Sphinx, estimate that it was constructed about 10,000 BCE. The ability of humans to harness the potential of the environment set the stage for technological advancements that continue to this day. The Egyptian civilization flourished for thousands of years and spawned a legacy that influenced their neighbors in the region, who benefited from their advancements.

Figure 8.4 Egyptian Pyramids of the Giza Plateau All Giza Pyramids

Wikimedia Commons – CC BY-SA 2.0.

The human activities that created the civilizations in Mesopotamia and along Egypt's Nile River gave humanity a rich heritage to help us understand our history. Many of our legends, stories, and myths have their origins in these cultures. Their cultural developments provided the basis for much of the Western world's religious beliefs and early philosophical ideas. The engineering feats needed to create the magnificent temples and pyramids have by themselves been studied and analyzed over the centuries to give modern scientists and scholars a reason to pause and recognize the high level of organization and structure that must have gone into developing and managing these civilizations. Various aspects of science and the arts were being developed by these ancient people. Writing, mathematics, engineering, and astronomy were becoming highly advanced. Artifacts such as clay tablets and hieroglyphs are still being discovered and interpreted and shed additional light on the advancements of these civilizations and their contribution to our collective human civilization.

Access to Fresh Water

Water is one of the necessities for human existence, and human settlements have long been based on the availability of water for human consumption and agriculture, navigation, and the production of energy. In North Africa and Southwest Asia, the availability of water has an even greater relevance because of the dominant type B climate. Methods used to address the shortage of water or to access fresh water have been nearly as diverse as the people who live here. Large populations of people can be found wherever there is fresh water. Water has historically been transferred from source to destination through canals, aqueducts, or special channels. Many ruins of extensive aqueducts from Roman times and earlier remain throughout the realm. The issues associated with water use continue to affect the lives of the people of this realm. Rapid population growth and industrialization have intensified the demand for fresh water.

Figure 8.1.58.1.5: Roman Aqueduct Near Caesarea. Photo by R. Berglee – CC BY-NC-SA.

Water can be found in the desert regions in a range of forms. For example, there are oases, springs, or noted wells from which people can draw underground water that is close to the surface. Mountainous regions such as the Atlas Mountains in North Africa or the Elburz Mountains in Iran trap moisture, which produces higher quantities of precipitation. The precipitation is then available in the valleys to irrigate crops. Discovering or developing other methods of acquiring fresh water is a requirement in areas without mountains.

Underground Water in Libya

In the Sahara region, Libya draws water from deep underground wells that tap into the vast aquifers beneath the desert that were charged with water when the region was tropical thousands of years ago. The water is referred to as fossil water. Extensive systems of canals and pipelines have been developed in Libya to extract fossil water and use it for agricultural production or for urban purposes. The man-made river project, one of the largest of its kind, has drawn fresh water from the desert to large cities such Tripoli and Benghazi. The local drinking water in Benghazi is contaminated by saltwater intrusion from the Mediterranean. Underground aquifers are underneath political boundaries, so the allocation and control of water are matters of political debate with the potential to lead to military conflict. The project's potential duration will be a function of how quickly the water is used and how many people use the aquifer system. The main problem with this system is that underground aquifers are not considered a renewable resource; as more countries tap into the aquifers, the available water will be depleted more quickly. As water is drawn from the aquifers for industrial irrigation, the water table goes down. Local settlements that rely on well water may in time have to dig deeper wells or move to locations where water resources are still available.

Nile Water in Egypt

Egypt draws water from the Nile to irrigate fields for extensive food production. For thousands of years, floods of the Nile annually covered the land with fresh silt and water. This made the land productive, but the flooding often caused serious damage to human infrastructure. The building of the Aswan High Dam in the 1970s helped control the flooding of the Nile Valley. The river no longer flooded annually, and water had to be pumped onto the land. Over time, the constant and extensive use of this type of irrigation causes the small quantities of salt in the water to build up in the soil to serious levels, thereby reducing the land's productivity. This process, called salinization, is a common problem in arid climates. To rid the soil of the salts, fresh water is needed to flood the fields, dissolving the salt and then moving the salty water back off the fields. High salinization in the soil and the reduction in agricultural productivity is a growing concern for Egypt. Egypt's growing population places a high demand on the availability of food. More than half of the eighty million people in Egypt live in rural areas, and many of them make their living in agriculture, growing food that plays a critical role in the country's economic stability

Water from the Tigris and Euphrates

The major source of water in the Fertile Crescent region comes from the Tigris and Euphrates rivers. Both have their origins in Turkey and converge at the Shatt al-Arab waterway that flows into the Persian Gulf. The Euphrates is the longest river in Southwest Asia and flows through Syria from Turkey before entering into Iraq. Turkey had developed large dams on both the Tigris and Euphrates for agricultural purposes and to generate hydroelectric power. As water is diverted for agriculture in Turkey there is less water flowing downstream for Syria or Iraq. Disputes over water resources continue to be a major concern in the Tigris-Euphrates Basin. The Atatürk Dam in Turkey is largest dam on the Euphrates, and it has a reservoir behind the dam that is large enough to hold the total annual discharge of the river. All three countries have dams on the Euphrates and both Turkey and Iraq have dams on the Tigris. The three countries signed a memorandum of understanding in 2009 to strengthen cooperation within the Tigris-Euphrates Basin. All three countries need the water for agriculture to produce food for a growing population. Agreements to share water have been difficult as a result of the Iraq War and the recent protests and demonstrations in Syria that have contributed to further political tension between the three countries.

Water Conservation in Israel

Israel has taken innovative steps to conserve water and use it efficiently. Drip irrigation mixed with fertilizers is called fertigation. Fertigation is used extensively in the area. Israel grows plantation crops such as bananas, which require large quantities of water. Banana groves are covered with material that allows sunlight to penetrate but reduces the amount of transpiration, which conserves water. Israel has worked to recycle water whenever possible. Gray water is water extracted from sewage that has been treated to be used in agriculture. Underground wells in the West Bank region provide water for a high percentage of people in both the Palestinian areas and Israel. The issue of control over the water is contentious at times. Just as the control of water may have been an important factor in the early Mesopotamian civilizations, it remains a point of political conflict in places such as Israel and the West Bank. The lack of fresh water and the heavier demand placed on water resources have caused countries that can afford it to desalinize seawater. This process is used extensively in the oil-rich states of the Arabian Peninsula. Israel is implementing a similar plan to accommodate their increasing population and fresh water requirements.

Figure 8.6: Banana Grove in Israel Near the Lebanese Border The grove is covered with material that allows sunlight to penetrate but helps reduce the loss of water through transpiration. Photo by R. Berglee – CC BY-NC-SA.

Mountain Water in Iran

Iran is largely a desert climate; thus most of the country does not receive copious amounts of precipitation. In an effort to redistribute the water available from the high mountains, Iranians have developed a system of qanats to collect water where it is available and channel it to the cities or urban areas for use. A qanat might include a system of shafts or wells along a mountain slope that allows water to collect into a common underground channel, which is then diverted to wherever it is needed. This system has been in use since ancient times in many arid regions of the realm and around the world. More than 2,700 years ago, a qanat was developed in what is present-day Iran. The qanat has a system of hundreds of wells and channels water for more than twenty-eight miles; it still provides drinking water to more than forty thousand people in the city of Gonabad. Thousands of qanats were developed over the centuries in this area. Persians used cold qanat water from the mountains to keep ice cool during the summer months. Agricultural production relies heavily on water from qanats, which in turn are dependent on climate conditions and local weather patterns.

Water Shortages in Arabia

The Arabian Peninsula almost always conjures up images of desert conditions; contrary to that image, there is water in the peninsula. Underground aquifers of fossil water are located beneath the sands similar to that in the Sahara Desert. Saudi Arabia taps into these water resources to irrigate agricultural land to grow food. This area faces the same issue as Libya and other areas of the world that rely on underground aquifers: eventually the water supply will be used up. The majority of the Arabian Peninsula is desert and water is always in short supply. Many coastal desert countries such as the United Arab Emirates (UAE), Bahrain, and Qatar have resorted to the desalinization of seawater for their needs, but this is viable only as long as the country can afford the energy cost required to operate the desalinization plants.

Yemen is a state on the Arabian Peninsula that is mountainous and has used terracing to more effectively benefit from what little precipitation they receive. Water from precipitation is trapped in terraces, and as it flows down the mountainside, each terrace uses what it needs and then passes the water to the next lower terrace. More land can be farmed to produce higher yields of agricultural crops such as sorghum and cereals. One of the main problems with Yemen's terrace system is that most of its trees are being cut down for firewood. Tree roots are essential for holding the soil together on the fragile mountain slopes. Another problem is heavy rains that cause serious erosion and damage to the terrace system. Additionally, the extremely rapid growth rate of Yemen's population is outstripping its agricultural production.

Diversion of Water in Turkestan

Fresh water is in short supply in many of the desert regions of Turkestan in Central Asia. Agricultural production has traditionally been dependent on water flowing in rivers and streams that originated with the precipitation from the mountains, but as humans have developed canals and irrigation systems, water from rivers has been diverted for agricultural use. Vast fields of cash crops such as cotton were developed during the Soviet era for economic reasons, and as discussed in more detail in Section 8.7 "Central Asia and Afghanistan", the result had devastating consequences for the Aral Sea, which depended on the water from these rivers for its survival. More than half the population of Central Asia depends on agriculture for their livelihood. The other half, of course, requires water and food for their existence.

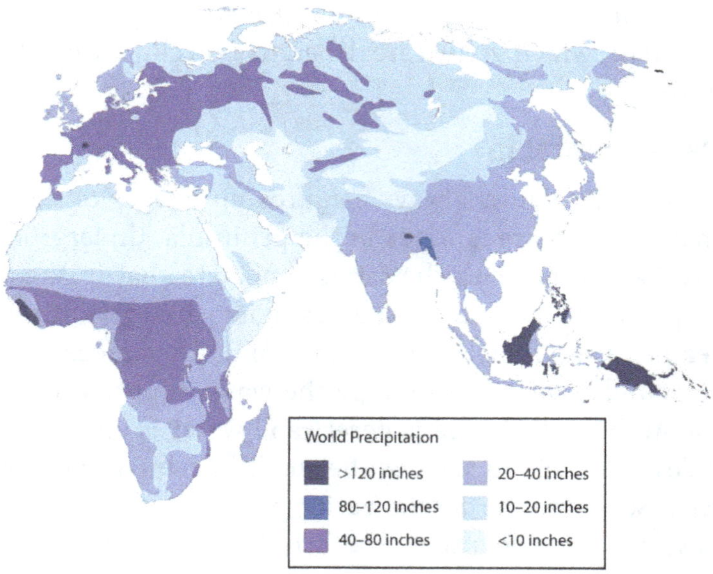

Figure 8.7: Precipitation Patterns for North Africa and Southwest Asia

Arab Spring of 2011

The year 2011 brought about important changes for the human geography of parts of this realm. The year ushered in a wave of human activity that awakened the power of the citizens to speak out against conditions in their country and actively protest against their governments. North Africa, the Middle East, and the Arabian Peninsula experienced the highest levels of protests and insurgency. Political leaders that had been in power for extended periods were challenged and removed from office. Democratic reforms were requested or demanded by citizens seeking more individual freedom and greater access to political power. Uprisings in some of the countries were internal; other countries received external support or intervention. Overall, demonstrations, protests, and outright revolution involved millions of people desiring improved living conditions and a better future for themselves and their families.

Protests emerged in North Africa in the beginning of 2011. Tunisia was the first country in which leadership felt the heat of civil resistance and open revolution. In January, the Tunisian president of more than twenty-three years was forced to flee to Saudi Arabia. In Egypt, millions of protesters demonstrated in the streets against political corruption and the lack of reforms. The revolution of Egypt's citizens was not an armed conflict, but it was an effective protest, because it eventually brought about the ouster of President Hosni Mubarak, who had been in office for almost thirty years. Demonstrations and protests continued against governments in Morocco and Algeria; the people voiced their concerns regarding issues such as high unemployment, poor living conditions, and government corruption. Libya's protests erupted into a full-scale armed revolution as antigovernment rebels took control of the city of Benghazi in an attempt to topple Muammar Gadhafi's forty-two years of authoritarian control of the government, oil revenues, and the people. The armed Libyan revolution was eventually successful in taking control of Tripoli and in removing Gadhafi and his family from power. The revolution in Libya was aided by North Atlantic Treaty Organization (NATO) air strikes and the implementation of a no-fly zone over the country.

The ripple effect that the Tunisian revolution had on North Africa was felt on the Arabian Peninsula in places such as Yemen and Bahrain. Mass public demonstrations in Yemen over government corruption, economic conditions, and high unemployment escalated into serious armed conflicts between government troops and opposition factions that wanted the president removed from office. In Bahrain, the protests and demonstrations were centered on the request for more personal freedoms and a greater role in leadership for the Shia population, who experienced discrimination by the Sunni-dominated government. Protests also occurred in Oman for greater reforms.

The Middle East did not escape the Arab Spring of 2011. Protests in Jordan forced King 'Abdullah II to reorganize his government. Israel and Lebanon were not as affected, as they have been addressing many of these issues on an ongoing basis. The country experiencing the greatest impact was Syria. Major mass demonstrations and serious protests against the government were staged in a number of cities across the country. In Syria, the long-term leadership of an Alawite minority continues to run the government and control the military. The al-Assad family—a father and then his son—has ruled Syria since 1971. The Syrian government has cracked down on the revolution with hard-line measures aimed at subduing the protests and demonstrations. By September 2011, more than two thousand protesters had been killed in Syria, and many more were detained or tortured. Countless others have tried to flee to neighboring countries for their safety. The protesters in Syria want democratic reforms as well as the end of the al-Assad family reign.

Other parts of the realm also felt the effects of the Arab Spring of 2011 with mixed results. Iran has had similar protests and demonstrations in past years, but there was not a major revolution or uprising as a direct result of the Arab Spring. Iran is not an Arab country but has experienced ongoing political friction between citizen factions and the government. The wars in Iraq and Afghanistan have tempered or isolated internal protests or revolutionary activities in those countries even though armed conflicts continue. Various Central Asian states have been working through similar issues but either have not had mass demonstrations or have not received the attention of news media regarding their situations. The wave of change that swept over the realm in the Arab Spring of 2011 is an example of how centripetal and centrifugal cultural forces act on a state or region. The political landscape was altered or drastically changed in many countries. The impact of these changes will be realized in the years and decades to come.

Key Takeaways

- The realm of North Africa and Southwest Asia extends from the Atlantic Ocean along the Moroccan coast to the western border of China. It includes the regions of North Africa, Southwest Asia, and Central Asia (often referred to as Turkestan).
- Three basic features that dominate this realm include the arid type B climates, Islam as the predominant religion, and the export of petroleum and minerals to gain wealth. There are exceptions to all three features, but these three are found within most countries of the realm.
- The two main cultural hearths in this realm are located along the rivers in Mesopotamia and in Egypt. Control of and access to water resources to grow excess food were the basis for the success of the empires that flourished in these two areas.
- Fresh water is a valuable resource that is not always available in North Africa and Southwest Asia because of the climate and physical geography. Each region within the realm has developed its own methods to draw from or extract the valuable resource of fresh water.
- The Arab Spring of 2011 was a massive wave of protests and demonstrations by citizens of the realm against their governments over such issues as poor living conditions, high unemployment, government corruption, and the lack of democratic reforms. Various leaders were removed from office and governments were pressed to reform their power structures to allow for more shared governance and reduced political corruption.

Discussion and Study Questions

1. Why does this realm include parts of three continents?
2. Which countries have territory on more than one continent?
3. What common traits are shared by most people in this realm?
4. What might have prompted the first rural-to-urban shift for human settlements?
5. Why is the Middle East called by that term? What is it in the "middle" of?
6. What did the two cultural hearths contribute to the advancement of human culture?
7. Why did the cultural hearths develop where they did?

8. What are some of the methods Israel has used to address their fresh water requirements?
9. What prompted the Arab Spring of 2011? Which country was the first to see change?
10. What were some of the common issues that people protested across the realm?

Geography Exercise
Identify the following key places on a map:

- Arabian Peninsula
- Asia Minor
- Atlas Mountains
- Elburz Mountains
- Empty Quarter
- Euphrates River
- Fertile Crescent
- Giza Plateau
- Levant
- Libyan Desert
- Mesopotamia
- Middle East
- Nile River
- Nubian Desert
- Tigris River
- Turkestan

8.2. MUHAMMAD AND ISLAM

Located in the mountains of western Saudi Arabia, the city of **Mecca** (also spelled Makkah) began as an early trade center for the region and a hub for camel caravans trading throughout Southwest Asia and North Africa. Mecca is about forty-five miles from the Red Sea coast at an elevation of 531 feet. South of Mecca, the mountains reach more than 7,200 feet in elevation. According to Islamic tradition, the patriarch Abraham came to Mecca with his Egyptian wife Hagar and their son Ishmael more than two thousand years before the birth of the Prophet Muhammad (born 571 CE). When Hagar died, Abraham and Ishmael built the **Kaaba** (or Ka'ba), a rectangular shrine that included a special stone, in Mecca. The shrine was destined to become one of the holiest sites for nomadic groups in Arabia. Abraham later died in Palestine in what is now the country of Israel. Centuries after Abraham's death, the Kaaba and the rituals associated with it deteriorated and mixed with other local traditions.

Figure 8.8: The Holy Mosque in Mecca, the Most Holy Site for Muslims The black rectangular structure in the photo is the Kaaba. Menj – The Hajj Makkah – CC BY-ND 2.0.

The Prophet Muhammad

The traditional groups in the region of Arabia were polytheistic and worshiped their own gods. By the time of Muhammad, Mecca is said to have been a center of worship to more than 360 deities or gods; the greatest of these was Allah (meaning "the god"). Allah was known as the chief of the Meccan pantheon of gods and was worshiped from southern Syria to Arabia. Mecca was full of idols, temples, and worship sites. Tradition states that the god Allah was the only god without an idol; he would become the sole entity of Muhammad's new Islamic religion.

Muhammad, born in Mecca 571 years after the birth of Christ and about 100 years after the fall of the Roman Empire, was orphaned at an early age, and was employed in a camel caravan when he reached his teens. His life and what has been reported about it changed the Middle East forever. Muhammad traveled throughout the region with the camel caravans. He was fortunate to have been able to live as he did, because most orphans in the region did not have many opportunities in life. His travels introduced him to many people, places, and issues. His situation changed when Muhammad and a widow many years his senior were married. Muhammad became a merchant, the leader of a camel caravan, and a respected member of his community. He was reported to have been intelligent and a wise businessman.

The traditional groups that traded in Mecca held many different religious beliefs. The city was a forum for the many vices and activities associated with trade, travel, and metropolitan business. To escape the activities of Mecca, Muhammad would often seek the solitude of the mountains, where he would contemplate and think. Tradition states that the angel Gabriel appeared to him while he was meditating in a mountain cave in 610 CE, when Muhammad was about forty. Muhammad was given words from Allah, which he recited from memory to his followers. According to tradition, Muhammad was illiterate; his supporters wrote down his words, compiled them into the Koran (Qur'an), the holiest book of Islam. Muhammad was the founder of the new religion, which he called *Islam* (meaning "submission to Allah"). The term *Muslim* (meaning "one who submits") refers to a follower of Islam.

After Muhammad returned to Mecca and related his visions and Allah's words from the angel Gabriel, he began to speak out against the city's vices and many gods. He stated that there was only one god: Allah, the same creator god of Abraham. He spoke out against gambling and drinking alcohol. He advocated the caretaking of widows and orphans. He also preached regarding family and community. His message was not well received: in 622 CE the people of Mecca forced Muhammad out. He fled to the safety of the nearby city of **Medina** in a journey known as the hejira (hijra). This historic journey became the start of the Islamic calendar, which is based on the lunar cycles. Muhammad found refuge in Medina and became a respected citizen.

Launching out from Medina, Muhammad and those loyal to him defeated the army of Mecca and converted the city into Islam's holiest place. They destroyed all Mecca's idols and temples except the Kaaba. Muhammad's teaching united the many Arabian groups under one religion. Since the Koran was written in Arabic, Arabic became the official language of Islam. The Kaaba and the mosque built at Mecca became the center of the Islamic world and a destination for Muslim pilgrims. Islam brought a new identity, a faith in one god, and a set of values to the Arab world. Islam made sense in a world with many traditional beliefs and few unifying principles.

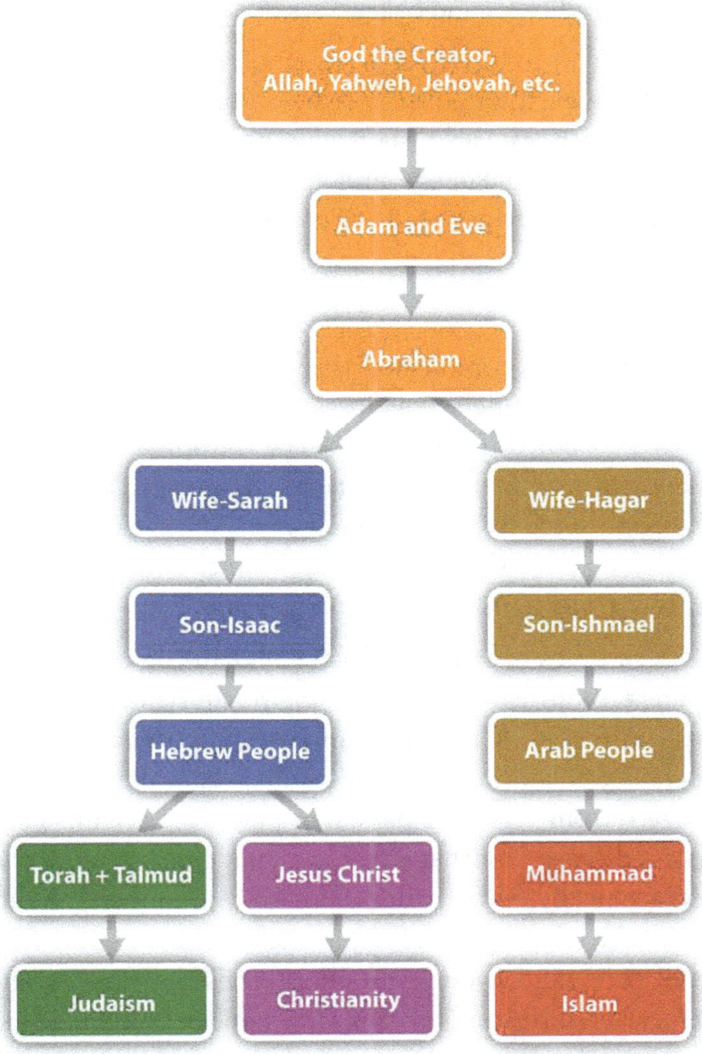

Figure 8.9 Traditional Succession of the Three Main Monotheistic Religions of the Middle East: Judaism, Christianity, and Islam

It's important to keep in mind that monotheist belief was not new: Christianity had been around for more than six hundred years. Judaism and Zoroastrianism in Persia had been around for centuries before Christianity. The principles of Islam and Muhammad's teachings are a continuation of Judaism and Christianity. All three traditions assert a faith in a divine creator, with important messages coming through prophets or holy messengers. All three religions acknowledge Abraham as a founding patriarch. Muslims believe that Moses and Jesus were major prophets and that Muhammad was the greatest and final prophet. All three religions have stories about creation, Adam and Eve, the flood, and other similar stories that have been adapted to the traditions and characters of each religion.

Religion is a part of culture. The religions that emerged out of the Middle East absorbed many of the existing cultural traits, traditions, or habits of the people into their religious practices. Early Islam adapted many Arab cultural traits, styles of dress, foods, and the pilgrimage and folded them into its principles. Early Christianity and Judaism also adopted cultural traits, holidays, styles of dress, and cultural traditions.

Spatial Diffusion

The spread of Islam was accomplished through trade and conquest. Mecca was a center of trade. When camel caravans left Mecca, they carried Muhammad's teachings with them. Islam diffused from Mecca and spread throughout the Middle East and into Central Asia and North Africa. The geographic principle of spatial diffusion can be applied to any phenomenon, idea, disease, or concept that spreads through a population across space and through time. The spatial diffusion of Islam outward from Mecca was significant and predictable.

There are two main types of spatial diffusion: expansion diffusion and relocation diffusion. Expansion diffusion has two main subtypes: contagious diffusion and hierarchical diffusion. A religion can spread from individual to individual through contagious diffusion when a religion starts at one point and propagates or expands outward from person to person or place to place in a pattern similar to the spreading of a disease. Another way a religion can spread through expansion diffusion is hierarchically, when rulers of a region convert to the religion and decree it as the official religion of their realm; the religion filters down the political chain of command and eventually reaches the masses. The second type of diffusion, relocation diffusion, takes place when the religion relocates to a new place from a central point. When Islam jumped from the Middle East to Indonesia, it diffused through relocation. Relocation diffusion also occurred when Islam spread to the United States.

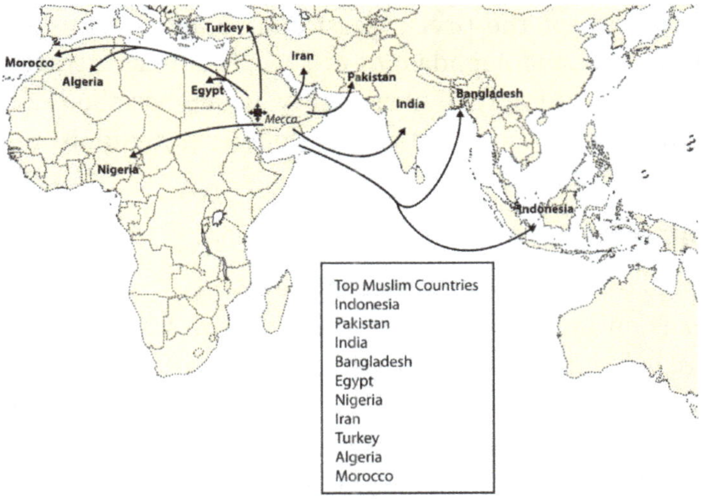

Figure 8.10: Diffusion of Islam and the Ten Countries with the Highest Muslim Populations. Data courtesy of Pew Research Center, Mapping the Global Muslim Population: A Report on the Size and Distribution of the World's Muslim Population.

Figure 8.11: Faithful Muslims Praying toward Mecca in Umayyad, Damascus. Wikimedia Commons – CC BY 3.0.

Early on, the unifying principles of Islam found their way into the regional groups of Arabia and into the minds of their leaders. By 700 CE, Islam had spread to the east, to the Mogul Empire of Pakistan and northern India. In India, the Emperor Shah Jehan, who built the famous architectural marvel of the Taj Mahal as a mausoleum for his wife, was a Muslim. The expansion of Islam fueled the Arab Empire of the Middle East. The capital city of the Arab Empire was first established at Medina and then moved to Damascus and later to Baghdad. While Europe was enduring the Dark Ages, Islam was experiencing a renaissance, expanding its knowledge of mathematics, architecture, and the sciences. The Arab institutions of higher learning kept the Greek classics alive and established universities in Toledo (Spain), Cairo, and Baghdad. As of 2010, Islam has attracted as many as 1.5 billion followers, second only to Christianity, which has about 2 billion followers. Hinduism is third, with about 900 million followers. Buddhism is considered the world's fourth-largest religion.

The Five Pillars of Islam

The basic tenets of the **Five Pillars of Islam** create the foundational structure of Islam. Prayer is an important part of the religion. A Muslim must offer prayers five times a day: before sunrise, at midday, at midafternoon, after sunset, and in the early evening. During prayer, Muslims face toward the compass direction of Mecca. Before clocks and time were well established, a mosque leader would climb a minaret (a tall tower next to the mosque, their place of public worship) and call the faithful to prayer at the required times of day. Muslims gather together for common prayer on Friday, which is a time to unite the community of believers. Mosques sprang up after Muhammad died, and they became the center of community activities in the Islamic world.

The Five Pillars of Islam can be translated as follows:

1. **Express the basic creed (*Shahadah*).** Profess that there is no god but Allah and his messenger and prophet is Muhammad.
2. **Perform the prayers (*Salat*).** Pray five times a day.
3. **Pay alms or give to charity (*Zakat*).** Share what you have with people who are less fortunate.
4. **Fast (*Sawm*).** During the month of Ramadan, abstain from personal needs, drinking, and eating from dawn to dusk (as one's health permits).
5. **Make the pilgrimage to Mecca (*Hajj*).** Conduct at least one pilgrimage to the holy city of Mecca (if within one's capacity).

The Death of Muhammad

Muhammad died at the age of sixty-two. He never claimed to be a god or anything other than a mere mortal. His tomb is located in Medina, the City of the Prophet. No provision was made to continue Muhammad's work after he died. One division thought his successor should be a blood relative. This division led to the **Shia** (or **Shi'ite**) branch of Islam, which makes up about 15 percent of Muslims. Others felt that the successor should be a worthy follower and did not need to be a blood relative. This branch became known as **Sunni**, which makes up about 84 percent of Muslims. Various smaller branches of Islam also exist, including **Sufi**, which approaches the Islamic faith from a more mystical and spiritual perspective.

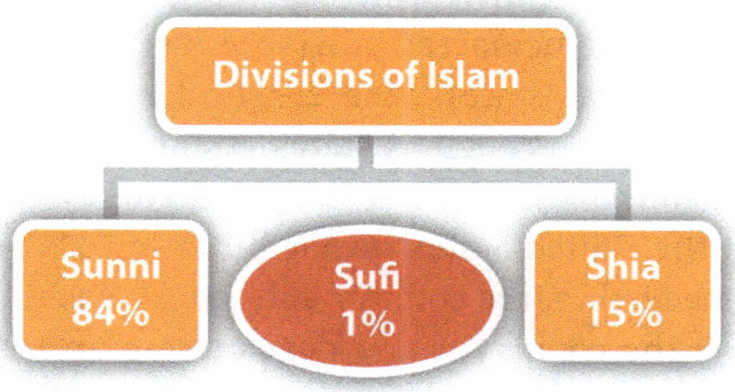

Figure 8.12: Three Main Divisions of Islam with Approximate Percentages

Sunni Muslims look to the family and community for direction; Sunni leaders are elected by the whole community. Shia Muslims look to their imams for the official source of direction. Imams hold the religious and political leadership in the Shia faith. Through the right of divine appointment, Imams are considered by many in the Shia division to hold absolute spiritual authority. Imams often have the final word regarding religious doctrine. Shia Muslims consider Muhammad's son-in-law and cousin Ali to be the origin of the direct line of succession for Imams. Ali is considered in Islam to have been the first person to accept and follow the words of Muhammad. There are many subgroups or branches in each of the Islamic divisions.

In understanding the Middle East, it is most critical to understand the Sunni and Shia divisions of Islam. The Shia and Sunni divisions of Islam have sometimes had divergent beliefs, resulting in conflicts. In the early sixteenth century, the Persian Empire, which is now Iran, declared the Shia branch its official religion. Its surrounding neighbors were predominantly Sunni. This divergence is part of the basis for the current civil unrest in Iraq. The two divisions of Islam currently vie for political power and control in Iraq. The majority of the Arab population in Iraq, about 60 percent in 2010, follows the Shia division of Islam, but the leadership under Saddam Hussein until 2003 was Sunni. Tradition states that Ali is buried in the Iraqi city of Najaf, which is considered by Shia Muslims to be one of the holiest sites in Islam. Just north of Najaf is the city of Karbala, which is also considered to be a holy place for Shia Muslims because it is the burial place of the Prophet Muhammad's grandson Husayn. The Shia majority in Iraq who are Arab share their faith with the Shia majority in Iran who are ethnically Persian.

Secular State versus Religious State

Islam has a code of law called the **Sharia** criminal code, which is similar to Old Testament law. The Sharia dictates capital punishment for certain crimes. For example, if a person is caught stealing, his or her arm would be severed. For more serious offenses, he or she would be beheaded or stoned to death. Some countries use the Sharia as the law of their country. Countries are called religious states

(Islamic states in this case) when religious codes take precedence over civil law. States in which people democratically vote on civil law based on common agreement are called secular states. Whereas secular states attempt to separate religious issues and civil law, religious states attempt to combine the two. Iran is a good example of an Islamic religious state, and Turkey is a good example of a secular state. When the United States invaded Iraq in 2003 and Saddam Hussein was removed from leadership, the country entered a transitional period in which they had to decide if the country would develop into an Islamic state with the Sharia or move to a democratically elected government with civil law. The debate on these issues continually surfaces in many of the countries in North Africa and Southwest Asia whenever transition occurs.

The cultural forces of democratic reforms and Islamic fundamentalism have been pushing and pulling on the Islamic world. Democratic reformers push for a more open society with equality for women, social freedoms for the people, and democratically elected leaders in government. Islamic fundamentalists pull back toward a stricter following of Islamic teachings; they oppose what they consider the decadent and vulgar ways of Western society and wish to restrict the influence of liberal, nonreligious teaching. A rift between militant Islamic fundamentalists and moderate Islamic reformers is evident throughout the Muslim world. Militant leaders strive to uphold the Sharia criminal code as law. Moderate reformers work toward a civil law based on democratic consensus. This rift adds to the conflicts that have been occurring in this realm. Islamic fundamentalists push for a more traditional and conservative society and express opposition to the United States' intervention in the realm. The Muslim world will continue to confront such arguments over the future direction of Islam in a globalized economy.

Key Takeaways

- Islam has its origins with Muhammad, who was born 571 years after the birth of Jesus, when Christianity was well established. Muhammad received his revelations through the angel Gabriel and passed them to his followers, who wrote down his words into what became the Koran.
- Islam is the youngest of the three major monotheistic religions of the realm: Judaism, Christianity, and Islam. All three trace their origins back to the patriarch Abraham. Other monotheistic religions, such as the Baha'i faith and Zoroastrianism, are also evident in the Southwest Asia.
- Spatial diffusion can be applied to any phenomenon, idea, disease, or concept that spreads through a population across space and through time. Islam has diffused through both expansion diffusion and relocation diffusion to become the second most followed religion in the world.

- Since the death of Muhammad, Islam has divided into a number of different factions. The two most prevalent divisions of Islam are Sunni (followed by about 84 percent of Muslims) and Shia (followed by about 15 percent of Muslims). Other minor divisions of Islam, such as Sufi, also exist.
- Religious states are structured around laws based on religious regulations that are usually determined by religious leaders. Secular states are structured around civil law, which is decided on by democratic consensus.

Discussion and Study Questions

1. What was Muhammad's early life like? Do you think his background influenced his teachings?
2. Who were Sarah and Hagar? How were their sons influential to the people of the Middle East?
3. Name the individual who is considered a patriarch to the three major monotheistic religions of the Middle East.
4. What event triggered the start of the Islamic calendar? What is this calendar based on?
5. List the Five Pillars of Islam. On which day of the week does the Islamic community gather for prayer?
6. Name the two main divisions of Islam. What percentage of Islam's disciples follow each of the divisions?
7. What are the differences between the two main divisions of Islam?
8. What are the two main types of spatial diffusion? Provide an example of each.
9. What is the difference between a religious state and a secular state? Which of these types of state does the United States fit into?
10. How has the conflict between Islamic fundamentalists and Islamic reformers manifested itself in the North Africa, Southwest Asia, and Turkestan realm in recent years?

Geography Exercise
Identify the following key places on a map:

- Baghdad
- Cairo
- Damascus
- Jerusalem
- Karbala
- Mecca
- Medina
- Najaf
- Toledo

Activity

1. On a map of this realm, indicate the type of government for each country. Determine which countries are religious states and which are secular states.

8.3. NORTH AFRICA AND THE AFRICAN TRANSITION ZONE

The principal link between North Africa and the Middle East, as well as Central Asia, is the diffusion of Islam from these regions to North Africa. Currently, it is a Muslim-majority region where Arabic is the predominant language. Historically, North Africa's ethnicity was primarily Berber, with the nomadic Tuareg and other indigenous communities intermingled. As Islam spread into North Africa, it became intertwined with Arab influence and culture. Contemporary Egypt has emerged as the epicenter of the Arab world; Cairo hosts a greater Arab population than any other metropolis globally. The primary regions of focus are the Maghreb in the northwest, the Nile River valley in the east, and the African Transition Zone, where the Sahara Desert merges with the tropical type A climates of Central Africa's equatorial region.

Figure 8.13: North Africa and the Maghreb The Maghreb traditionally includes Morocco, Algeria, and Tunisia, but Libya is also considered part of the Maghreb by many inhabitants of the region. Map courtesy of University of Texas Libraries.

Islam diffused through North Africa to the Berber people of the Maghreb and entered Europe across the **Strait of Gibraltar** to the **Iberian Peninsula.** The Arab-Berber alliance, called the Moors, invaded Spain in 711 CE. The Islamic influence thrived in Iberia and would have continued into mainland Europe if not stopped by Christian forces such as Charles Martel's army in the famous Battle of Tours. Islam was eventually pushed out of the Iberian Peninsula and held south of the Strait of Gibraltar. Islamic architecture and influence remain part of the heritage of Iberia.

The historical geography of North Africa is not complete without an understanding of the European influences that have dominated or controlled this region for centuries. The Roman Empire controlled much of the coastal area of the

Mediterranean during its zenith. The Romans built ports, aqueducts, roads, and valuable infrastructure. After the fall of the Roman Empire, common bonds of religion and language were created for the people through the invasion of the Arabs, who introduced the Islamic faith. North Africa was later dominated by European colonialism. France controlled and colonized the region of the Barbary Coast along North Africa's western waterfront, including Algeria, Tunisia, and parts of Morocco. Italians colonized the region that is now Libya. The Barbary Coast of the Mediterranean was once a haven for pirates and a danger to shipping during the colonial era. Even the United States involved itself with wars against the pirates off the coast of the Berber states of North Africa during the early 1800s. After the fall of the Ottoman Empire, Britain controlled Egypt and parts of the Sudan. The Spanish colonized parts of Morocco and Western Sahara. In due time, resistance movements were successful in defeating the colonial powers and declaring independence for all the countries of North Africa. However, the European influence remains through the region's dependence on trade and economic partnerships with Europe.

North Africa is separated from Subsaharan Africa by the African Transition Zone, a transitional area between Islamic-dominated North Africa and animist- and Christian-dominated Subsaharan Africa. It is also a transition between the Sahara Desert and the tropical type A climates of Africa's equatorial region. This is a zone subject to shifting boundaries. The region was once a major trade route between the Mali Empire of the west and the trade centers of Ethiopia in the east. Camel caravans have crossed this sector of Africa for centuries, and camel caravans from Mecca might have traveled across this zone. Many nomadic groups continue to herd their livestock across the region in search of grazing.

The Maghreb: "Isle of the West"

The **Maghreb** is a region extending from Morocco to Libya that is distinguished by the main ranges of the **Atlas Mountains**, which reach elevations of near thirteen thousand feet. The main Atlas range is often snow-covered at higher elevations. The name *Maghreb*, which in Arabic means "Isle of the West," receives between ten and thirty inches of rainfall per year. This is substantially more rainfall than what is received in the Sahara Desert to the south. The Atlas Mountains extract precipitation from the air in the form of rain or snow, which allows fruits and vegetables to be grown in the fertile mountain valleys of the Maghreb. To the south of this region is the vast Sahara Desert with lower precipitation and warmer temperatures. Libya is actually outside the range of the Atlas Mountains but is associated with the Maghreb by most local inhabitants.

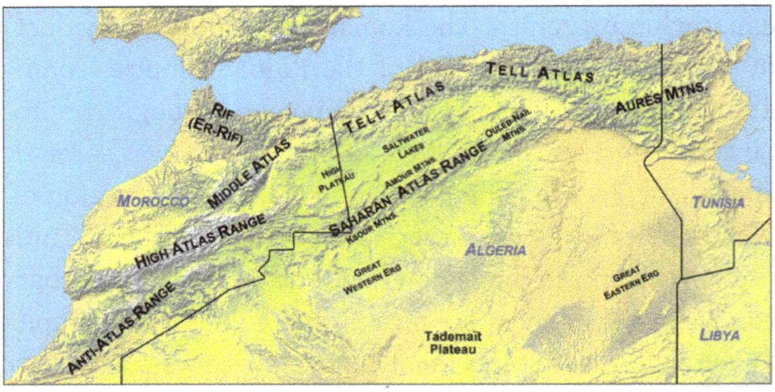

Figure 8.14: The Main Mountain Ranges of the Maghreb The Atlas Mountains of the Maghreb extend to the east and west of the main ranges. Wikimedia Commons – public domain.

Aided by a moderate type C climate, the northern coastal region of the Maghreb and the mountain valleys are a center for agricultural production, including grapes, dates, oranges, olives, and other food products. Think about how geography affects population: Which climate type do most human groups gravitate toward? What conditions will you find when you combine this climate type and generous quantities of water and food? As you fit the pieces of the geographic puzzle together, you can understand why populations centralize in some places and not in others. The Maghreb is an attractive place for human habitation, but it borders on the inhospitable vastness of the Sahara Desert. Most of the Maghreb's residents live in cities along the Mediterranean coast. There are few people in the vast desert interior of these countries. The exceptions are groups such as the Tuareg that are found in the Sahara.

The Maghreb is an expansive region with countries of size. Algeria, Libya, and Morocco are large countries in terms of physical area. Algeria is Africa's largest country as of 2011. Algeria is similar in area to the entire United States east of the Mississippi River, Libya is larger than the state of Alaska, and Morocco is the size of the state of California. A large percentage of Algeria south of the Maghreb and a large percentage of the area south of the Mediterranean coastline in Libya fit the classification of desert conditions. The largest cities of Libya are along the coast, but other Libyan cities are positioned in the desert region. Tunisia, the smallest of the countries on the Mediterranean, is about the same size as the US state of Wisconsin and has mountains in its north and desert areas to the southwest.

All the countries of the Maghreb have former connections to Europe. These ties have strengthened in recent years because of an increase in trade and the economic dependencies that have been created between Europe and the Mediterranean. North Africa can grow fruit and vegetable crops that are not as plentiful in the colder northern latitudes. In the last half of the twentieth century, an enormous amount of oil was discovered in the Maghreb, and Europe has a growing need for energy. The discovery of oil changed the trade equation: oil and natural gas revenues subsequently advanced past agricultural goods as the main export products. Oil and natural gas exports now make up 95 percent of the export income for Algeria and Libya.

Economic Geography of the Maghreb

Europe, which is in the higher stages of the index of economic development, has small families with fewer young people to fill entry-level service jobs, and North Africa has a burgeoning population of young people seeking employment. Many people from North Africa speak the languages of their former colonial masters, and when they leave North Africa seeking employment, they find the transition to a European lifestyle relatively straightforward. Migration from the shores of North Africa to Europe is not difficult; the Strait of Gibraltar, for example, is only about nine miles across from Morocco to Spain.

Figure 8.15: The Strait of Gibraltar North Africa is separated from the Iberian Peninsula by the Strait of Gibraltar, which connects the Mediterranean Sea with the Atlantic Ocean. The distance from Morocco to Spain across the strait is about nine miles, making immigration to Europe from North Africa only a matter of a short boat ride. Satellite image courtesy of NASA SRTM Team – public domain.

European countries have attempted to implement measures to halt the tide of illegal immigration into their continent from North Africa but have not been successful. The need for cheap labor in European countries is a major economic factor in this equation. The core-periphery spatial relationship creates the push-pull forces of migration based on opportunities and advantages. Europe needs cheap labor and more energy, provides employment opportunities, and has an advantage in its higher standards of living: these forces attract immigration and pull people toward

Europe. North Africa can supply labor and oil, has high levels of unemployment, and suffers from poor economic conditions: these factors push people to emigrate from North Africa to places where conditions are more attractive. Europe is the core economic region, and North Africa is the peripheral economic region. People usually shift from periphery to core in their migration patterns, and this is the case across the Mediterranean.

European and American influences have been strong in North Africa. Western culture continues to compete with fundamental Islamic tendencies in the region, especially in urban centers. In Morocco, which is the farthest country from Mecca, democratic reforms have opened the country to more opportunities for women and have led to exposure to Western fashions, ideas, and products. Tunisia has a supportive Westernized society. In Libya, birth control or family planning products have not always been supported by the government, so family size remained relatively high until about 1985 when it began to decline. Women are allowed to go to college in Libya, but a smaller percentage are enrolled compared to men. More than 90 percent of Libyans live in urban areas. Exposure to Western concepts, along with urbanizing and industrializing of the economy, has caused a drop in family size in Libya—from 7.5 children in 1975, to about 4.0 children in 1990, and down to 3.0 children in 2010 (Central Intelligence Agency). More women are trading traditional dress for Western-style clothing and are entering the workforce to the extent they are allowed.

Political Geography of the Maghreb

Economic and political pressures are building across North Africa. Toward the end of 2010 and the spring of 2011, activists called for the governments in North Africa to implement change and address the push of North Africans for stronger democratic openness, less government corruption, and the sharing of wealth accumulated by those in power. Leaders have been reluctant to relinquish power and are being challenged by protests and revolution. The driving forces for the people have been high unemployment rates, high food prices, and the lack of adequate housing.

Western Sahara and Morocco

Western Sahara is the region south of Morocco. In the latter half of the 1970s, Morocco annexed and took control of this region following the withdrawal of Mauritanian control. A local resistance group called the Polisario Front challenged Morocco's claim to Western Sahara and staged a guerilla war to fight for the independence of Western Sahara. By 1991, the United Nations (UN) had stepped in, brokered a halt to the fighting, and worked to resolve the matter. Both sides have offered proposals to the UN; Morocco's plan advocates for annexation, and Polisario's plan is for independence. Western Sahara is mainly desert terrain and only has about three hundred thousand people. The UN delegation has indicated that independence is not realistic. However, talks continue between the factions to work out a solution.

Figure 8.16: Morocco and Western Sahara

Algeria

Algeria has also experienced its share of violent clashes; in this case, the clashes are between Islamic fundamentalist groups and the democratically elected government. In the 1990s, the Islamic Salvation Front, which advocates for a fundamentalist Islamic state in Algeria, challenged the secular political mainstream. The electoral process was interrupted, and the government found itself fighting an Islamic insurgency within the country. By 1998, more than one hundred thousand people had been killed. The horror of the violence received international attention. Islamic extremists widened their attacks and massacred entire villages to send a message to support their cause. By the end of the decade, government forces gained control of the country, and the Islamic Salvation Front officially disbanded. Smaller extremist groups continued to operate. They joined forces with Osama bin Laden's al-Qaeda group to create an insurgent group called *al-Qaeda in the Lands of the Islamic Maghreb*, which has continued a campaign of terror and violence against the Algerian government and people in the region with Western interests. Coinciding with similar protests across North Africa, during 2010 and 2011, there was an insurgence of protests and demonstrations against the government of Algeria by its people, who were asking for better living conditions. The government made some concessions to address the issues, but the political climate in Algeria continues to be tense as the government struggles to find ways to satisfy the needs of the people.

Tunisia

Tunisia was once an outpost of the Roman Empire, and well-preserved Roman ruins can be found throughout North Africa. In 1956, at the end of the colonial era, France recognized Tunisia's independence, and the country established its own government. The political problems in Tunisia today stem from the fact that little has changed in the government since the time of independence. Since independence

Tunisia has had few changes in government leaders. The country's second president, **Ben Ali**, dominated the country for twenty-four years (1987–2011). Ben Ali worked to stem the tide of Islamic fundamentalism, opened the country to establishing rights for women, and allowed more tolerance of religious diversity; however, under his leadership the government struggled to find a balance between democratic openness and authoritarian measures to keep the country moving forward with economic development and social services for a growing population.

Figure 8.17: Egyptian Demonstration Protests in Cairo's Tahrir Square on March 4, 2011, were in support of demonstrations in Egypt and in Libya, which followed the protests in Tunisia. Tarek – Egyptian and Libyan Flags – CC BY-ND 2.0.

Beginning in 2010, corruption, unemployment, and a lack of personal freedoms prompted Tunisians to take to the streets in massive protests, which grew into a revolution for change. By the end of the year, President Ben Ali was removed from power and an interim government was formed. Protests continued as the government shifted to adapt to the conditions. In 2011, a new constitutional assembly was formed to address changes in the government. Economic conditions have remained thorny during the transition.

Libya

Muammar Gadhafi came to power in Libya in 1969 by overthrowing the king and never held an election for political control. He used Libya's oil revenues to build up the country's infrastructure, enhance his military, and create an anti-Western terrorist network with weapons production. Personal freedoms have been limited in Libya, and there has been no free speech, freedom of the press, or freedom of assembly. Gadhafi supported various Islamic terrorist groups and nationalized all the oil assets of international corporations in Libya in the early 1980s. This action of nationalism directly opposed and angered many corporate colonialists. His policies caused the international oil companies, which operated much of the oil industry there, to pressure the US government to conduct a series of devastating military bombing raids on Libya in 1986 in an attempt to kill Gadhafi.

Gadhafi deployed his military along Libya's southern border with Chad in a zone called the **Aozou Strip**. This border territory was the site of a boundary dispute over land that was reported to hold uranium reserves. Libyan forces controlled the Aozou Strip for a number of years before Chad forced them out. In 1994, the UN ruled that the zone belonged to Chad, and the dispute was finally resolved. In other activities, Libyan agents were accused of planting bombs on passenger airlines, such as the 1988 explosion on Pan Am flight over Scotland that killed 270 people and the 1989 explosion of UTA flight 772 over Niger that killed 170 people. After the 9-11 tragedy in New York, Gadhafi backed down on his aggressive anti-Western activities and agreed to dismantle some of his weapons production facilities and comply with international trade agreements. Libya then denounced terrorism, and the United States lifted economic sanctions against Libya and exchanged ambassadors.

Unrest in Libya surfaced again in 2011, coinciding with protests under way in neighboring Tunisia, Algeria, and Egypt. Factions in Libya that opposed Gadhafi's regime advocated for his removal from office and for more democratic freedoms. Based in Libya's second-largest city of Benghazi, in the eastern sector of the country, the resistance movement gained momentum and pushed for an all-out civil war that split the country in half. Forces loyal to Gadhafi in the capital city of Tripoli in the western part of the country faced off against the separatists under the structure of the National Transitional Council (NTC). In 2011, Gadhafi had been in power for more than forty-two years. His family and clan controlled much of the country's wealth from the export of oil. Gadhafi's control of Libya did not allow for the establishment of strong administrative divisions within the government to share power.

The UN, through the North Atlantic Treaty Organization (NATO), implemented a no-fly zone over Libya and allowed military strikes against Gadhafi's forces to stop the massacre of civilians in the civil war. By the end of the summer of 2011, the resistance forces headquartered in Benghazi had pushed across to the western part of Libya and took control of Tripoli. Gadhafi was no longer in control of the country, as the NTC became the transitional power base. Muammar Gadhafi was killed in October 2011. Post-Gadhafi Libya will most likely be quite different from the status quo of the past four decades. Nation building will become a major focus for North Africa, which may be more difficult for Libya than its neighbors because of the lack of an administrative structure.

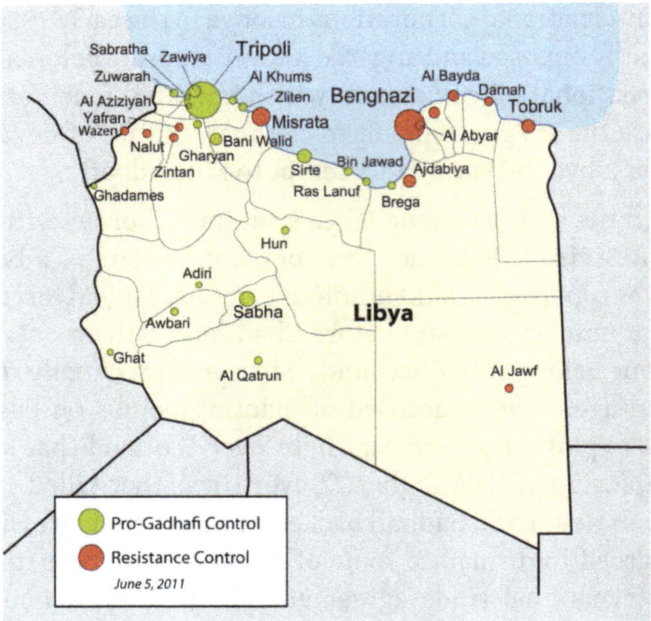

Figure 8.18: Control of Cities in Libya Based on the Civil War as of June 5, 2011. Wikimedia Commons – CC0 – public domain.

The Nile River and Egypt

Figure 8.19: Nile River System

The Nile River originates in East Africa in Lake Victoria and in Ethiopia in Lake Tana. The White Nile flows north from Lake Victoria through Uganda and into Sudan, where it converges with the Blue Nile at the city of Khartoum, Sudan's capital. The Blue Nile originates in Lake Tana in Ethiopia. From Khartoum, the Nile River flows north through the Nubian Desert into Egypt, where it eventually reaches the Mediterranean Sea. The fresh water of the Nile is a lifeline that enables agriculture and transportation and supports a growing human population in the region.

The White Nile originates in Lake Victoria, and the Blue Nile originates in Lake Tana. They converge at Khartoum.

Until the Aswan High Dam was completed in 1968, the river flooded its banks yearly, depositing silt and nutrients onto the soil and causing enormous damage to infrastructure. As far back as when the pharaohs ruled Egypt, the people used flood irrigation to grow their crops. Today, water is pumped from the controlled Nile River onto the fields to water crops. This change has increased the number of crops that can be grown per year. However, it has also caused a buildup of salt in the soil, resulting in declining soil quality. Without annual flooding, the salts cannot be dissolved away but remain in the soil, reducing yields. Almost a third of Egypt's population works in agriculture; about half the population is rural.

Population Dynamics

Cairo, Egypt's capital, lies at the northern end of the Nile River. With a population of more than ten million, it is the largest North African city and home to more Arabs than any other city in the world. It is considered the cornerstone city of Arab culture. Cairo is so crowded that more than a million people live in its old cemetery, the City of the Dead. Cairo's residents, and the millions of people in Egypt, depend on the Nile River for their survival. About 95 percent of Egypt's population lives within fifteen miles of the Nile River. As the population has grown, urban expansion has encroached on the farmland of the Nile Valley. Egypt can no longer produce enough food for its people; about 15 percent of its food comes from other countries, mainly the United States.

Conflicts between democratic reforms and Islamic fundamentalism are evident in Egypt. The growing population of about eighty million in 2010 is a major concern. In Egypt's case, democratic reformers were able to promote a strong program of family planning and birth control to help reduce family size, which in 2008 was at 2.8 children per woman and declining. The government even created a popular Egyptian soap opera to promote the concept that it was appropriate in an Arab culture to use family planning and have a small family. The prime-time program, called *And the Nile Flows On*, told the story of a young village bride dealing with the issues of pregnancy and life complicated by the interjection of a progressive sheik and a meddling female doctor. The drama addressed many family planning and religious issues regarding the acceptability of breaking with tradition to address the growing population problem in Egypt.

Television programming is popular in Egypt, and even reruns of old American shows such as *Bonanza* and *Dallas* are dubbed into Arabic and shown on Egyptian television. Egypt is a cultural mix with a strong heritage steeped in Arab history with a secular side that is open to the outside world. The cultural forces that create this paradox have not always been in unison. Egypt has a major connection to Western society because of tourism. The Pyramids of Giza and the Great Sphinx are major attractions that pull in millions of people per year from around the world. Tourism opens up Egypt to outside elements from various cultural backgrounds, most of which are secular.

Political Dynamics

On the political front, the democratically elected government has received strong opposition from the Islamic Brotherhood, which advocates a more fundamentalist Islamic lifestyle and government structure. The democratic reformers that vie with the Islamic Brotherhood for political power support a more open and democratic civil government. These two elements are what drive Egyptian culture and society.

Figure 8.20: Protesting in Egypt On January 25, 2011, the "Day of Revolt" protests were held in Egypt. Tens of thousands of Egyptians went onto the streets to peaceably protest against the Mubarak government. Muhammad Ghafari – CC BY 2.0.

The political situation changed in Egypt with the Arab Spring of 2011. Student protests against government corruption and the lack of democratic reforms emerged with an intensity that gained the support of the Egyptian people and forced the Egyptian government to respond. Egyptian president Hosni Mubarak had been in power since 1981 after the assassination of the previous president, Anwar Sadat. President Mubarak was able to win every election for president that had been held since that time. Millions of Egyptians took to the streets in early 2011 in civil protests against the Mubarak government. Massive protests and demonstrations continued until February 11, when President Mubarak announced his resignation. The transition to an interim government has been complicated by continued protests and calls for justice from the people, who have called for Mubarak to stand trial for stealing billions of dollars from the state and concealing it in his own bank accounts. The people and the government continue to search for progressive opportunities to address their issues. What started out as the Arab Spring turned into the Arab Year as all three long-term leaders in Tunisia, Libya, and Egypt were removed from positions of power.

The African Transition Zone

Figure 8.21: Camel and Tuareg Rider in the Southern Sahara Desert The Tuareg are masters of the desert and camels. They often lead camel caravans on long trips through the desert. Marco Bellucci – Tuaregs with their dromedaries – CC BY 2.0.

Stretching across the widest part of Africa on the southern edge of the Sahara Desert is the African Transition Zone. Known as the Sahel, meaning "border or margin," this zone is where the dry arid conditions of the desert north meet up with the moister region of the tropics. For thousands of years, the seasonal grazing lands of the Sahel have been home to nomadic groups herding their livestock across the zone and eking out a living held together by tradition and heritage. Changing climate conditions and overgrazing has enhanced the desertification process, and the region is slowly turning into desert. The Sahara Desert is shifting southward, altering the economic activities of the millions of people who live in its path. Ten thousand years ago, North Africa and the Sahara Desert were tropical environments filled with all the biodiversity and wildlife now found in Subsaharan Africa. This desertification process has been occurring for centuries; it is not a new process. Human factors and climate change may be accelerating this process, but they did not create it.

Political stability is complicated to achieve in the African Transition Zone. The political borders established by European colonialism during the Berlin Conference of 1884 remain basically intact and create barriers that hamper the nomadic groups from traveling through the Sahel in search of grazing land for their livestock. Political boundaries now restrict movement and keep people divided and separated into national identities. The African Transition Zone is also in transition from a rural, traditional agrarian culture to a society confronting the information age and modern technology. Camel caravans that once transported goods and materials across wide expanses of desert terrain are being replaced with motor vehicles and aircraft. The many traditional groups across this zone are adapting to the conditions of the modern world but work to retain their values and the traditions of their heritage.

The colonial political borders have impeded progress in the region's effort to establish stable governments and control the land and resources within its borders. Postindependence governments have thus far been unable to establish stable economic conditions within many of the countries in the Sahel. Natural resources are being exploited for economic profit, which is changing the natural environment. Security and safety have become significant issues. Today this region is unstable, with political and cultural conflicts between the local groups and governments. The current conflicts in Sudan are examples of the instability.

Sudan: Slavery and Genocide

Comparable in size to the entire United States east of the Mississippi River, Sudan is the largest country in Africa. The capital city of Khartoum lies where the Blue Nile River converges with the White Nile. Khartoum's government has a black Arab majority and follows Islam, complete with Sharia laws. The African Transition Zone crosses Sudan and separates the Arab-Muslim north from the mainly African-Christian south. There has been a civil war between the north and the south for decades. Before a peace agreement brokered in 2005, military soldiers from the north would raid the villages in the south, taking women and children as slaves. Though the Sudanese government denied the slave trade, thousands of Africans were owned by northern black Arabs in Sudan, and many still are. The world community has made little effort to intervene. The price for a slave in Sudan is about fifty US dollars.

The differences in religion, ethnicity, and culture have always divided southern Sudan from the north. Additional economic considerations might fuel the debate because of an increase in oil production in the region. In January 2011, the southern region of Sudan voted on a referendum that would allow the south to break away and become an independent country called the Republic of South Sudan. The acceptance of this new republic will change the map of the region and the dynamics between South Sudan and North Sudan. The new Republic of South Sudan was formalized in July of 2011. Juba is designated as the capital with talk of creating a new forward capital in the center of the newly formed country in the future. The many clans and indigenous groups make it difficult for unity and cohesiveness in the new country. Armed groups in the various states continue to cause internal division, while at the same time boundary disputes continue to be worked out with North Sudan.

In 2003, various groups in **Darfur** complained that the Khartoum government was neglecting them. A militia group calling itself the Janjaweed was recruited by the local Arabs to counter the resistance in Darfur. The Janjaweed began an ethnic cleansing campaign that pushed into the Darfur region, burning villages, raping women, and killing anyone who opposed them. Refugees began to flee into the neighboring country of Chad.

South Sudan has elected to break away and become independent. The Darfur region has been

Figure 8.22: Sudan, the Region of Darfur and the Republic of South Sudan

experiencing genocide by Janjaweed militias backed by the Arab majority in northern Sudan.

In this particular case, the campaign was not based on religious divisions, because both sides were Muslim. This was an ethnic conflict in that the people of Darfur are of a traditional African background and the people of northern Sudan consider themselves Arab, even though they may have dark skin. Accurate numbers have been difficult to verify, but as of 2010 an estimated 300,000 people have died in this conflict. There are more than 2.7 million refugees, many of them in Chad (BBC News). Just as the government of Sudan denied the slave trade, it denies that it supports the Janjaweed. The African Union provided a modest number of peacekeeping troops before the UN stepped in to provide security. It has been up to the world community and Sudan to take more action and provide more assistance. Food, water, and care for the refugees have taxed the region's aid and support system.

The African Union

Former Libyan leader Muammar Gadhafi was instrumental in the development of the African Union (AU) in the mid-1990s. The Sirte Declaration (titled after Gadhafi's hometown of Sirte in Libya) was issued by the Organization of African Unity, which outlined the need for the creation of the AU. The AU was launched in Durban, South Africa, on July 9, 2002. Fifty-three countries formed this intergovernmental organization. The focus of the AU is on the health, education, economic development, political stability, environmental sustainability, and general welfare of the people of Africa. The organization strives to integrate the socioeconomic and political stability of its members and promote a continent-wide effort for security and peace. The AU is working to create a proper political climate, one that helps its member states engage in the global economic marketplace by negotiating international issues and policies that affect Africa.

The dominating activities of colonialism and neocolonialism (corporate colonialism) are big concerns for the AU. The AU's objective is to bring more unity to the political and economic arena between the African countries to address the transition to a globalized world. It faces many challenges within its realm, including health care issues such as HIV/AIDS and malaria that have devastated much of Africa. The AU is working to bring political stability to countries such as Sudan and other countries experiencing civil unrest because of political turmoil or civil war, such as the Congo, Somalia, Sierra Leone, Ivory Coast, and Liberia. AU peacekeeping troops are assisting in this process. The legal issues regarding border disputes or territorial disputes such as that of Western Sahara are problems that the AU attempts to address.

In the global scale of economic and political supranationalism, the AU will be up against three main powerhouses: the European Union, the North American Free Trade Agreement, and the East Asian Community. Regions across the globe are working on trade associations to create economic networks to bring about greater cooperation and commerce between nations. The AU is one part of that network that represents a growing percentage of the world's population and the second-largest continent on Earth.

Gadhafi was the AU chairman when the uprisings and demonstrations of the Arab Spring began in January of 2011, at which time his chairmanship ended. At the beginning of the civil war in Libya, the AU attempted to meet with Gadhafi to mediate the situation. The NATO no-fly zone declaration and intervention restricted AU activities in Libya. After Gadhafi was no longer in power in Libya by the fall of 2011, the AU continued to not recognize the NTC (National Transition Council) as the legitimate government of Libya.

Key Takeaways

- Three main physical features of North Africa are the Atlas Mountains, the Sahara Desert, and the Nile River. Most of North Africa's population lives along the Mediterranean coast or along the Nile River. The ethnic majority in the Maghreb are Berber, with Arabs dominating in Egypt.
- The Maghreb centers on the Atlas Mountains, which traditionally has provided for a diversity of food production. Oil has been found in North Africa, the export of which has surpassed the export of food products.
- Europe has had a strong influence on the region, ranging from the Roman Empire, to colonial activity, to becoming a destination for immigrants looking for employment and opportunities.
- North Africa has experienced serious political conflicts. Political leaders in Tunisia, Libya, and Egypt remained in power for decades. People have protested for economic and political reforms. The presidents of Egypt and Tunisia have stepped down, but Libya's longtime leader, Muammar Gadhafi, did not step down but instead engaged the country in a civil war.
- The African Transition Zone creates the southern boundary for North Africa. This zone serves as the transition between the arid type B climates and the tropical type A climates. It is also the transition between the dominance of Islam and the dominance of Christianity and animism.
- The African Transition Zone cuts through the center of Sudan and divides the country along religious and ethnic distinctions. Civil war has been waged in the south and in the Darfur region, which has split the country

into separate regions. Southern Sudan has initiated the process for independence.

Discussion and Study Questions

1. What is the name of the majority ethnic group that resides in the Maghreb? What is the main physical feature there?
2. What have been the main ties between North Africa and European continent?
3. How did the construction of the Aswan High Dam change the way crops are grown along the Nile River?
4. What political changes have been occurring in North Africa? Why are they occurring?
5. Why has there been a civil war in Libya? How was the country divided in this civil war?
6. Where is the African Transition Zone? What does it transition between?
7. What centripetal force brought villagers together in Timbuktu?
8. Name the country in the Sahel where slavery has become prominent. Why did no one step in to stop it?
9. How has the African Transition Zone divided Sudan? What are the main groups on each side?
10. What are the most serious problems in the Darfur region of Sudan? Who is conducting ethnic cleansing and why?

Geography Exercise
Identify the following key places on a map:

- African Transition Zone
- Atlas Mountains
- Blue Nile
- Darfur
- Iberian Peninsula
- Lake Tana
- Lake Victoria
- Maghreb
- Nile River
- Strait of Gibraltar
- Western Sahara
- White Nile

References

BBC News, "Q&A: Sudan's Darfur Conflict," http://news.bbc.co.uk/2/hi/africa/3496731.stm.

Central Intelligence Agency, "The World Factbook," https://www.cia.gov/library/publications/the-world-factbook; "Demographics of Libya," Wikipedia, en.Wikipedia.org/wiki/Demographics_of_Libya.

8.4. ISRAEL AND ITS NEIGHBORS

Israel is situated in a complex geopolitical landscape, bordered by several countries, each with its own historical and political dynamics. To the north lies Lebanon, which shares a contentious history with Israel, particularly regarding conflicts and territorial disputes. To the east, Syria has also been a focal point of tension, especially over the Golan Heights. To the south, Egypt has a peace treaty with Israel, but relations can be fluctuating, influenced by broader regional developments. The Palestinian territories of the West Bank and Gaza Strip, located to the east and west respectively, are central to the Israeli-Palestinian conflict, significantly impacting Israel's relationships with its neighbors and shaping regional politics. This intricate web of relationships involves issues of security, identity, and national aspirations, making the region one of the most challenging and dynamic in the world.

The State of Israel

At the center of the Middle East, on the shores of the Mediterranean in the Levant (the area bordering the eastern Mediterranean Sea), lies the country of Israel. Israel is bordered by Lebanon to the north, Syria and Jordan to the east, and Egypt to the south. Covering an area of only 8,522 square miles, Israel is smaller than the US state of Massachusetts and only one-fifth the size of the state of Kentucky. The coastal region, which has a moderate type C climate, receives more rainfall than the dry interior and the **Negev Desert** in the south, both of which have arid type B climates. The **Sea of Galilee**, also called Lake Kinneret or the Sea of Tiberias, is a major fresh water supply. The **Jordan River** flows from the Sea of Galilee to the **Dead Sea**. The Dead Sea is 1,300 feet below sea level, so it has no outlet. Over time, salts and minerals have built up, creating an environment that does not support fish or aquatic life. South of the Negev Desert is the **Gulf of Aqaba**, which provides access to the Red Sea for both Israel and Jordan. Israel does not have substantial oil resources but has a potential for natural gas in offshore locations along the Mediterranean Sea.

Though most of the population in the Middle East is Islamic, there are exceptions, such as in Israel, which has a Jewish majority. Israel was established in 1948. Before that time, the country was called **Palestine**. The region went through a series of tumultuous transitions before it became the nation of Israel. Before 1948, most people in Palestine were called Palestinians and consisted primarily of Arab Muslims, Samaritans, Bedouins, and Jews. Most Jewish people were dispersed throughout the world, with the majority in Europe and the United States.

The Division of Palestine

Palestine was a part of the Turkish Ottoman Empire before the end of World War I. Britain defeated Turkish forces in 1917 and occupied Palestine for the remainder of the war. The British government was granted control of Palestine by the mandate of the Versailles Peace Conference in 1919 at the end of World War I. Britain supported the Balfour Declaration of 1917, which favored a Jewish homeland. The British Mandate included Palestine and Transjordan, the area east of the Jordan River, which includes the current country of Jordan.

Between 1922 and 1947, during British control, most of the population of Palestine was ethnically Arab and followed Islam. In 1922, Jews made up less than 20 percent of the population. The Jewish settlements were mainly along the west coast and in the north. Jewish people from other countries—including primarily Jews escaping German oppression in the 1930s—migrated to the Israeli settlements. Palestine was turned over to the control of the newly created United Nations (UN) in 1945 at the end of World War II.

The United Nations Special Committee on Palestine (UNSCOP) was created by the UN in 1947. To address the Palestine region, UNSCOP recommended that Palestine be divided into an Arab state, a Jewish state, and an international territory that included Jerusalem. Palestine was divided by the UN. About 44 percent of the territory was allocated to the Palestinians, who consisted of about 67 percent of the population, which was mainly Arab. Approximately 56 percent of the territory was allocated to the minority Jewish population, who only consisted of about 33 percent of the population. The country of Jordan was created out of the region east of the Jordan River and the Dead Sea. The city of Jerusalem was to remain under the administrative control of the UN as an international city. The Jewish State of Israel was officially recognized in 1948. The Palestinians, who were a majority of Israel's total population at the time and who owned about 90 percent of the land, denounced the agreement as unacceptable (Knowledgerush). One of the consequences of the territorial partition was that thousands of Palestinian Arabs were forced off the land that was allocated to the Jewish state. These Palestinians became refugees in the Palestinian portion or in neighboring countries.

Palestine's Arab neighbors—Syria, Saudi Arabia, Lebanon, Iraq, and Egypt—sided with the Arab Palestinians and declared war on Israel. The war did not end favorably for the Arabs. With support and aid from Britain and the United States, the Jewish State of Israel defeated the attacking Arab armies and took control of a larger portion of the land, including some of the land designated by the UN as a portion of the Arab half.

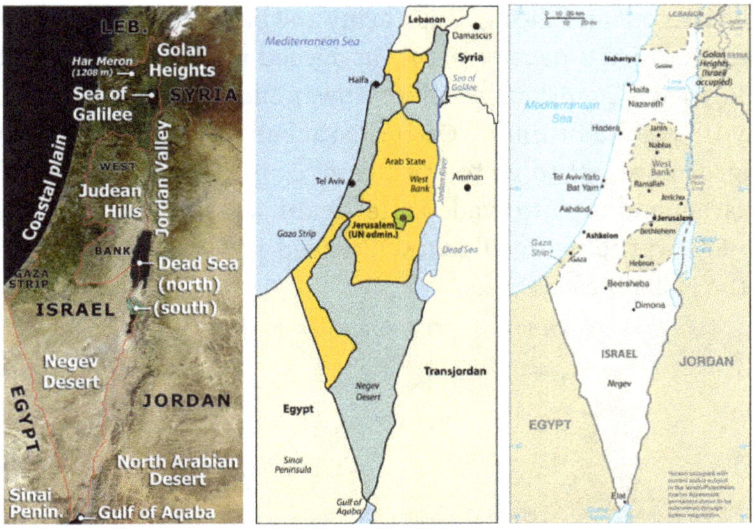

Figure 8.23: Satellite Image of Palestine (*left*); 1948 UN Division of Palestine into Half Jewish State and Half Arab State (*center*); Political Map of Israel in 2011(*right*). Wikimedia Commons public domain; Wikimedia Commons - public domain; CIA World Factbook - public domain.

After the Arabs lost the first war against the Jewish State of Israel, the Palestinians' problems increased. By 1950 over 750,000 Palestinians living in the Jewish-controlled regions of Israel were forced out of their homes and into refugee camps. According to the UN in 2010, about one-third of the registered Palestine refugees, more than 1.4 million, live in fifty-eight recognized refugee camps in Jordan, Lebanon, the Syrian Arab Republic, the Gaza Strip, and the West Bank, including East Jerusalem. Jordan has the highest number of Palestinian refugees: almost two million, with more than 330,000 living in refugee camps (United Nations Relief "Palestine Refugees"; United Nations Relief "Statistics"). Jordan has allowed most of its Palestinian refugees to receive Jordanian citizenship.

By 1967, the Arab armies had regrouped and were willing to attack Israel again. The 1967 war was short lived, lasting only about a week. The Arab armies were devastated once again, and Israel gained even more territory. Israel took the Sinai Peninsula and the Gaza Strip from Egypt, the Golan Heights from Syria, and the **West Bank** from Jordan. The entire city of Jerusalem came under Israeli control. The 1967 war solidified the control of the region of Palestine under the Israeli government and placed Israel at greater odds with its Arab neighbors. Syria wanted Israel to return the Golan Heights, which has a strategic military advantage in overlooking northern Israel, and Egypt wanted Israel to return control of the Sinai Peninsula.

Egypt and Syria attacked Israel again on October 6, 1973, which was Yom Kippur, the most solemn holiday in the Jewish religion. The Israeli army counterattacked, driving the Syrians out and the Egyptian army and back across the **Suez Canal**. After a few weeks of conflict, a peace treaty was agreed upon. In 1977, Israeli president Menachem Begin and Egyptian president Anwar Sadat were invited to Camp David, Maryland, by US president Jimmy Carter. Israel and Egypt signed the Camp David Accord, an agreement not to go to war again. Egypt agreed to officially recognize the State of Israel and to not invade Israel again. Israel agreed to return the Sinai Peninsula to Egypt; the peninsula was returned in 1982. Each participant in the accord won the Nobel Peace Prize.

Figure 8.24: The Western Wall in Jerusalem A remnant of the Jewish temple built by Herod the Great and destroyed in 70 CE, the Western Wall is the most holy place for the Jewish people. The Dome of the Rock mosque in the background is the third-holiest site for Muslims. Photo by R. Berglee – CC BY-NC-SA.

In 1980, Israel passed the Jerusalem Law, which stated that greater Jerusalem was Israeli territory and that Jerusalem was the eternal capital of the State of Israel. The UN rejected Israel's claim on greater Jerusalem, and few if any countries have accepted it. Israel moved its capital from Tel Aviv to Jerusalem to solidify its claim on the city even though most of the world's embassies remain in Tel Aviv. The move of the capital was designed to create a *forward capital*, the purpose of which is usually either to protect a nation's territory or to spur the development of the country. In this case, it was to protect valuable territory.

Palestinians were left with only the regions of the Gaza Strip and the West Bank, which is controlled by the Israeli government and is subject to Israel's national jurisdiction. As of 2010, about 1.5 million Palestinians live in the Gaza Strip and 2.5 million live in the West Bank. A number of cities in the West Bank and Gaza Strip have been turned over to the **Palestinian Authority (PA)** for self-governing. The PA was established between the Palestine Liberation Organization (PLO) and the Israeli government to administer internal security and civil matters. The PLO and the PA are two separate entities. The PLO is the internationally recognized governing body of the Palestinian people. It is legitimately recognized by the UN to represent the area known as Palestine in political matters. There are two main political parties within the PLO: Hamas and Fatah. The Hamas party is the strongest in the region of the Gaza Strip, and the Fatah party is more prominent in the West Bank.

The Palestinians, Israel, and Possible Solutions

The future of the West Bank and the Gaza Strip has been the focus of talks and negotiation for decades. There are various ways to approach this issue; a one-state solution and a two-state solution have been proposed. The one-state solution proposes the creation of a fully democratic state of Israel and the integration of all the people within its borders into one country. Integration of the Gaza Strip and the West Bank into the Jewish State of Israel is part of this plan; in other words, "Take the walls down and create one state" (Masri, M., 2011). Many Palestinians support the one-state solution, but most of the Jewish population does not. Family size is much larger in the Palestinian side, so it would be only a matter of time before the Jewish population would be a minority population and would not have full political control with a democratic government. To have the Jewish State of Israel, the Jewish population needs to keep its status as the majority.

In a two-state solution, Palestinians would have their own nation-state, which would include the Gaza Strip and the West Bank. The rest of former Palestine would be included in the Jewish State of Israel. The two-nation concept (Israel and a Palestinian state) has been proposed and supported by a number of foreign governments, including the United States. Implementation of a two-state solution is, of course, not without its own inherent problems. At the present time, the West Bank and the Gaza Strip are subjects under the Jewish State of Israel without full political or economic autonomy. The two-state solution would buy more time for the Jewish population with smaller families to retain power as a majority political voting bloc.

Parties to the negotiations have acknowledged that the most likely solution is to create a Palestinian state bordering Israel. However, it is not clear how to make this happen. Palestine is now divided between the Jewish State of Israel (with 7.3

million people) on one side and the Palestinians (with 4.0 million people) in the West Bank and the Gaza Strip on the other side. About 75 percent of Israel's population of 7.3 million people are Jewish, and about 25 percent are Arab. Travel between Israel and the Palestinian areas is heavily restricted and tightly controlled. A high concrete and barbed wire barrier separates the two sides for much of the border. The West Bank provides fresh water used on the Israeli side for agriculture and industrial processes. The industries also employ Palestinians and support them economically.

Jewish people from various parts of the world continue to migrate to Israel, and the Israeli government continues to build housing settlements to accommodate them. Since the West Bank region is under the Israeli national jurisdiction, many of the new housing settlements have been built in the West Bank. The Palestinians who live there strongly oppose the settlements. In 1977, only about five thousand Jews lived in the West Bank settlements. As of 2010 there were more than two hundred thousand. The Palestinians argue that if they were to have their own nation-state, then the Jewish settlements would be in their country and would have to be either resettled or absorbed. Israel responds by indicating that the two-state solution is indefensible because the Jewish settlements in the West Bank cannot be protected if the West Bank is separated from Israel.

The issues in Israel are generally complicated. After a series of wars and considerable negotiations, the central problems remain: Jews and Palestinians both want the same land, both groups want Jerusalem to be their capital city, and neither group can find a compromise. Support for the Jewish State of Israel has primarily come from the United States and from Jewish groups external to Israel. There are more Jews in the United States than there are in Israel, and the US Jewish lobby is powerful. Israel has been the top recipient of US foreign aid for most of the years since 1948. Through charitable donations, US groups provide Israel additional billions of dollars annually. Foreign aid has given the Jewish population in Israel a standard of living that is higher than the standard of living of many European countries.

In the past decade, most of the PLO's operating budget has come from external sources. Arab neighbors provide millions of dollars annually. Though Iran is not Arab, they have provided aid to the Palestinian cause in support of fellow Muslims against the Jewish State of Israel. The PLO has received the bulk of its funding from the European Union. Russia has also provided millions of dollars in aid. The United States provides millions in direct or indirect aid to the Palestinians annually.

The Gaza Strip is one of the most densely populated regions on Earth, and it is located in a desert. There are few jobs and no real methods of gaining wealth.

Palestinians living in the Gaza Strip or the West Bank need to rely on outside support to survive. These are small areas with lots of people. The West Bank is only about thirty miles wide by seventy-five miles long, yet more than 2.5 million people call it home. The Gaza Strip is a desert region about six miles wide by twenty-three miles long and is home to more than 1.5 million people. In 2010, family size in the West Bank was about 3.2, and in the Gaza Strip, it was about 5.0. Unemployment rates averaged about 40 percent in the Gaza Strip and over 20 percent in the West Bank (Central Intelligence Agency). Underemployment is also a major issue in that there may be few employment opportunities for professionals with specialized skills or a university degree in a specific field of interest.

Figure 8.25: A Street in the West Bank City of Nablus Cell phones are ubiquitous in Nablus. Photo by R. Berglee – CC BY-NC-SA.

In 2006, both Israel and the PLO held democratic elections for their leaders. In 2006, a candidate from the Hamas party won the election for the leadership of

Figure 8.26: Security Wall between Israel and the West Bank. Photo by R. Berglee – CC BY-NC-SA.

the PLO, which concerned many of the PLO's external financial supporters. The Israeli government characterizes Hamas as a terrorist organization that supports the destruction of the State of Israel. Hamas has advocated for suicide bombers to blow themselves up on populated Jewish streets. The Jewish State of Israel has been fighting a low-level war against Palestinian resistance groups such as Hamas since 1948. In 2008, the leaders of the Fatah party, which are more prominent in the West Bank region, took matters to the PLO Central Council, which chose Mahmoud Abbas as president of the State of Palestine.

The problems between Israel and Palestinians are far from settled. The region has plenty of interconnected concerns. The biggest supporter of Israel, the United States, invaded Iraq in 2003, an invasion that raised the concern level of Islamic groups in the Middle East, including the Islamic leaders that control the government of Iran. Israel has nuclear weapons, and Iran has worked at developing nuclear weapons. US involvement in the region has heightened tensions between Iran and Israel. Oil revenues are driving the economies of most of the Arab countries that support the Palestinians. Oil is an important export of the region, with the United States as a major market. The difficulties between Israel and the Palestinians continue to fuel the conflict between Islamic fundamentalists and Islamic reformers. Some Islamic groups have accepted Israel's status as a country and others have not. The Israel-Palestinian problem drives the geopolitics of the Middle East. The US war in Iraq has complicated the situation but has not superseded it. The situation in Palestine is predicted to continue long after the problems in Iraq have stabilized.

Figure 8.27: West Bank Settlements and Palestinian-Controlled Areas. Updated from map courtesy of CIA World Factbook – public domain.

Jordan

North of the Arabian Peninsula are three Arab states that surround Israel: Jordan, Syria, and Lebanon. Each country possesses its own unique physical and cultural geography. The country of Jordan was created through the British Mandate after World War I, when Britain defeated the Turks in Palestine. The area east of the Jordan River became the modern country of Jordan in 1946. From 1953 to 1999, during the most volatile period of the region, the country was ruled by a pragmatic leader, King Hussein, who was able to skillfully negotiate his way through the difficult relationship with Israel and yet keep his country stable. When Palestine was divided by the UN to create the State of Israel, the region of Jordan received more than a million Palestinian refugees from the West Bank and Israel. Refugees make up a large portion of the more than six million people who live in Jordan today; about a half million refugees from the US war in Iraq are included in that total.

Jordan is not large in physical area. Natural resources such as oil and water are not abundant here, and the country often has to rely on international aid to support its economy. Inflation, poverty, and unemployment are basic issues. The government of Jordan is a constitutional monarchy. King Hussein's son 'Abdullah II took power after the king's death in 1999. Economic reforms were implemented by King 'Abdullah II to improve the long-term outlook of the country and raise the standard of living for his citizens. The king allowed municipal elections to be conducted, which allowed for 20 percent of the positions to be dedicated to women candidates. Parliamentary elections were held by a democratic vote.

Figure 8.28: King 'Abdullah II of Jordan Visits US President Barack Obama in the White House in 2011 Jordan has had good political relations with the United States. King 'Abdullah II has worked to maintain a stable government in Jordan and maintain civil stability in spite of Jordan's lack of economic opportunities. Photo courtesy of the White House – public domain.

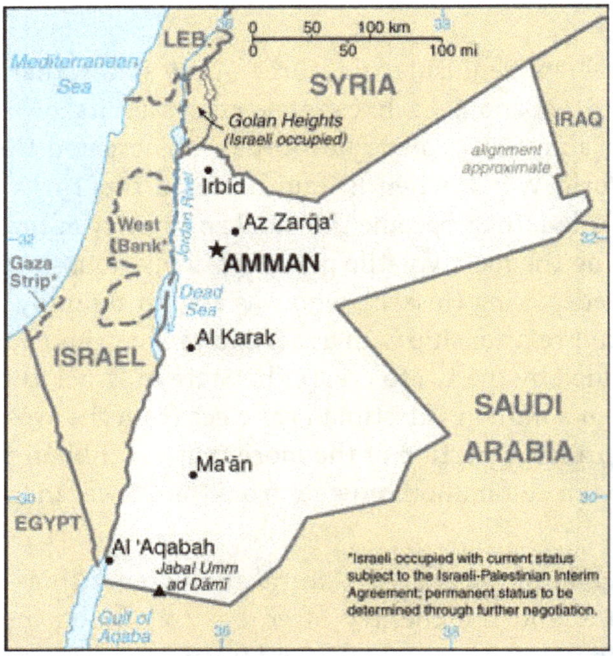

Figure 8.29: Jordan. Map courtesy of CIA World Factbook – public domain.

Jordan has demonstrated how a country with few natural resources in a volatile region of the world can proceed down a progressive path despite difficult circumstances. Jordan has developed a positive trade relationship with Europe and the United States while at the same time working with its Arab neighbors to access oil and to maintain a civil state of affairs. Jordan is not without its challenges but has managed to confront each issue yet retain a sense of stability and nationalism.

Syria

The strategically located country of Syria is at the center of the Middle East's geopolitical issues. Syria gained its independence from the French Mandate in 1946, the same year as Jordan. Syria has strived to work out and stabilize its political foundation. In a move to create greater Arab unity in the realm, Egypt and Syria joined forces and created the United Arab Republic in 1958. This geopolitical arrangement lasted until 1961, when the partnership was dissolved. Syria returned to its own republic. The Arab Socialist Baath Party gained strength, and in 1970 Hafiz al-Assad, of the Alawite minority (an offshoot branch of Shia Islam making up about 10 percent of the Syrian population), took over leadership in a coup that stabilized the political scene. It was during this era that the Golan Heights was lost to Israel in the Arab-Israeli War of 1967. This strategic geographical location is a point of contention in the peace negotiations between Syria and Israel.

Hafiz al-Assad served as the leader of Syria for twenty-nine years without having been democratically elected to the office by the people. His son Bashar took the reins of leadership after Hafiz died in 2000. The Alawite sect held power in Syria through the Assad family under military control. Syria has been accused of using its military power to influence conditions in Lebanon, where it brokered a peace deal in its civil war (1975–1990). Syria has also been accused of supporting the anti-Israel groups headquartered in Lebanon.

Syria is located in an ancient land with a long history of empires and peoples. The region of Syria was once part of the cradle of civilization that sprung up in Mesopotamia. Damascus claims to have been continually inhabited longer than any other capital city on Earth. The largest city and the center of industrial activity is Aleppo, which lies in the north of Syria. Syria's physical area is slightly larger than the US state of North Dakota. Overall, Syria's climate is characterized as an arid type B climate; some regions receive more rain than others. The western region, because it borders the Mediterranean Sea, is an area that receives more rainfall. The additional rainfall translates into extensive agricultural production. The northeast area of Syria is also productive agriculturally through water resources provided where the Euphrates River cuts through the country. Oil and natural gas have been the country's main export products. The petroleum reserves are being depleted, and few new fields are being developed. Eventually, the wealth generated by the sale of petroleum reserves, which are finite resources, is projected to diminish, even as the population continues to increase.

Figure 8.4.308.4.30: Female Protesters in Douma, a Suburb of Damascus, in 2011. Syriana2011 – Syria Damascus Douma Protests 2011 – 05 – CC BY 2.0.

The Syrian government has exerted strict control over the economy. The country will face serious economic issues in the future. There is a high rate of unemployment. Because oil production has not been increasing, the government has been forced to take on additional national debt. The arid climate and the need to supplement agriculture production have placed additional pressure on precious fresh water supplies. The Euphrates River provides fresh water, but it originates in Turkey, where large dams restrict the flow. Water rights for the region are therefore an issue. One third of Syria's population is under the age of fifteen, which indicates a rapid population growth pattern that will tax future resources at an increasing rate. In 2010, Syria had about twenty-two million people. The country holds political significance; its strategic location between Iraq and Israel makes it is a vital player in any solution for lasting peace in the Middle East.

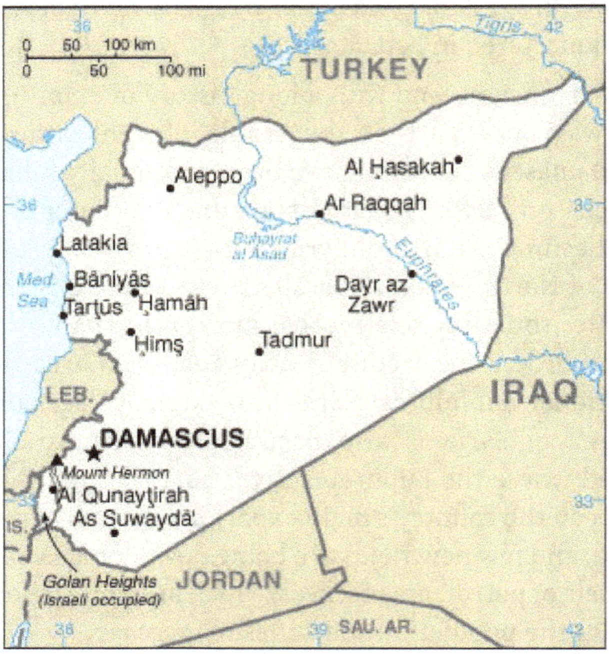

Figure 8.31: Syria. Map courtesy of CIA World Factbook – public domain.

Syria has experienced protests and demonstrations similar to those that swept through North Africa in the Arab Spring of 2011. Citizens expressed dissatisfaction with the government because of the lack of democratic reforms, high unemployment, and the loss of civil rights, which had been taken away when the government declared a state of emergency in 1963. Student protests escalated to massive citizen demonstrations that emerged in various Syrian cities in the spring of 2011. The government cracked down on protesters, killing some. After extensive demonstrations on March 15, the government arrested more than three thousand people. Hundreds have been killed in violent clashes between the people and government security forces.

The lack of democratic processes by President Bashar al-Assad's government has continued to prompt protests and demonstrations in Syria. The US government eventually placed sanctions against Assad and a number of high-level Syrian officials. A new cabinet was installed in Syria, and the decades-old state of emergency was rescinded. Unrest and protests by the people continue, and the government responds by cracking down on them with harsher methods. Syria's neighbors and various European nations have increased sanctions and denounced the Assad regime, which has increased the pressure on Syria.

The uprising in Syria that started as a part of the wave of the Arab Spring continues. The Arab Spring protests and demonstrations in Syria have often been compared to those in Libya. Both Syria and Libya have had long-time leaders that ruled for decades without democratic reforms or wide-spread personal freedoms for their citizens. The difference is that in Syria the Assad regime has held total control over a large military and comprehensive control over political and economic activities. This is unlike Libya under Muammar Gadhafi, whose modest military forces included hired mercenaries from other countries and the government had a fragmented or weak political structure. Speculation regarding changes in the government and the future of Syria is more difficult.

Lebanon

Phoenicians created an empire along the Mediterranean coast of present-day Lebanon four thousand years ago, and many armies fought over the strategically located region. After the demise of the Ottoman Empire, the area became a European protectorate under the French Mandate. Independence was granted by the French in 1943.

Lebanon is smaller than the US state of Connecticut with a population of about four million. The country's high central mountain chain, the **Lebanon Mountains**, reaches as much as ten thousand feet in elevation. At these elevations, precipitation turns to snow and allows the operation of ski resorts. To the east of the central range is the fertile **Bekaa Valley**, which plays a vital role in the country's agriculture. On the eastern side of the Bekaa Valley is another shorter mountain range that borders Syria.

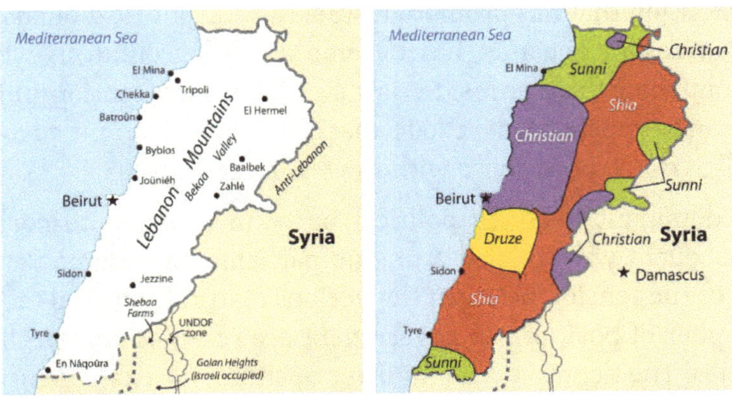

Figure 8.32: Lebanon (*left*); Majority Religious Factions in 2006 (*right*). Maps courtesy of CIA World Factbook – public domain.

Following World War II, Beirut, the capital city of Lebanon, became known as the "Paris of the Middle East," complete with Western-style night clubs and a jet-setting business class. In the past, of Lebanon was called the "Switzerland of the Middle East" because of its capabilities in banking and finance, which were supported by a relatively stable political climate. Unfortunately, stable and progressive conditions were not enough to keep the country from escalating into division and civil war in the 1970s.

In the latter half of the twentieth century, internal tensions were building between the many religious and cultural factions competing for power in Lebanon. By the early 1970s, the minority Christian government clashed with a majority Muslim population. Many factions entered the arena on both the Christian and Muslim sides. On the Christian side are Orthodox, Catholic, and Protestant factions. The Islamic side includes the Sunni, the Shia, and the Druze, a semi-Islamic offshoot group that incorporates Gnostic or neo-Platonic philosophies.

Lebanon was experiencing a full-scale civil war by 1975 that continued until 1990, when the Syrian military brokered a peace deal. By the time it ended, the bitter civil war had destroyed the infrastructure of the country. Only one-third of Beirut's population remained. The former thriving city had been reduced to a collection of bullet-ridden empty buildings. It took more than a decade, but through the resiliency of the people, Beirut rebounded and continues to recover. A massive rebuilding program has resurrected the city of Beirut and stimulated the economy.

Still, conflicts linger, and discord between Israel and Syria has violent results. Israel has taken military action against anti-Israel factions within Lebanon on a number of occasions. In 1982, Israel attacked PLO strongholds, which were operating out of Lebanon in the Bekaa Valley and West Beirut. In 1993, Israel conducted air raids and military strikes against guerilla bases in Southern Lebanon. Anti-Israel groups such as Hezbollah operate out of Lebanon and receive aid from other Arab states, a source of contention that has prompted Israel to confront Hezbollah on Lebanon's territory. As a result, Lebanon has become a battleground on which factions try to work out their differences. Syria's continual intervention in Lebanon has sometimes been unappreciated; in 2004, massive demonstrations advocated for the removal of all Syrian troops. Syria withdrew its forces in 2005.

There is no dominant majority political party in Lebanon to coalesce power. Lebanon consequently developed a unique parliamentary democratic system to relieve some of the tension between the various cultural-political factions. In this system, a number of positions in government are reserved for specific religious/political parties. The deputy prime minister position, for example, is reserved for an Orthodox Christian; the prime minister is a Sunni Muslim; the speaker of the parliament is a Shia Muslim; and the president can only be a Maronite Catholic Christian.

Lebanon's cultural and religious factions within its small state clash with political fervor and conviction, at times creating chaotic conditions that interrupt economic growth and discourage international investments. Add the dynamic cultural conditions to Lebanon's attractive physical features—the beautiful Mediterranean coast, the attractive interior mountains, and the cosmopolitan city of Beirut—and it is easy to see why Lebanon is such a fascinating geographic study. Lebanon holds a unique location and position in the Middle East that will remain a focus of interest to the rest of the world.

Figure 8.33: Beirut, Lebanon Located in the middle of an upscale neighborhood in Beirut, this food shop is in a building that has not been fully repaired from damage resulting from the 1975 civil war. The restaurant offers diverse foods, including Chinese food and pizza. Steven Damron - mini chinese, pizza papa, faysal - CC BY 2.0.

Key Takeaways

- The current Jewish State of Israel was recognized in 1948. Before this time, the region was called Palestine and the people who lived there were called Palestinians.
- Victorious in war against their Arab neighbors, Israel acquired the Golan Heights, the West Bank, the Gaza Strip, and the city of Jerusalem. The West Bank and the Gaza Strip are considered Palestinian territory. Two plans have been proposed to address the division but have not been agreed upon.
- Jordan is a constitutional monarchy led by King 'Abdullah II, who has worked to implement reforms to maintain a country that has few natural resources.
- The government of Syria is led by Bashar al-Assad, a member of a minority ethnic group called the Alawites. Assad and his father have ruled Syria for more than forty years under a state of emergency. Massive protests and

demonstrations against the government have resulted in hundreds of deaths and thousands imprisoned.

- Lebanon is a diverse country with a large number of religious groups that dominate the cultural and political scene. The physical geography includes the Lebanon Mountains, the Bekaa Valley, and the Mediterranean coast.

Discussion and Study Questions

1. How did the UN divide Palestine? Why was the division a problem for the Palestinians?
2. What happened to the Palestinians that lived in the Jewish-controlled areas after Israel was divided?
3. What was the Camp David Accord about? Which country's leaders were involved?
4. What was so important about the Golan Heights? What is the issue with this territory?
5. Why would Israel move its capital to Jerusalem and establish it as a forward capital?
6. How are the governments of Israel, Jordan, Syria, and Lebanon different?
7. Why were there massive protests and demonstrations in Syria in 2011?
8. Why was the United Arab Republic created in 1958? Why was it dissolved?
9. Where was the "Paris of the Middle East"? What happened to the city in 1975?
10. How has Lebanon attempted to satisfy all the political factions in their government?

Geography Exercise
Identify the following key places on a map:

- Allepo
- Bekaa Valley
- Dead Sea
- Euphrates River
- Gaza Strip
- Golan Heights
- Gulf of Aqaba
- Jordan River
- Lebanon Mountains
- Negev Desert
- Sea of Galilee
- Sinai Peninsula
- Tel Aviv
- Transjordan
- West Bank

References

Central Intelligence Agency, "The World Factbook," https://www.cia.gov/library/publications/the-world-factbook

Knowledgerush, "Division of Palestine by United Nations," www.knowledgerush.com/kr/encyclopedia/Division_of_Palestine_by_United_Nations. Masri, M., Nablus—West Bank 2011 (Paraphrased quote). United Nations Relief and Works Agency for Palestine Refugees, "Palestine Refugees," http://www.unrwa.org/etemplate.php?id=86. United Nations Relief and Works Agency for Palestine Refugees, "Statistics," http://www.unrwa.org/etemplate.php?id=253.

8.5. ARABS, ISLAM, AND OIL

The intersection of Arabs, Islam, and oil has profoundly influenced global politics and economics over the past century. The discovery of vast oil reserves in the Arab world, particularly in countries like Saudi Arabia, Iraq, and the Gulf states, transformed these nations into key players on the international stage. This newfound wealth has not only fueled rapid economic development but has also allowed Arab states to wield significant political influence, particularly in organizations like OPEC. Islam, as the dominant religion in the region, shapes cultural identities and social norms, often intertwining with the political landscape. The interplay of oil wealth and Islamic identity has also played a role in both fostering unity and exacerbating divisions within the Arab world, as varying interpretations of Islam and economic interests can lead to differing national agendas and regional rivalries. This complex relationship continues to impact global energy markets, geopolitical alliances, and the quest for stability in the region.

States of the Arabian Peninsula

The Arabian Peninsula is a desert environment surrounded by saltwater bodies. The **Persian Gulf**, the **Arabian Sea**, and the **Red Sea** border the peninsula on three sides. Arid type B climates dominate the region. Saudi Arabia only receives an average of four inches of precipitation per year. The southern portions of the peninsula are some of hottest places on Earth. Summer temperatures can reach more than 120 ºF. In the south is the **Rub' al-Khali** (Empty Quarter), which is mainly desert and comprises about 25 percent of Saudi Arabia. It is extremely dry and virtually uninhabited, though oil discoveries have brought temporary settlements to the region. There are no natural lakes or major rivers on the peninsula. Agricultural activity is dependent on the availability of water by rainfall, underground aquifers, oases, or desalinization of seawater.

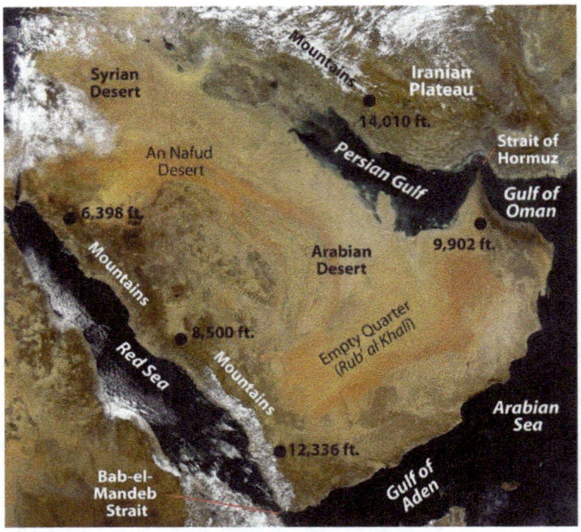

Figure 8.34: Satellite Image from 2008 of the Arabian Peninsula Illustrating the Mountainous Regions, the Uninhabited Empty Quarter Desert Region, and the Surrounding Bodies of Water. Satellite image courtesy of NASA's Sea-viewing Wide Field-of-view Sensor project and John Nevard – public domain.

Most of the people living on the peninsula are Arabs, and most of the peninsula's countries are ruled by monarchs who rely on oil revenues to gain wealth. Minerals are mined in the mountains that dominate the peninsula's western and southern regions. The highest peaks reach more than twelve thousand feet in elevation in northern Yemen. Of the countries on the peninsula, Yemen has the fewest oil resources and has had the sole democratically elected government. Saudi Arabia dominates the region in size and in oil resources. Islam, the major religion, infiltrates all aspects of Arab culture.

Saudi Arabia

The holy cities of **Medina** and **Mecca** are in Saudi Arabia, the birthplace of Islam. Islam first united the many traditional groups of Arabia with religion and then with the Arabic language. The region was further united after 1902, when Abdul Aziz Al-Sa'ud and his followers captured the city of Riyadh and brought it under the control of the **House of Sa'ud**. In

Figure 8.35: Political Map of the Arabian Peninsula. Map courtesy of University of Texas Libraries.

1933, the lands under the control of the king were renamed the Kingdom of Saudi Arabia. Saudi Arabia is an absolute monarchy. In 1938, US oil corporation Chevron found large quantities of oil in the region, which has sustained the royal family ever since. Aramco is the state-run oil corporation. Controlling about one-fifth of the world's known oil reserves, the Saudi royal family claims considerable power.

The Saudi royal family gave safe haven to thousands of Kuwaitis, including the emir and his family, during the First Persian Gulf War (1991). Saudi Arabia allowed US and Western military forces to use bases on its soil during Operation Desert Storm. Acquiescence to non-Muslims operating military bases on the same soil as the holy cities of Mecca and Medina gave extremist groups a reason to engage in terrorist activities. Out of the nineteen hijackers in the 9-11 attack in New York, sixteen were from Saudi Arabia. The Saudi government has been forced to step up its efforts against terrorism and domestic extremist groups.

The entire economy of Saudi Arabia is based on the export of oil, and more than 20 percent of the known oil reserves in the world are located in Saudi Arabia. The country is a key member of the Organization of Petroleum Exporting Countries (OPEC) and has been the world's number one oil exporter. Millions of foreign workers in the petroleum industry make up a vital component of the country's economy.

A high rate of population growth has been outstripping economic growth in Saudi Arabia. In 2010, more than one-third of the population was younger than fifteen years old, and family size was about 3.8 children. The unemployment rate is high, and there is a shortage of job skills in the workforce. The government has been working to shift its focus away from a petroleum-based economy and increase other economic opportunities; it plans to heavily invest in the necessary infrastructure and education to diversify its economy.

Saudi Arabia has made several efforts to move forward and put the country more in line with globalization efforts that are modernizing the other Persian Gulf States. The World Trade Organization accepted Saudi Arabia as a member in 2005. In 2008, the king implemented the initiative for interfaith dialogue in an effort to address religious tolerance and acceptance. The first woman was appointed to the cabinet, and municipal councils held elections for its members.

Figure 8.36: Modern Infrastructure Illustrated on Medina Road in Jeddah, Saudi Arabia. Wikimedia Commons – public domain.

Saudi Culture

The royal family and most of the people in Saudi Arabia are Sunni Muslims. The country has a strong fundamentalist Islamic tendency. The law of the state is strict and supports conservative Islamic ideals. The **Wahhabi** branch of Sunni Islam has a major influence on culture. Activities such as gambling, alcohol consumption, and the promotion of other religions are outlawed. Alcohol and pork products are forbidden in accordance with Islamic dietary laws. Movie theaters and other Western-style productions are prohibited but can be found in areas where workers from other countries live in private compounds. Though movie theaters are restricted, movies on DVDs are not prohibited and are widely available. The dress code in Saudi Arabia strictly follows the Islamic principles of modesty. The black abaya (an article of clothing that looks like a cloak or robe) or modest clothing is appropriate for women. Men often wear the traditional full-length shirt and a headcloth held in place by a cord.

In Saudi Arabia, human rights organizations, legal associations, trade unions, and political parties are banned. The country maintains a tight censorship of all local media. The press is only allowed to publish what the government permits it to report. Communication with foreigners, satellite media, and Internet access are highly controlled. Those who speak out against the government can be arrested or imprisoned.

Figure 8.37: Women in Saudi Arabia Former First Lady Laura Bush meets medical staff in Saudi Arabia. Note the women's attire. Photo courtesy of the White House – public domain.

The Sharia is the basic criminal code in Saudi Arabia, along with whatever law is established by the king. A wide range of corporal and capital punishments—from long prison sentences to amputations (arm or foot), floggings, and beheadings—are proscribed for legal or religious offenses. Trials are most often held in secret without lawyers. Torture has been used to force confessions that are then used to convict the accused. Torture techniques—including the use of sticks, electric shocks, or flogging—can be applied to children and women as well as men. Executions are usually held in a public place every Friday.

Role of Women

Men hold the dominant roles in Saudi society. Under strict Islamic law, women do not have the same rights as men, so Saudi women do not have the opportunities that women in many Western countries have. For example, it is not customary for a woman to walk alone in public; traditionally, she must be accompanied by a family member so as to not be accused of moral offences or prostitution. The mutawa'een (religious police) have the authority to arrest people for such actions. The punishment could be as many as to twenty-five days in prison and a flogging of as many as sixty lashes.

As of 2011, the following restrictions have been made on women:

- Women are not allowed to drive motor vehicles.
- Women must wear modest clothing such as the black abaya and cover their hair.
- Women can only choose certain college degrees. They cannot be engineers or lawyers, for example.
- Women cannot vote in political elections.
- Women cannot walk in a public spaces or travel without a male relative.
- Women are segregated from men in the workplace and in many formal spaces, even in homes.
- Women need written permission from a husband or father to travel abroad.
- Marriages can be arranged without the woman's consent, and women often lose everything in a divorce.

Saudi Arabia is a country steeped in tradition based on the heritage of its people, and many of the traditions regarding women were implemented to protect and care for them. However, as the forces of globalization seep into the fabric of society, many of these traditions are evolving and changing to adapt to the times and to a more open society. Women are asserting themselves in the culture, and many of these long-standing traditions are starting to break down. In late 2011, there were a number of women who organized to defy the ban on driving. One woman was arrested for driving a car and sentenced to ten lashes. Saudi King Abdullah then overturned the sentence and promised to support or protect women's rights. There is no law stating that women cannot drive a vehicle. The taboo is based on tradition and religious views. More women have taken to the roadways in spite of the taboo against it (Huffington Post). Saudi Arabia is an example of how Islamic fundamentalism is being challenged by modernity and democratic principles.

Kuwait

Kuwait, a small country located on the Persian Gulf, is a monarchy ruled by an emir from the royal family. Immense oil reserves have made Kuwait attractive to international oil investors. In 1961, **Zapata Oil Company** (now Pennzoil), owned by former US president **George H. W. Bush**, drilled the first offshore Kuwaiti oil well in the Persian Gulf. Thanks to ample oil revenues, the small Kuwaiti population (about three million people) has adequate social services. The country has a high standard of living. Education is free, and much of the labor base comes from non-Kuwaiti migrants. Petroleum exports account for most of the government's income.

Kuwait has an excellent port at Kuwait City. However, one of the environmental problems with building a large city in the desert is the shortage of fresh water. To solve this problem, Kuwait has turned to the desalinization of seawater to provide for its domestic, agricultural, and industrial needs.

The United States and an international coalition fought the First Persian Gulf War in 1991 to "liberate" Kuwait from the grip of Saddam Hussein. It is compelling to note that the war was not about democracy.

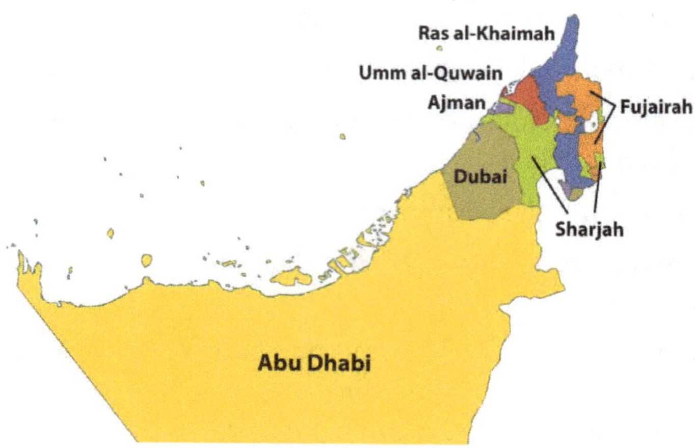

Figure 8.38: Desert Storm US fighter jets in Operation Desert Storm fly over burning oil well fires in Kuwait. The fires were set by retreating Iraqi forces. Wikimedia Commons – public domain.

The war was about the control of oil resources. Under Hussein, Iraq invaded Kuwait and took over its enormous oil industry and port facilities. By taking over the oil assets, Hussein was in actuality taking over the oil assets of various international oil corporations. With the support of United Nations (UN) resolutions demanding that Hussein leave Kuwait, President George H. W. Bush organized an international military coalition to remove Hussein from Kuwait. The US mission was called **Operation Desert Storm.** The war started on bases in Saudi Arabia and pushed the Iraqi army out of Kuwait. When Hussein realized that he could not benefit from the oil in Kuwait, he had approximately 750 oil wells in Kuwait dynamited, which caused serious well fires and large lakes of oil flowing out onto the desert sands. The fires and spilled oil caused extensive environmental damage.

Kuwait was not a democracy during the Persian Gulf War and is not a true democracy today. It is considered a constitutional emirate. The emir, or head of the royal family, is the head of state. He appoints the prime minister and has a high level of control over the government. The emir has the authority to dissolve the National Assembly, which has members that hold seats by election. A number of groups wish to have a political voice in the government, including Islamists, business merchants, secular liberals, Shia activists, and a small number of local groups. Islamist groups are usually those who support an Islamic religious state as the desired type of government.

Kuwait had to invest nearly five billion dollars to reestablish the oil industry after the Persian Gulf War, but the emirate has recovered, and its economy is growing with the increased sale of exported oil. Kuwait has about 104 billion barrels of oil in known reserves. In 2010, the four largest export partners were Japan, South Korea, Taiwan, and Singapore, which are all economic powers in East Asia that have to import almost all of their energy and raw materials. The US has traditionally been Kuwait's number one source of imported goods.

Bahrain

Bahrain is a small archipelago (group of islands) in the Persian Gulf. The country received its independence from Great Britain in 1971. Iran has made claims on the islands to no avail. Similar to other small monarchies in the region, Bahrain has lots of oil and a small population. Though more than 50 percent of the population is Shia, the country is opening up to democratic reforms. In 1999, elections were approved for a parliament, all political prisoners were released, and women were allowed to vote. The royal family, ruled by the king, has had an enormous degree of power over its government. Officially, Bahrain is a constitutional monarchy, but the king appoints the members of the upper house in its bicameral legislature. The first female was appointed to a cabinet position in 2004, which was an indication of the move toward openness to the globalization process and modernization. Some in the country think the implementation of these measures is still too slow.

Figure 8.39: Development in Bahrain This Hardee's franchise in Bahrain is a clear example of Americanization of the Arabian Peninsula. Claus Wolf – Fast Food in Dammam – CC BY-NC-SA 2.0.

Most of Bahrain's wealth is gained through the extraction of natural resources. Enormous natural gas reserves are located in Bahrain's coastal waters, and oil now makes up about 60 percent of the export profits. The small land area size of the country, lack of sufficient supplies of fresh water, and few other natural resources has prompted a shift for Bahrain to expand into the financial sector. Islamic banking and financial services for the global marketplace have been an expanding sector of the economy. The objective in diversifying the economy is to reduce the dependency on oil as a future source of national wealth. In addition, the United States has entered into a free-trade agreement with Bahrain, which has attracted multinational corporations to do business in the region. Bahrain has been supportive of a US military presence for both protection and cooperation and is the permanent headquarters for the US Fifth Fleet navel operations. In a mutual defense agreement, some one thousand American navel officers and personnel are stationed on the island. Bahrain has been a frontline state for the US military in the Iraq War and the war in Afghanistan.

The country is also opening up development in the service sectors. Tourism is not what usually comes to mind when one thinks about the Arabian Peninsula, but Bahrain has been attracting millions of visitors yearly. The country's authentic heritage is attractive to tourists from neighboring Arab states and the global community. The country boasts of nearly five thousand years of human activity. UNESCO has designated the Qal'at al-Bahrain castle as a World Heritage Site. The country has invested heavily in modern shopping malls and international sports facilities in an effort to modernize its country and attract more international events.

The citizens of Bahrain have had to work to balance the shift toward modernization and globalization with the strong Arab heritage and Islamic beliefs that have been the foundation of their culture. The term *Middle East Lite* has been applied to Bahrain because Bahrain has been investing in modern infrastructure but has worked hard to maintain its Arab heritage with a Persian Gulf identity that is more accepting and open to the outside world. The growing and prosperous middle class is more tolerant and liberal than many of its Middle East neighbors.

The same level of tolerance toward outsiders has not been witnessed within the country. The 2011 protests and demonstrations that swept across North Africa and the Middle East also occurred in Bahrain. The king, the royal family, and the majority in government follow the Sunni branch of Islam; however, most of the population follows the Shia branch of Islam. Many within the Shia community felt that they were being discriminated against and protested the lack of democratic reforms. Protests and demonstrations in Bahrain have prompted the government to call in military support from Saudi Arabia to help quell the uprising. A number of Shia mosques were reported to have been destroyed, and hundreds of people were detained by police. The protests and demonstrations in Bahrain are more than just a conflict between Shia and Sunni, though this split has been a major concern for years. Many Sunni have participated in the demonstrations because they are in support of more democratic reforms as well.

Qatar

The small peninsula jutting out from Arabia into the Persian Gulf is an Arab land in transition. Ruled by an emir who has supported democratic reforms, Qatar is moving forward with a globalization policy similar to other Westernized nations. Many of these reforms are similar to those in Bahrain and the United Arab Emirates (UAE). Oil and gas exports have fueled a building boom that has produced shopping malls, wide boulevards, and even a large US military base. Women are allowed to vote, Western clothing and products are permitted, and rap music can be heard in the streets. Though still politically restrictive in many ways, Qatar is more open than many of its neighbors. Qatar is also home to the Al Jazeera news organization, which often balances out Western news programming. Al Jazeera is also allowed to report critically on its home country.

In the past few years, oil revenues have provided Qatar with a rapidly growing economy and a high standard of living. Proven reserves of oil and natural gas are enormous for such a small country. Qatar's natural gas reserves are the world's third largest. Qatar has been pursuing development of private and foreign investments in non-energy-related businesses, including banking and financial institutions.

Figure 8.40: Building Boom in Qatar City Revenues from oil and natural gas are fueling the rapid development of this small peninsula that is rivaling core economic countries. Amina Tagemouati – Doha Cityscape – CC BY 2.0.

Modernization efforts have supported Qatar's push for a greater emphasis on education. Infrastructure and financial support have been allocated to support educational reform, and university opportunities are expanding rapidly. Qatar University was founded in 1973, and in the last decade many more universities from Western countries have opened up branch campuses in **Education City**, which was established to advance Qatar's educational reform goals. The emir's second wife has actively promoted educational reforms and has encouraged women to pursue higher education to excel in their careers. She has also created greater visibility for women in public roles and has broken through some of the cultural barriers and taboos that have restricted women in other conservative Islamic Arab countries.

United Arab Emirates (UAE)

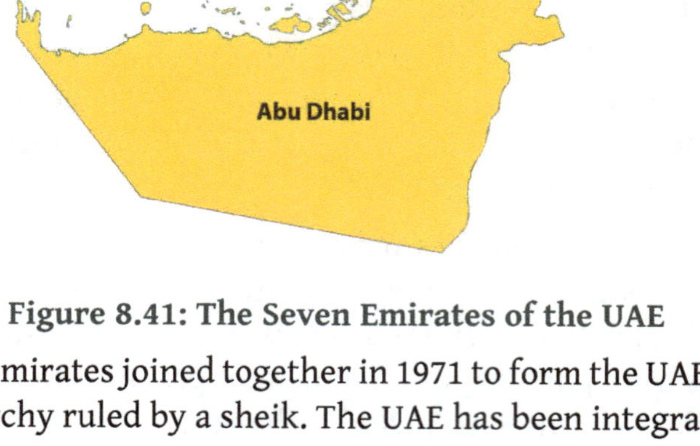

Figure 8.41: The Seven Emirates of the UAE

Seven small Arab emirates joined together in 1971 to form the UAE. Each emirate is an absolute monarchy ruled by a sheik. The UAE has been integrating its economy with the global marketplace and has established a high standard of living for its people. Two of the emirates—Abu Dhabi and Dubai—possess most of the oil reserves. Abu Dhabi is the capital city and consists of 87 percent of the land area in the UAE. The head of the royal family in this emirate is considered the head of state for the UAE.

It is evident that one emirate, Abu Dhabi, dominates in terms of square miles of physical area.

Dubai has turned its small emirate into an international trade center. The emirate has used its oil reserves to promote trade and commerce. Dubai built itself a world-class port facility equaling that of Hong Kong or New York. As a **free-trade zone,** there are no taxes or tariffs, so international corporations use the location as a trade center to bring high-volume buyers and sellers together. Dubai has been looking ahead to its future when the oil runs out. The creation of an international trade center would be a means to gain economic income when the revenue from the sale of oil diminishes.

Oil wealth and the need for workers have opened up economic opportunities in the UAE that have attracted laborers and businesspeople from many parts of the world. Noncitizens make up about 80 percent of the population; about half the noncitizens are from South Asia, and many are Muslims from India. The large number of laborers that are required to develop the infrastructure has created an imbalance between the percentage of men and women. There are about twice the number of men than there are women in Dubai. This has created an interesting dynamic for women in Dubai, who have more rights and opportunities than those in more conservative Islamic countries such as Saudi Arabia. The UAE presents an excellent example of people migrating from peripheral countries to a core region in search of opportunities and advantages in a globalized economic community.

The UAE has invested its oil income in building up its infrastructure to compete in a global economy. Hundreds of billions of dollars of construction projects are under way in Dubai alone. Dubai has the world's tallest structure, the most expensive hotel, the world's most expensive airport (when completed), and the world's largest artificial islands. Dubai is even home to an indoor downhill ski resort complete with real snow. Other ventures such as the Dubailand entertainment complex and Dubai Sports City have also been proposed. The downturn in the world's economic situation has slowed the development in this emirate but has not diminished its perspective on the future.

Figure 8.42: Free-Trade Zone of Dubai The skyline of Dubai includes the tallest building in the world. Wikimedia Commons - public domain.

The Sultanate of Oman

Ruled by a sultan, the absolute monarchy of Oman on the southern end of the Arabian Peninsula also controls the tip of land next to the **Strait of Hormuz.** All oil tankers leaving the Persian Gulf must pass through this vital choke point. Mountains reach more than nine thousand feet in the eastern region of Oman, and rugged arid central plains cover the central region. The country gets plenty of sunshine and has some excellent beaches. Annual rainfall varies from four inches or fewer in the eastern sector to as much as twenty-five inches in the southwest. The climate is generally hot: temperatures can reach 120 °F from May to September.

Oman has been using its oil income to build infrastructure to benefit its people. The sultan of Oman has widespread support from his people and has built up goodwill from the international community for his investments in his country. He has built a free-trade zone with a giant container port facility, luxury tourist hotels, a good road system, and a first-rate international airport. He has also provided clean drinking water to the rural areas. Though Oman is not a democracy, the sultan has been positive role model for other monarchs. He has used Oman's oil wealth to help his country develop and modernize. The mountains of Oman have additional natural resources such as gold, marble, and copper.

A lack of fresh water is a concern for Oman. The nation has limited renewable water resources. More than 90 percent of the water available is used in agriculture, and the rest is used for industry and domestic consumption. Fresh water is piped throughout most of the country, but shortages occur at times because of droughts and limited rainfall. Environmental problems have also arisen in Oman. For example, irrigation operations have caused soil conditions such as a salt buildup. Oil tankers traveling through the Strait of Hormuz and along the coast in the **Gulf of Oman** have leaked oil, which has washed up on coastal areas where attractive beaches are located. The higher level of exploitation of the environment by a growing population has exacted a toll on the organisms that live in the fragile desert ecosystems. Mammals, birds, and other organisms are in danger of extinction, including the Arabian leopard, the mountain gazelle, and the Arabian oryx. The country may lose its biodiversity unless action is taken toward preservation.

Yemen

Yemen is a mountainous country bordering Oman and Saudi Arabia. The tallest mountains on the peninsula—reaching more than twelve thousand feet in elevation—are located here. The four main regions of Yemen are the eastern desert region of the Rub' al Khali; the Eastern Highlands south of the Rub' al Khali; the Western Highlands, which have the highest peaks; and the western coastal plains. The Western Highlands receive about thirty inches of rain per year, while the eastern desert received almost no rainfall. A number of volcanic islands are located off the coast. A volcano on one of the islands erupted as recently as 2007. This region is an extension of the rift valley system coming out of East Africa.

Figure 8.43: Four Physical Regions of Yemen The highest peak in the Western Highlands is 12,028 feet in elevation. Map courtesy of CIA World Factbook – public domain.

The economy has traditionally been based on agriculture. Most of the farmland is in the form of terraces cut into the mountainsides that trap rainwater as it flows down the slope from one terrace to the next. Food production is a primary concern because in 2010 the population was more than twenty-four million and increasing rapidly. The arid land has few trees, but firewood is in high demand for cooking. The demand for firewood has caused deforestation, which in turn has caused serious soil erosion and damage to the mountain terraces that produce the food. Yemen is facing serious environmental concerns. The fast-growing population will only put more pressure on the environmental systems. On the positive side, oil and natural gas reserves are being found in some quantities, which will assist with the economic conditions and help supply the energy needed in the future.

Women in Yemen do not have the opportunities available in some of the more urbanized and modernized Gulf States. The contrast between the rural dynamics of Yemen and the urban culture of Dubai provides an excellent example of a rural versus urban or a core versus peripheral spatial relationship. The basic formula of family size and income levels applies to Yemen. In 2010, the fertility rate declined to 4.8 from more than 7.0 in 2000 (IndexMundi). Rural-to-urban shift is causing cities within Yemen to grow at an increasing rate. Large family sizes are forcing young people to seek out opportunities and advantages in the cities or in other countries. The situation is only going to intensify, as the population of Yemen is projected to reach sixty million by 2050 at the current rate of increase.

Yemen has a democratically elected government that came about when North Yemen and South Yemen merged into one country to create a democratic republic in 1990. The population of Yemen is about 40 percent Shia and 60 percent Sunni. Yemen sided with Iraq and Saddam Hussein in the First Persian Gulf War in Kuwait, which resulted in Saudi Arabia expelling thousands of Yemeni workers. Yemen and Saudi Arabia have had a long-standing territorial dispute and only recently agreed on the desert border between the two countries. As the lone democracy on the peninsula, Yemen contrasts with the more conservative Islamic states and monarchies such as Saudi Arabia that are more common in the Middle East. Poor, rural, and agriculturally based, Yemen does not fit the mold of the typical oil-rich sheikdom of the region.

The cultural forces within Yemen demonstrate the dichotomy between modernization with democratic reforms and fundamentalist Islamic tendencies. Stability in Yemen is critical for the security of the regional waterways. In the past decade, piracy against ships off the coast of Somalia has increased, and many ships have been boarded and taken hostage by pirates demanding high ransoms for the release of the ship and crew. Security in the region is critical to support safe passage for international shipping activity through the **Gulf of Aden** and the **Bab-el-Mandeb Strait**.

Yemen experienced civil unrest and citizen protests in the spring of 2011 similar to those in the other Arab countries. The protests and demonstrations targeted political corruption, economic conditions, and high unemployment. During this time, the government was also looking to modify its constitution. The protesters shifted their focus to call for the president of Yemen to resign after twenty-one years in power. The president and government resisted, and the conflicts turned more serious. Clans not loyal to the president sided with the protesters, and the Yemeni president was seriously injured in a military clash. He was flown to Saudi Arabia for medical treatment. When the president recovered and returned to Yemen he still refused to step down. The country continues to adjust to the situation.

Figure 8.44: The Capital of Yemen Sana'a, an ancient city, is the capital of Yemen and the largest city, with a population of more than two million people. Step – From atop the Gate, Sana'a Yemen – CC BY 2.0.

Key Takeaways

- The Arabian Peninsula is a desert region. The Rub' al-Khali (Empty Quarter) provides an example of desert extremes. The mountains along the western and southern edges receive the most rainfall. There are no rivers or major lakes on the peninsula. Water, a resource that is vital to human activity, is scarce throughout the region.
- The export of oil and natural gas is what drives the economies of the region. Many of the states are working to diversify their economies with banking, free-trade zones, and even tourism.
- Yemen is the only country with a truly democratically elected government. Leaders of royal families rule in the other countries as heads of state with varying degrees of shared governance.
- Family size varies widely in the region, from more than 4.5 in Yemen to around 2.5 in the progressive Gulf States of Qatar, the UAE, and Bahrain. Women's rights and opportunities have an inverse relationship with family

size; that is, when women's rights and opportunities increase, family size usually decreases. In the Gulf States, the smaller countries in land area are more open to promoting women's rights and responsibilities in the public and private sectors.

Discussion and Study Questions

1. Where are the mountains located on the Arabian Peninsula? How high do they reach in elevation?
2. Why type of government does Saudi Arabia have? What is the law based on?
3. What are some things that women are not allowed to do in Saudi Arabia that they can do where you live?
4. Who are the mutawa'een? When would one encounter them?
5. Who drilled the first offshore Kuwaiti oil well? Why is this significant?
6. What measures have the governments of Bahrain and Qatar taken to modernize their countries?
7. How has the emirate of Dubai been able to promote a globalized economy?
8. Why does the term *Middle East Lite* apply to Bahrain?
9. How has the current sultan of Oman used his country's oil revenues?
10. Outline five differences between the emirate of Dubai and the country of Yemen.

Geography Exercise

Identify the following key places on a map:

- Arabian Sea
- Bab-el-Mandeb Strait
- Gulf of Aden
- Gulf of Oman
- Mecca
- Medina
- Persian Gulf
- Red Sea
- Rub' al-Khali
- Strait of Hormuz

Activities

1. Check the news for world affairs and follow up on what is happening with the protests in Bahrain and Yemen. Analyze how those two countries have been addressing their internal issues.
2. Find a way to access the English version of Al Jazeera news broadcasts and compare its coverage with the coverage from other international news agencies.

References

Huffington Post, "Saudi Arabia: Woman Driver Verdict Reportedly Overturned By King," http://www.huffingtonpost.com/2011/09/28/saudi-arabia-woman-driver-verdict_n_985857.html.

IndexMundi, "Total Fertility Rate (Children Born/Woman)," http://www.indexmundi.com/g/g.aspx?c=ym&v=31.

8.6: IRAQ, TURKEY, AND IRAN

Iraq

Iraq is situated in the Fertile Crescent, positioned between the Tigris and Euphrates Rivers, where the ancient civilizations of Mesopotamia emerged. Historic cities including Nineveh, Ur, and Babylon were situated in this region. Modern Iraq and Kuwait were formed from the British Mandate area acquired after Britain's victory over the Ottoman Empire. Britain delineated linear political borders between Iraq and Jordan, Syria, and Saudi Arabia. Such borders are termed geometric boundaries as they do not adhere to any physical characteristics. In 1961, upon Britain's withdrawal from the region, the emir governing the southern territory adjacent to the Persian Gulf petitioned for the establishment of his oil-rich kingdom as an independent nation. This territory transformed into Kuwait, while the remainder of the area evolved into Iraq. The Baath Party ascended to power in Iraq in 1968, facilitating Saddam Hussein's rise to authority in 1979.

Figure 8.45: The Tigris and Euphrates Rivers and the Shatt al-Arab Waterway between Iraq and Iran. Map courtesy of CIA World Factbook – public domain.

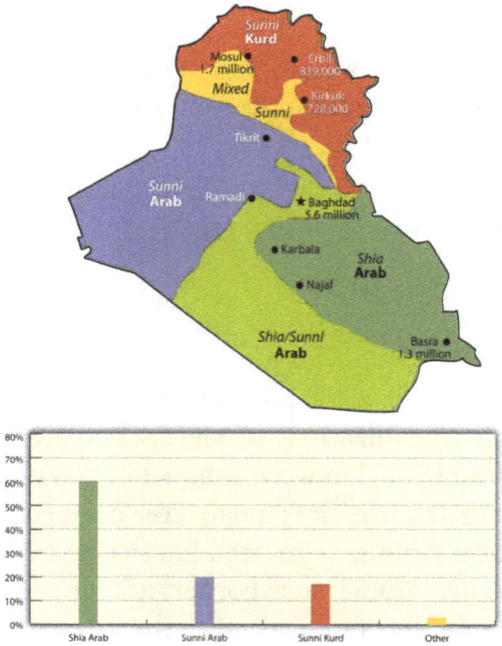

Figure 8.46: Iraq's Divisions of Islam as of 2008

The two ethnic divisions are Arab and Kurd. The two religious divisions of Islam are Shia and Sunni. Karbala and Najaf both have holy sites for Shia Muslims.

Iran-Iraq War (1980–88)

In 1980, a disagreement arose over the **Shatt al-Arab** waterway in the Persian Gulf on the border between Iraq and Iran, and the feud led to war between the two countries. The people of Iran are not Arabs; their ethnic background is Persian. Most Iranians are Shia Muslims. Saddam Hussein and his Baath party were Arabs and Sunni Muslims. Ethnic and religious differences thus fueled the conflict. The Shatt al-Arab waterway was quickly filled with wrecked ships. The local battle escalated into an all-out war, which ended in 1988 without anyone declaring a victory. The Iran-Iraq War was as close to World War III as the world has ever seen, with more than a million casualties and a cost of more than one hundred billion dollars. World powers aligned themselves with one side or the other. Before the war, the Iranian government had been taken over by Islamic fundamentalists who opposed the US intervention in the region; therefore, in the Iran-Iraq War, the United States supported Hussein and provided him with industrial supplies and materials.

The Persian Gulf War (1990–91)

After the Iran-Iraq War, Saddam Hussein looked to Kuwait to gain new oil wealth and expand access to the Persian Gulf. By taking over Kuwait, Iraq would gain an excellent port on the Persian Gulf and earn more income from oil reserves. Hussein accused Kuwait of slant drilling oil wells along the Iraqi border and removing oil that was legally Iraq's. It was common knowledge that both sides were engaged in this practice, but it was the excuse Hussein needed to invade Kuwait and reclaim it as the nineteenth province of Iraq.

In 1990, the Iraqi military invaded and occupied Kuwait. Though the world community opposed this action, it was not until Hussein nationalized all the oil assets of the international oil corporations that resistance was organized. Under the leadership of US president George H. W. Bush, the United Nations (UN) organized a military coalition to remove Hussein from Kuwait. On January 16, 1991, Operation Desert Storm began. After forty-five days of fighting, Iraq was overwhelmingly defeated and its military was ousted from Kuwait. This was a major victory for the coalition. It was during this time that President Bush publicly announced the emergence of the potential **New World Order.** Kuwait was not a democracy but a monarchy ruled by an emir. Clearly, the war was not a war over democracy; it was a war over the control of resources.

When it became evident that Hussein would lose Kuwait, his forces dynamited all the oil facilities and set all the oil wells in Kuwait on fire. His position was that if he could not have the oil, then nobody would. This was one of the worst environmental catastrophes regarding oil on record. Oil flowed into the Persian Gulf and covered the water's surface up to three feet thick. Most mammals, birds, and organisms living on the water's surface died. Oil flowed out onto the desert sand into large petroleum lakes. The air pollution caused by burning oil wells dimmed the sun and caused serious health problems.

Ethnic and Cultural Divisions

To keep Iraq from breaking apart after Operation Desert Storm, coalition forces allowed Hussein to remain in power. Ethnically and religiously, Iraq is divided into three primary groups that generally do not get along. Sunni Arabs dominate central Iraq in a region often referred to as the **Sunni Triangle**, which includes the three cities of Baghdad, Tikrit (Hussein's hometown), and Ramadi. Sunnis were the most loyal to the Hussein government. Southeastern Iraq is dominated by Arabs who follow the Shia division of Islam, which is also followed by most of Iran's population. A group that is ethnically Kurdish and follows the Sunni division dominates northern Iraq. Kurds are not Arabs or Persians; rather, they originated from somewhere in northern Europe centuries ago with their own religion, language, and customs. Many have converted to Islam.

Hussein was a Sunni Muslim, and when he was in power, he kept the other two groups in check. He used chemical weapons on the Kurds during the Iran-Iraq

War. In 1988, he used chemical weapons on the Kurdish town of Halabja and killed about 10 percent of the eighty thousand who lived there. Thousands of Kurds died in other attacks, and thousands more continue to suffer serious health effects. After Operation Desert Storm, Hussein pushed the Kurds north until the UN and the United States restricted him at the thirty-sixth parallel, which became a security zone for the Kurds. The Arab Shia population in the south often clashed with Hussein's military in an attempt to gain more political power, and Hussein subjected them to similar harsh conditions and treatment.

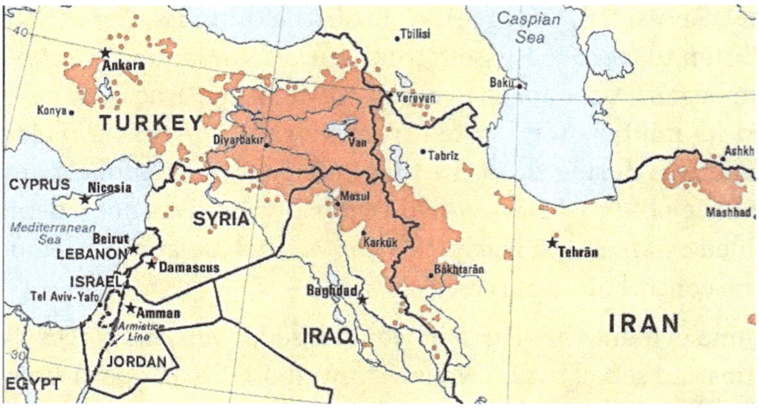

Figure 8.47: Homelands of the Kurdish People, Indicated by the Shaded Areas Future Kurdistan would be the main portions in Iran, Iraq, and Turkey and a corner of Syria. The city of Diyarbakir in Turkey would be the Kurdish capital city. Map courtesy of University of Texas Libraries.

The Arab Shia population in the southeast makes up most of the Iraqi people. The two main cities of **Karbala** and **Najaf** contain holy sites for Shia followers worldwide. The Shia population is three times larger than the Kurdish population in Iraq; more Kurds live outside Iraq than live in Iraq. The Kurds are the largest nation of people in the world without a country. About twenty-five million Kurds live in the Middle East, and most—about fourteen million—live in Turkey. About eight million Kurds live in Iran, about seven million live in Iraq, and a few others live in neighboring countries. At the 1945 conference of the UN, they petitioned to have their own country called Kurdistan carved out of Iran, Iraq, and Turkey, but they were denied. You will recall from Section 8.4 "Israel and Its Neighbors" that Israel was approved to become a nation at the same UN conference.

The Iraq War (2003–11)

After the September 11, 2001, attack on the United States, there was a renewed interest in weapons of mass destruction (WMD). Knowing that Hussein had used chemical weapons on the Kurds, the Iranians, and the Shia, there was a concern that he would use them again. UN Weapons inspectors in Iraq never could confirm that

Hussein retained WMD. They had been destroyed, moved out of Iraq, or hidden. US president George W. Bush decided to invade Iraq in 2003 to remove Hussein from power. In the invasion, Hussein's two sons were killed and Hussein was captured. One aspect of the invasion plan was to use Iraq's vast oil reserves to help pay for the cost of the war, which quickly ballooned to more than a billion dollars a week. Fundamentalist Islamic insurgents made the war difficult.

The US invasion of Iraq brought about the removal of the Baath Party from power and Iraq came under a military occupation by a multinational coalition. An Iraqi Interim Government was formed that assumed sovereignty in 2004. A new constitution was drafted and approved by vote of the Iraqi people. Elections were held and a new government was formed under the newly drafted constitution. Occupying troops continued to remain as the country struggled to adapt to the reforms. Insurgencies developed that brought about an increase in violence that peaked in about 2007. By 2010, the combat operations by occupying troops were ending and the country worked to sustain stable political conditions. Under an agreed upon mandate all combat troops were to withdraw as of the end of 2011. A number of US troops will remain in Iraq in an advisory capacity.

Figure 8.48: Soldiers from the 926th Engineer Brigade Combat Team and the Army 432d Civil Affairs Battalion Patrol in Baghdad's District of Sadr City The United States invaded Iraq in 2003 and removed Saddam Hussein from power. US Department of Defense - public domain.

Resources and Globalization

In geographic terms, the Persian Gulf War and the Iraq War were wars over resources—namely, oil. Wars have historically been fought over territory and resources. When the United States invaded in 2003, Hussein was contracting billions of dollars worth of projects to oil companies in France, Russia, and China. Other support projects were contracted out to other European countries. When the United States invaded, the contracts with were summarily canceled, and British and US oil contractors took over the projects.

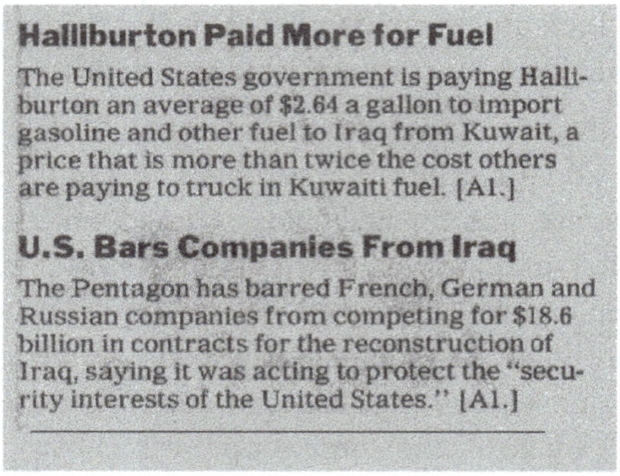

Halliburton Paid More for Fuel

The United States government is paying Halliburton an average of $2.64 a gallon to import gasoline and other fuel to Iraq from Kuwait, a price that is more than twice the cost others are paying to truck in Kuwaiti fuel. [A1.]

U.S. Bars Companies From Iraq

The Pentagon has barred French, German and Russian companies from competing for $18.6 billion in contracts for the reconstruction of Iraq, saying it was acting to protect the "security interests of the United States." [A1.]

Figure 8.49: News Reports of Iraqi Contracts at the Time of the US Invasion of Iraq After the US invasion, many of the contracts Iraq had with other countries were canceled. Source: *New York Times*, December 10, 2003.

Iraq is an example of the second wave of globalization. Neocolonialism has been the dominant force in Iraq's economy since before the **Persian Gulf War**. Industrialization requires high energy demands; therefore, the industrialized countries of the world consume energy on a massive scale. Iraq is not a core economic country, but it holds vast petroleum reserves, making it vulnerable to exploitation by industrialized core countries. It is interesting to note that the Persian Gulf War, initiated in response to Iraq's invasion of Kuwait, can be traced directly to globalism. It was Britain that established the straight-line borders separating Kuwait and Iraq. The war over the control of Kuwait in 1991 was a war over the control of resources, just as the ongoing competition between Sunni Muslims and Shia Muslims, for example, is a competition for the control of political power or resources and not a competition related to ethnic principles.

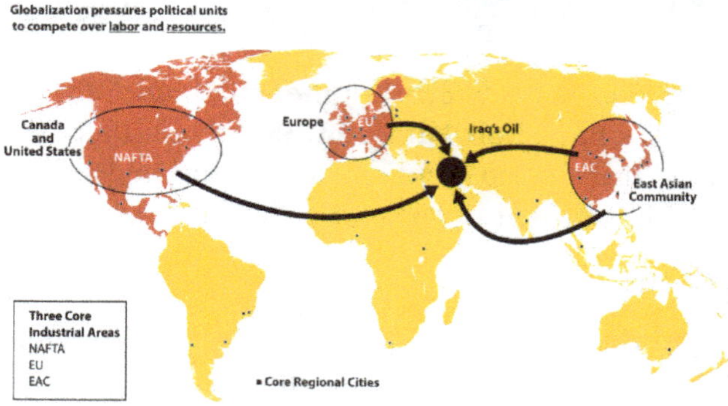

Figure 8.50: The Three Core Economic Areas of the World That Consume High Levels of Energy

The East African Community (EAC) is not an official organization but a core economic area.

Hussein proved to be a destabilizing force and a potential threat to the established so-called

New World Order of global security and trade as outlined by US president George H. W. Bush. The removal of Hussein from power brought to the surface the competition between the trilateral powers of North America, Europe, and East Asia. It remains to be seen who will have control of or access to Iraq's natural resources.

Politics, Oil Companies, and the Administration of US President George W. Bush

There are many connections between the administration of former US president George W. Bush and the international oil industry. Bush once owned the failed Texas oil company Arbusto Energy and was president of Spectrum 7 Energy Corporation, which bought out Arbusto. Bush was also on the board of Harken Energy Corporation when it bought out Spectrum 7 Energy. Bush's father, former US president George H. W. Bush, owned the Zapata Oil Company, which drilled the first offshore Kuwaiti oil well and later merged to become Pennzoil.

Figure 8.51 Vice President Dick Cheney and Secretary of State Condoleezza Rice During the US Presidential Administration of George W. Bush Photo courtesy of the White House – public domain.

US vice president Dick Cheney was CEO of Halliburton, the world's most extensive oil service company. Commerce Secretary Donald Evans was former CEO of Tom Brown Inc., an independent oil and gas company based out of Denver. Secretary of State Condoleezza Rice was a Chevron Corporation board member and had an oil supertanker named after her—the *Condoleezza Rice*. The former US secretary of state under former president George H. W. Bush (Sr.) was James Baker III, who had been a central attorney for an oil consortium building one of the largest oil pipelines in Central Asia. President George W. Bush appointed Zalmay Khalilzad, a former aide to the American oil company UNOCAL, as special envoy to Afghanistan, then as ambassador to Iraq, and then to the UN.

The globalized economy forces political units to compete over valuable resources, which often results in the blending of those in leadership roles in the corporate world with those in positions of power in the political arena. The top personnel in the George W. Bush administration (2001–9) is an example of the relationships that develops between corporate leaders and political leaders. In this case, the relationship centered on the oil industry.

Turkey

Turkey is the only remaining country of the vast Ottoman Empire, which ruled the region for seven hundred years (1299–1923 CE). When the empire was at its peak in the sixteenth and seventeenth centuries, it controlled parts of Europe, North Africa, the Middle East, and Arabia. Located on the **Bosporus**—the straits that connect the Black Sea with the Mediterranean Sea—the ancient city of **Constantinople** was the capital of the Ottoman Empire. Called **Istanbul** today, this city is the largest in Turkey, but it is not the current capital. Turkey moved its capital to Ankara on the **Anatolian Plateau**, which is centrally located in Turkey.

Turkey has a small portion of its land area on the western side of the Bosporus to claim its connection to Europe. Most of Turkey's land area is a part of the Asian continent, and Turkey has often been referred to as Asia Minor. Mountains on Turkey's eastern border with Armenia include Mount Ararat, which is the highest peak in the country at 16,946 feet in elevation. Legend has it that Mount Ararat was the resting place of the legendary Noah's ark. The people of Turkey are neither Arab nor Persian; they are Turkish and speak the Turkish language. As much as 90 percent of the Turkish population is Sunni Muslim, which is similar to many of the other Muslim countries in the Middle East.

Figure 8.52: Turkey The map shows the portion of Turkey in Europe to the west, the central Anatolian Plateau, the Euphrates and Tigris Rivers, and Mount Ararat on the Armenian border. CIA World Factbook – public domain.

Turkey is an established secular democracy with a democratically elected political leadership. To maintain its democracy, it has had to deal with Islamic fundamentalists, who often have supported a shift to an Islamic religious state. Turkey has also had to negotiate with its neighbors, Syria and Iraq, over water rights to the Tigris and Euphrates Rivers, which originate in Turkey but flow through the other countries. Turkey has built dams on these rivers, much to the dismay of its neighbors, who want to use more of the water.

Turkey borders northern Iraq and was home to fifty-six million Turks and fourteen million Kurds as of 2010. The Kurdish claim of a homeland in eastern Turkey has not been recognized by the Turkish government. Open rebellion has been expressed by Kurdish groups wishing to become independent of Turkey, but the Turkish government has oppressed any movement toward independence. Many Kurds have migrated to Istanbul in search of work. They live and work in Istanbul and send money back to their families in eastern Turkey. The large city of **Diyarbakir** in eastern Turkey is predominantly Kurdish and is considered by many Kurds as the city that would be their capital if they had their own country of Kurdistan.

Figure 8.53: A City Park outside the Blue Mosque in Istanbul The capital city of Turkey is Ankara, which is located in the center of the country. Istanbul remains the primate city of the country and is home to various world-class mosques. Jeremy Vandel – Blue Mosque – CC BY-NC-ND 2.0.

Turkey is a member of the North Atlantic Treaty Organization (NATO) and has been a US ally. The United States has been allowed to have military bases in Turkey, which were helpful in Operation Desert Storm during the First Persian Gulf War. In spite of being allies, Turkey did not allow the US military to use these bases as direct invasion points when the United States invaded Iraq in 2003. The Iraq War is a major concern for Turkey, since Turkey has a major stake in the stability of the Middle East. Possessing a small portion of its land area in Europe and being a member of NATO provides Turkey with an opening to join the European Union (EU), but so far it has been denied membership, mainly because of Turkey's controversial human rights record, conflicts with the Kurds, disagreements with Greece over Cyprus, and low economic indicators. Conditions are improving, and at some future point Turkey may have an opportunity to join the EU.

To date, the EU does not have an Islamic country in its membership; Turkey may be the first. The country has various ways of gaining wealth that could help support Europe's other industrial activities. Turkey grows large quantities of food in the central Anatolian Plateau; vast fields of grain extend across this central plateau. Turkey also has some oil resources in the east and minerals in the mountains bordering Armenia and Georgia.

Cyprus

The island of Cyprus in the eastern Mediterranean Sea is officially a part of Europe and a member of the EU. The island is separated by the Green Line, monitored by the UN, which divides the Greek-dominated south from the Turkish-dominated north. The two sides have been separated since a civil war in 1974. Turkish groups in the north have declared their half of the island the independent Turkish Republic of Northern Cyprus. Turkey is the only country that recognizes this proclamation; the rest of the world considers the whole island the Republic of Cyprus.

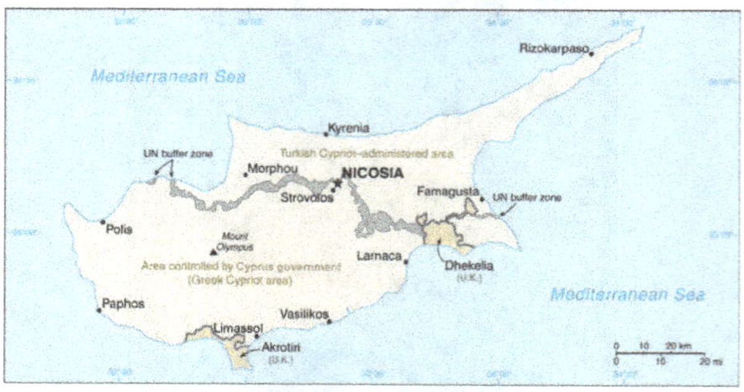

Figure 8.54: Divided Cyprus The island of Cyprus is divided by the Green Line, which is monitored by the UN. Turkish Cypriots control the north and have declared it the Turkish Republic of Northern Cyprus. Greek Cypriots control the southern half of the island. The whole island is considered one country by the UN, which does not recognize the Turkish Cypriots' claim to independence. CIA World Factbook – public domain.

Iran

Iran covers a physical area larger than the US state of Alaska. It is a land of mountains and deserts: Iran's central and eastern regions are mainly desert with few inhabitants, and the northern and western regions of the country are mountainous. The **Elburz Mountains** in the north around the Caspian Sea reach as high as eighteen thousand feet in elevation near the capital city of Tehran. The **Zagros Mountains** run along the border with Iraq and the Persian Gulf for more than nine hundred miles and can reach elevations greater than fourteen thousand feet. Similar to the Atlas Mountains in the Maghreb, Iran's mountains trap moisture, allowing minor agricultural activities in the valleys. Most of Iran's population lives in cities along the mountain ranges. Qanats—systems of shafts or wells along mountain slopes—bring water from the mountains to the valleys for irrigation and domestic use.

Persian Empire to Islamic Republic

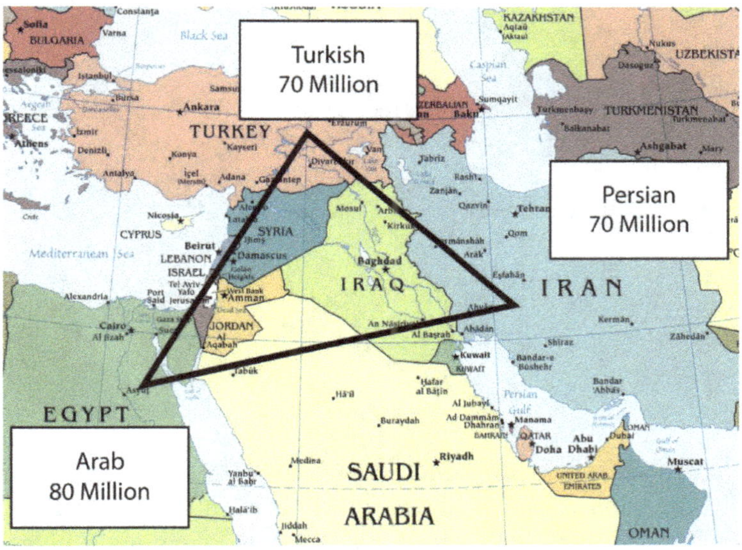

Figure 8.55: 2010 Population Estimates for the Ethnic Triangle of the Middle East, with Egypt, Turkey, and Iran Anchoring Each Corner. Updated from map courtesy of University of Texas Libraries.

Iran was once the center of the **Persian Empire**, which has its origins as far back as 648 BCE, and the country was called Persia until about 1935. The Ethnic Triangle of the Middle East consists of Persians in Iran, Turks in Turkey, and Arabs in Arabia (see Figure 8.55 "2010 Population Estimates for the Ethnic Triangle of the Middle East, with Egypt, Turkey, and Iran Anchoring Each Corner"). Most of the seventy million people in Iran are Persian. Iran has a long history with the ancient Persian Empire and the various conquering armies that followed it. During the rise of Islam, Iran had major contributions to the arts, mathematics, literature, philosophy, and science. The highly advanced carpet-weaving traditions from centuries past are but one example of the advancements in design and the technical expertise of the people. The country's Persian identity and culture continued throughout the centuries under different ruling powers. The Persian language remains and is a branch of the Indo-European language family. Arabic is widely used as a second language and a language used in science, which was helpful in reaching a broader audience and reaching out to the regional community with academic and scientific findings.

Figure 8.56: Main Mountains and Desert Areas in Iran. Updated from map courtesy of University of Texas Libraries.

In 1971, Iran celebrated the 2,500th anniversary of Persia's first monarchy. The monarchy was ruled by a shah, which is a title for the sovereign leader in Iran similar to a king. The shah's royal family ruled Iran from 1923 to 1979, when Islamic fundamentalists took control of the government. It was during this revolution on November 4, 1979, that **Ayatollah Ruhollah Khomeini**, who had previously been exiled by the shah, urged Iranians to oppose US activities in Iran. Iranian students stormed the US embassy and took US citizens hostage. Fifty-two Americans were held for 414 days during the US presidency of Jimmy Carter. The hostages were released the day that US president Ronald Reagan took office. Khomeini indicated he had not been aware of the students' plan but supported it. This is one reason the United States backed Saddam Hussein in the Iran-Iraq War of the 1980s. Khomeini ruled Iran until his death in 1989. Since then, Iran has been an Islamic state with an ayatollah as the Supreme Leader. An ayatollah is a high-ranking Shia cleric that is an expert in the Islamic faith and the Sharia (Muslim code of law based on the Koran). There are cases where women have reached the same status as an ayatollah and known as *Lady Mujtahideh*. Approximately 90 percent of Iran's population follows the Shia division of Islam.

Political Challenges

On the international front, Iran's leadership has indicated a drive to develop nuclear weapons and use them against Israel, which has caused concern in the global community. The government of Iran does not recognize the legitimacy of the nation of Israel. US president George W. Bush included Iran, along with Iraq and North Korea, in his 2002 Axis of Evil speech, the focus of which was the lack of democratic freedoms and the restrictive measures of the Islamic fundamentalist government. Iran is similar to Saudi Arabia in its restrictions of civil rights. A number of countries including the United States have placed trade sanctions against Iran regarding any materials associated with nuclear weapons or missiles. The US sanctions extend to an almost total trade embargo against Iran stemming from the 1979 revolution.

Open protests expressing a need for change have periodically erupted in the streets of Tehran. Ultimately, protesters are seeking personal freedoms and a more open society.

Figure 8.57: Twitter Revolution in Iran During the 2009 election in Iran, students and other individuals used the Internet, Twitter, and cell phones to organize a protest against the current president and in support of opposition candidates. The demonstrations were called the Twitter Revolution. Wikimedia Commons – CC BY 2.0.

The Arab Spring of 2011 was a phenomenon that spread across North Africa, the Arabian Peninsula, and the Middle East. Iran was not immune from the impact of the protests and demonstrations that occurred in their neighboring countries, but Iran was different. First of all, Iranians are not Arab but Persian in their ethnicity. Their history and heritage creates a distinct identity that separates them from the rest of the region. Many Iranian citizens want the same outcomes that the protesters and demonstrators want in the countries experiencing the Arab Spring revolutions. The difference is that Iranians have been demonstrating and protesting issues with their government in the years before 2011. Political tensions in Iran have been high since the 2009 elections and even earlier. During the 2009 election in Iran, students and other individuals used the Internet, Twitter, and cell phones to organize a massive protest against the current president and in support of opposition candidates. The demonstrations were called the Twitter Revolution.

Iran is at a crossroads in the conflict between conservative Islamic fundamentalists and Islamic reformers. The government of the Islamic state is controlled by Muslim clerics who tend to be more conservative in their rulings, but the young people are mainly on the side of the democratic reformers. Young people are becoming more familiar with Western culture. For example, the unofficial holiday of Valentine's Day has become extremely popular in Iran and is celebrated by a large sector of the population, mainly young people. In an effort to curb the influence of Western culture, on February 13, 2011, the government of Iran officially banned all symbols or activities associated with Valentine's Day. One claim was that the day was named after a Christian martyr and therefore was not supportive of Islam.

Economic Resouces

Iran has abundant oil and natural gas reserves that are being exploited to form the base of their economy. Iran holds about 15 percent of the world's reserves of natural gas, which is second only to Russia (Infoplease). In 2010, the country was the fourth-largest oil exporter in the world and held about 10 percent of the world's known oil reserves (Opec.org). The UN has classified Iran's economy as semideveloped. The government has taken control of the oil and natural gas industry and implemented a type of central planning over many major businesses. Small-scale agriculture and village trading activities are not usually owned by the government. The **Caspian Sea** provides for fishing and has oil reserves under the seabed. Oil and gas revenues make up most of the state's income. However, fluctuations in commodity prices have resulted in a more volatile income stream, and Iran's manufacturing base has been increasing to support a more diversified economy.

Key Takeaways

- Iraq was ruled by Saddam Hussein from 1979 to 2003. Iraq invaded Kuwait in 1990, which prompted the First Persian Gulf War. Hussein's use of weapons of mass destruction provoked the US invasion in 2003, which eventually removed Hussein from power.

- The majority Arab population in Iraq shares the country with the Kurds in the north. Sunni and Shia groups have divided the country and opposed each other since the US invasion of 2003.
- Turkey has a relatively stable, democratically elected government. Turkey's physical geography provides for large supplies of food crops and hydroelectric energy.
- Iran is a religious state that has often been at war or clashed with its Middle Eastern neighbors. Reformers and many young people in Iran would like to see the country become more democratic and more open to personal freedoms.

Discussion and Study Questions

1. Name three reasons Iraq gave for invading Kuwait in 1990. Why did the United States get involved in this war?
2. What are the three main geographic regions of Iraq? What groups are the majorities in each region?
3. Why was Saddam Hussein allowed to stay in power after Operation Desert Storm?
4. Why did the United States invade Iraq in 2003? What was one plan to finance the war?
5. Where are the two main mountain chains located in Iran? How do they support agriculture?
6. What three ethnic groups form an Ethnic Triangle that dominates the Middle East?
7. Why were protesters demonstrating in the streets of Tehran?
8. Where do Kurds hope to establish the proposed country of Kurdistan? When was the hope of establishing Kurdistan first proposed to the UN?
9. What economic group does Turkey want to join? What factors are keeping Turkey from joining this group?
10. Why are there conflicts in Iran over such things as Valentine's Day? What was the Twitter Revolution about?

Geography Exercise

Identify the following key places on a map:

- Anatolian Plateau
- Bosporus
- Caspian Sea
- Cyprus

- Diyarbakir
- Elburz Mountains
- Ethnic Triangle
- Euphrates River
- Istanbul
- Karbala
- Kurdistan
- Najaf
- Shatt al-Arab
- Sunni Triangle
- Tigris River
- Zagros Mountains

References

Infoplease, "Greatest Natural Gas Reserves by Country, 2006," http://www.infoplease.com/ipa/A0872966.html. http://www.opec.org/opec_web/static_files_project/media/downloads/publications/ASB2010_2011.pdf.

8.7. CENTRAL ASIA AND AFGHANISTAN

Central Asia and Afghanistan are regions rich in history, culture, and diversity, marked by their unique geographical landscapes and historical significance along the ancient Silk Road. Central Asia, comprising countries like Kazakhstan, Uzbekistan, Turkmenistan, Kyrgyzstan, and Tajikistan, boasts a blend of nomadic traditions and settled agricultural societies. Afghanistan, often seen as a crossroads of civilizations, has a complex tapestry of ethnic groups, languages, and historical influences due to its strategic location. Both regions face challenges, including political instability and economic development, but they also hold immense potential for growth and collaboration, particularly in trade and cultural exchange. The interplay between their histories and contemporary dynamics continues to shape the broader narrative of this pivotal area of the world.

Central Asia (a.k.a. Turkestan)

Central Asia is a region in the Asian continent that extends from the mountains of western China to the shores of the Caspian Sea. Pakistan and Iran create the southern border of the region, and the vast expanse of Russia is to the north. Afghanistan is considered a part of the region even though it was never a formal part of the Soviet Union. Central Asia was located on what was known as the Silk Road between Europe and the Far East and has long been a crossroads for people, ideas, and trade.

Central Asia has an extremely varied geography, including high mountain passes through vast mountain ranges, such as the **Tian Shan**, **Hindu Kush**, and the **Pamirs**. The region is also home to the vast **Kara Kum** and **Kyzyl Kum** Deserts, which dominate the interior with extensive spans of sand and desolation. The expansive treeless, grassy steppes that surround the desert regions are considered an extension of the steppes of Eastern Europe. Some geographers think of the Eurasian Steppes as one single, homogenous geographical zone.

Under the sand and prairie grasses lay the some of the most extensive untapped reserves of gas and oil on the planet. Natural resources are the main attraction of the region driving the economic forces that determine the development patterns of individual countries. Multinational corporations have vigorously stepped up their activity in the region.

The political systems are adjusting from the old Soviet Union's socialist policies to new democratic systems that are subject to high levels of authoritarian rule and corruption in business and politics.

The five countries of **Kazakhstan**, **Uzbekistan**, **Turkmenistan**, **Tajikistan**, and **Kyrgyzstan** were part of the former Soviet Union until its breakup in 1991. Today, with **Afghanistan**, they are independent countries that make up the region called Central Asia. The term *stan* means "land of," so, for example, Uzbekistan is the land of the Uzbeks. Central Asia is also referred to as Turkestan because of the Turkish influence in the region. The people of Turkey did not originate from the Middle East; they originated from northern Asia. They swept through Central Asia and dominated the region on their way to the Middle East. The Turkish language and heritage have had the most significant impact on the people of Central Asia. Turkmenistan's name is another reminder of the Turkish connection; it means "the land of the Turkmen."

Most of the groups of Central Asia were nomadic peoples who rode horses and herded livestock on the region's vast steppes. This way of life continued until the 1920s, when the Soviet Union forced many of the groups to abandon their lifestyle and settle on collective farms and in cities.

Figure 8.58: Central Asia, Formerly Part of the Soviet Union Afghanistan is also usually included as part of Central Asia, though it was never officially part of the Soviet Union. Updated from map courtesy of University of Texas Libraries.

Most of the people of Central Asia continue to identify culturally with their nomadic past. Central Asians who live in cities often demonstrate a mix of local and Russian culture in terms of dress and food because of the large influx of Russian populations in the region. More than six million Russians and Ukrainians were resettled into Central Asia during Soviet rule. Russian is often used as a lingua franca.

One of the primary ways in which people distinguish themselves culturally is through religious practices. Despite the area being part of the Soviet Union, where religious activities were discouraged, Islam was and still is the dominant religion. Most Central Asian Muslims are Sunnis.

Kazakhstan

The traditional people of Kazakhstan, who share a Mongol and Turkic heritage, moved into the Central Asia region sometime after 1200 CE. The expansion of the Russian Empire under the tsars integrated Kazakhstan and its neighbors, which eased their transition when the tsarist system of Russian government gave way to the Soviet Union. The influx of Russian people and culture had a major influence on Kazakhstan.

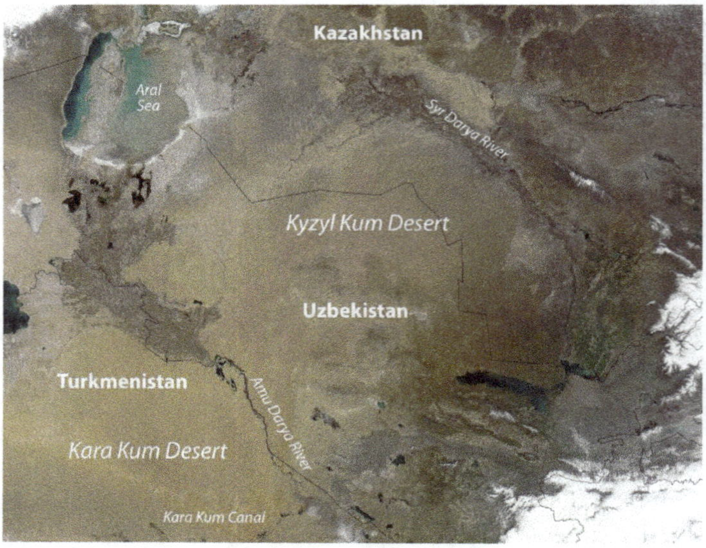

Figure 8.59: Deserts of Central Asia The Aral Sea is decreasing in volume because of water being diverted from the two rivers to be used for irrigation of desert lands to grow enormous crops of grains and cotton. Updated from map courtesy of NASA – public domain.

Kazakhstan gained independence from the Soviet Union in 1991, thereby creating the world's largest landlocked nation. It is the ninth-largest state on the planet in terms of square miles and is larger in physical area than of all of Western Europe. This vast land is host to a wide variety of physical landscapes, including the high, snow-capped peaks of the ranges on the Chinese border. The western portions are lowlands bordering the Caspian Sea. The seemingly endless grasslands of the interior are one of the largest steppe regions in the world. The steppe region has a semiarid type B climate. A large portion of southern Kazakhstan is desert, including the northern regions of the Kyzyl Kum Desert. Colder type D climates are found in the northern regions of the country.

The steppe produces grain in large quantities and other agricultural products, while the productive mining of minerals adds to the national wealth. Kazakhstan ranks high in the mining of many metals and uranium. Even diamonds are found here. Oil and natural gas extraction accounts for the largest sector of the country's economy and generates the largest export income. The **Tengiz basin** around the northeast shores of the Caspian Sea is home to extensive petroleum reserves. Oil pipelines are expanding to transport the oil to port locations and other countries, including China. The economy of Kazakhstan has been larger than the economies of all the other Central Asian states combined.

Figure 8.60: Cultural Traditions in Kazakhstan Kazakh performers demonstrate a cultural tradition of a game called "Catch a Girl," in which the two riders (male and female) take off on horseback and try to catch each other. The female rider whips the male until she is caught. If the male catches the female, he is rewarded with a kiss. US Air Force – public domain.

Kazakhstan also has a forward capital. During the Soviet era, the capital was located in the southeast at **Almaty**, but after gaining independence in 1991, the capital was moved north to Astana to ensure that the Russian-dominated northeast would be monitored against devolutionary forces that desiring to secede and become part of the Russian Republic.

Uzbekistan

Uzbekistan physically borders all the Central Asian countries. It is the most populous Central Asian country, with a population that exceeds twenty-seven million. Uzbekistan's eastern boundary extends deep into Kyrgyzstan territory. The boundary lines were created during the Soviet era to provide the central government with more control over its republics by politicizing enclaves to their benefit. Geographers call Uzbekistan a doubly landlocked nation because all the countries that surround it are also landlocked. The main source of fresh water comes from the Eastern Highland regions. The main rivers have been heavily diverted for irrigation and are often depleted before reaching their destination at the **Aral Sea.**

Cotton is the main agricultural crop. Uzbekistan is one of the top producers of cotton in the world and is a major exporter to world markets. The central and western regions have mainly arid desert climates and rely heavily on the fresh water flowing in from the mountains. Agriculture employs a full one-fourth of the population and accounts for one-fourth of the gross domestic product (GDP). The extraction of gold, minerals, and fossil fuels are also prime economic activities. The country has been transitioning from the old Soviet Union's command economy, which was controlled by the central government, to a market economy competing in a global marketplace.

Uzbekistan is a country of young people: about one-third of the population is under the age of fifteen. Education was heavily emphasized during the Soviet era; as a result, about 99 percent of the population is literate—though about one-third of the people still live in poverty. Islam emerged in this country after Uzbekistan won its independence in 1991 after the fall of the Soviet Union. In a culture of openness, Islam has risen in prominence to the point that approximately 88 percent of Uzbeks profess Islamic beliefs. The most commonly spoken language is Persian/Farsi.

Samarkand and the country's capital city, **Tashkent**, are located in the eastern core region, which is home to most of the population. Tashkent has an unofficial population of more than three million people. The city, which sits on the confluence of a local river and its tributaries, started as a **caravanserai**, or **oasis** for trade, along the Silk Road. Samarkand is Uzbekistan's second-largest city and is most noted as the central city of the Silk Road as well as an important historical city for Islamic scholars. In 2001, UNESCO declared this 2,750-year-old city a World Heritage Site. It is one of the oldest continually inhabited cities in the world and has been one of the more important cities in Central Asia. The historical architecture is heavily influence by Islamic styles from Iran. The region around Bukhara, Uzbekistan's fifth-largest city, has been occupied for at least the last five thousand years. Bukhara was another important city on the Silk Road and is known for its Asian carpet and textile industry. This region has been an important cultural, economic, and scholarly center for most of its known existence.

Figure 8.61: Bibi Khanym Mosque, Samarkand, Uzbekistan Samarkand, called the crossroads of culture, is located on the ancient Silk Road to China. The city is a UNESCO World Heritage Site. Fulvio Spada - Bibi Khanum Mosque, Samarkand - CC BY-SA 2.0.

Aral Sea Environmental Disaster

Central Asia's shrinking Aral Sea is shared by Uzbekistan and Kazakhstan. The sea was once the fourth-largest body of water in the world, but it has been reduced to a fraction of its original area. In 1960, the Aral Sea covered about 26,254 square miles, an area larger than the size of the US state of West Virginia. By 2009, the sea covered less than 10 percent of the same area. The entire eastern portion of the sea has become a sand desert, complete with the deteriorating hulls of abandoned fishing vessels. The loss of water is approximately equivalent to the complete draining of both Lake Erie and Lake Ontario in North America.

The water loss escalated when the **Syr Darya River**, which flowed into the northern part of the sea, and the **Amu Darya River**, which flowed into the southern side of the sea, were diverted for the irrigation of cotton and other crops. At about 1,500 miles long, the Amu Darya is the region's longest river. Its source is the high mountain streams and lakes of the **Pamir Mountains**. Environmental problems were further exacerbated by the extensive use of pesticides and fertilizers in agricultural processes. The chemicals contaminated the water flowing into the Aral Sea. Once the water dried up in the sea, the winds carried the buildup of chemicals and salt from the dry seabed over the land, causing serious health-related problems in the nearby human population. Cancer and respiratory illness rates continue to be higher than normal. Water and land pollution is a serious problem. Even the climate around the Aral Sea has changed gradually because of the loss of water from evaporation for precipitation. The climate is getting warmer in the summer and colder in the winter. The moderating affect that this large body of water had on its surrounding area is no longer as prominent as it once was.

The decline of the Aral Sea has destroyed habitats and the local economy. The fishing industry, which employed more than sixty thousand people, has been devastated. The remaining western portion of the sea has a rising salt content that is contributing to the decline of the fish population. Adding to the environmental devastation, the Soviets conducted biological weapons experiments on an island that was once in the middle of the Aral Sea. Hazardous wastes such as anthrax and toxic chemicals contaminated the land and found their way into the sea. Efforts have been made to marginalize the environmental damage of the contamination, but the damage has not been completely ameliorated. The sea has historically been an important environmental location for wildlife. It is located in a major flyway for migratory waterfowl in Central Asia and served as an important habitat. The deterioration of the Aral Sea and the destruction of habitat for waterfowl and other organisms is one of the world's worst environmental catastrophes. The fact that the Aral Sea is located in a region that is not part of the core economic area of the global community has rendered it "out of sight and out of mind" by entities that could potentially provide economic support.

Figure 8.62: Orphaned Ship in a Dry Seabed of the Former Eastern Region of the Aral Sea in Kazakhstan The eastern half of the sea is completely dry. Wikimedia Commons – public domain.

In the northern portion of the Aral Sea, called the **Little Aral Sea**, there has been some success in abating the deterioration of this once-thriving environmental habitat. A major dam has been constructed that partitions off the Little Aral Sea, causing water from the Syr Darya River to increase the water level of the Little Aral Sea and reduce the salt content. Canals, dikes, and irrigation processes have been updated to reduce the loss of water and increase the amount that flows into the northern section. The development efforts have caused the water level to rise and therefore have reinvigorated the once-struggling fishing industry. The efforts have been undertaken by Kazakhstan's government, which controls the Little Aral Sea.

A major part of the southern portion of the once thriving sea is located in Uzbekistan, which has not contributed the economic attention necessary to halt the sea's continued deterioration. The remaining western portion of the Uzbekistan side of the Aral Sea will continue to shrink if measures are not taken to address the loss of water from the Amu Darya River. The eastern side was completely dry by 2009. Uzbekistan has responded to the situation by contracting out to various multinational oil companies from Korea, China, and Russia to explore for oil beneath the dry seabed.

Figure 8.63: Aral Sea These satellite images compare the Aral Sea between 1989 and 2008. The eastern region was completely dry by 2009, whereas the northern portion was responding favorably to Kazakhstan's conservation efforts. NASA – public domain.

The demise of the Aral Sea was caused in part by the diversion of water from its northern inlet, the Syr Darya River. At the other end of the Syr Darya River an additional factor augmented the lack of water flow: the Soviet Union placed a dam on the river and allowed the overflow from the dam to flow into low-lying dry pans, creating artificial lakes. As a result, **Aydar Lake** was created and became the second-largest lake in Uzbekistan. Various species of fish were introduced and the lake became a major source for commercial fishing. Hundreds of tons of fish are harvested annually. Just as fishing was declining in the Aral Sea, the fishing industry was growing at Aydar Lake.

Turkmenistan

To the south of the Amu Darya River is the desert country of Turkmenistan, which extends from the Caspian Sea to Afghanistan in the east. Turkmenistan is slightly larger in physical area than the US state of California. Roughly 80 percent of the country is covered by the Kara Kum Desert, which is among the driest in the world. The southern mountains along the Iranian and Afghan border reach as high as 10,290 feet in elevation. Water from the Amu Darya River has been diverted by the seven-hundred-mile-long **Kara Kum Canal** through Turkmenistan to help grow cotton and other agricultural products.

The transition from a Soviet republic to an independent state in 1991 brought many changes. The former leader of the Turkmen Soviet Socialist Republic, Saparmurat Niyazov, who went by the name *Turkmenbashi* (leader of the Turkmen people), dominated the presidency for fifteen years. Through his authoritarian rule, he promoted a traditional culture—with Islam as the predominant religion—and was notorious for developing a cult of personality. For example, he changed all the names of the days of the week and the months of the year to his name, the names of his family members, and the names of Turkmen heroes or famous people. Turkmenbashi's image was printed on the currency, and large posters of him could be seen throughout the country. His book on important concepts, the *Ruhnama* (*The Book of the Soul*), was to be read by all schoolchildren and the public. After his death in 2006, many of his actions were reversed. The country continues to transition to a stable democratic state, though many of the same dynamics of corruption and authoritarian rule remain.

Turkmenistan is blessed with the fourth-largest natural gas reserves in the world; the top three are Russia, Iran, and the United States. The income from natural gas exports has become the country's greatest means of gaining wealth. Because Turkmenistan is landlocked, its government has been forced to partner with Russia to use of Russia's pipelines to export the natural gas. Not wishing to rely on Russia's monopoly on the pipelines, Turkmenistan developed an additional pipeline to China to help boost income and profits. Many international corporations are seeking to do business in Turkmenistan and Central Asia to corner a piece of the vast natural resources. Corporate colonialism is extremely active and has contributed to a high level of corruption in the government and the business sector. It is unclear how much of the country's wealth filters down to most of the population. Over the past decade, unemployment rates have exceeded 50 percent, and more than half the population lives below the poverty line.

The administrative center and largest city of Turkmenistan is its capital, Ashgabat, which has a population of about one million. Ashgabat lies between the Kara Kum Desert and the mountains near the former Silk Road. In the historic region of Central Asia, it is comparatively a very young town, having grown out of a small village founded in 1818. Ashgabat is primarily a government and administrative center, although it has thriving cotton, textile, and metalworking industries. Ashgabat is also a popular stop along the Trans-Caspian Railway.

Kyrgyzstan

Local groups in the mountains of Central Asia make up the population of Kyrgyzstan. The forty rays of sun on the country's flag symbolize of the legendary forty tribes of Manas that represent the nation. The rugged landscape of this mountainous land includes the high ranges of the Tian Shan Mountains, which can reach elevations as high as 24,400 feet and cover about 80 percent of the country. Snowfall from the mountains provides fresh water for agriculture as well as hydroelectric energy. Food crops can be grown in the valleys and the few lowland areas. Half the population works in agriculture, and self-sufficiency in food production is a major objective for survival. The mountains hold deposits of metals and minerals that have a strong potential for adding to the national wealth. Oil and natural gas reserves are also available for exploitation. The government is seeking foreign aid and investments to help develop these resources.

Figure 8.64: Flag of Kyrgyzstan The flag of Kyrgyzstan has a symbol of the sun and forty rays of light indicating the legendary forty tribes of Manas that represent the nation. Wikimedia Commons – public domain.

In 2009, Kyrgyzstan had a population of about 5.4 million in a land area about the size of the US state of South Dakota. About 30 percent of the population is under the age of fifteen, and about 36 percent of the population is urban. The western boundary with Uzbekistan and Tajikistan is winding and creates various small enclaves and exclaves of people from one country surrounded by people of another country and separated from their home nations.

Kyrgyzstan's transition from a Soviet republic to independence was not smooth. The loss of the state social safety net pushed the economy further to the informal sector, where trading and small transactions for personal survival are common. Shortages of consumer goods occur in rural areas and small towns. Kyrgyzstan is an isolated country that has been working to integrate itself into the global economy through technology and modernization. In 2010, clashes between Uzbeks and Kyrgyz brought about riots in the streets of major cities, resulting in more than two hundred casualties and three hundred thousand displaced citizens. After the situation cooled down, the government worked to stabilize itself with new leadership.

Figure 8.65: Base Camp on South Inylchek Glacier in the Tian Shan Mountains, Kyrgyzstan The mountain in background is Khan Tengri (22,949 feet). Wikimedia Commons – public domain.

A form of improvisational oral poetry, which allegedly dates back to more than one thousand years ago, is an aspect of traditional culture that has been preserved. While common throughout the region, it is mainly found in Kyrgyzstan and Kazakhstan. Practitioners will often engage in "lyrical battles" of folklore. These poets, often accompanied by two- or three-stringed instruments, will recite the *Manas*, an epic poem of Kyrgyzstan that details the life of the Kyrgyzstan hero Manas. This epic tale is a renowned part of the culture and festivals of Kyrgyzstan.

Figure 8.66: Storytelling Tradition A manaschi—a traditional storyteller—in Karakol, Kyrgyzstan, who has memorized entire epic legends of his history, recites the stories to others at a local festival. Wikimedia Commons - public domain.

Tajikistan

The eastern region of Central Asia has some of the highest mountain ranges in the world; about 90 percent of Tajikistan is mountainous, and more than half the country is 10,000 feet in elevation or higher. Ranges of the **Himalayas** extend from the south all the way to the western border with China. The **Pamirs** is a mountain range located where the **Tian Shan**, **Karakorum**, and **Hindu Kush** mountain ranges meet in Tajikistan, an area referred to as the **Pamir Knot**, or the roof of the world. Elevations in the Pamirs often exceed 24,500 feet. The Pamirs is the source of the Amu Darya River and is home to the longest glacier outside the polar regions (forty-eight miles long in 2009). There is great potential for hydroelectric power generation, and Tajikistan is developing the world's highest dam.

Tajikistan has the smallest physical area of any country in Central Asia but has a population of about 7.3 million. Only about one-fourth of the population is urban, and one-third of the population is younger than fifteen years of age. There is less ethnic or religious diversity; 80 percent of the people are ethnically Tajik and are Sunni Muslims. Though it has natural resources similar in quantity to those in Kyrgyzstan, Tajikistan's economy is not advanced enough to fully take advantage of its economic potential. Half the labor base works abroad and sends remittances back to their families for economic support. Unemployment is high, and job opportunities have not been able to keep up with demand.

Dushanbe, the capital and largest city of Tajikistan, is situated on the confluence of two local rivers and is famous for its Monday markets (Dushanbe means "Monday" in Tajik). Dushanbe, like Ashgabat in Turkmenistan, was originally a small village; it became an administrative center for the region when the Soviet army conquered the area in 1929. Similar to many of the other cities and regions in Central Asia, the Soviets transformed the political and economic landscape and made Dushanbe a center for cotton and silk production. The Soviets also transformed the cultural and ethnic makeup of the city by relocating tens of thousands of people from Russia and other regions of Central Asia to Dushanbe.

The transition from a Soviet Republic to an independent country in 1991 was difficult for Tajikistan. From 1992 to 1997, a bitter civil war between regional factions killed more than fifty thousand people. Political instability and corruption has hampered the growth of a market economy, and political power remains in the hands of the economic elite. Debt restructuring with Russia and an infusion of development loans from China have aided the ailing economy. Aid from the US helped fund a thirty-six-million-dollar bridge linking Tajikistan and Afghanistan, which opened in August 2007. US aid has also contributed to infrastructure development designed to help US military operations in Afghanistan and in the region as a whole. Countries such as Russia, China, and the United States are all looking to gain an advantage with their ties to Tajikistan to exploit the region's natural resources.

Afghanistan

Present-day Afghanistan has been conquered by the likes of Genghis Khan, Alexander the Great, and the Mogul Empire and was a buffer zone for colonial feuds between Russia and British India. The high central mountain range of the Hindu Kush dominates the country and leaves a zone of well-watered fertile plains to the north and a dry desert region to the south. Afghanistan is a remote region without access to the sea and acts as a strategic link between the Middle East and the Far East.

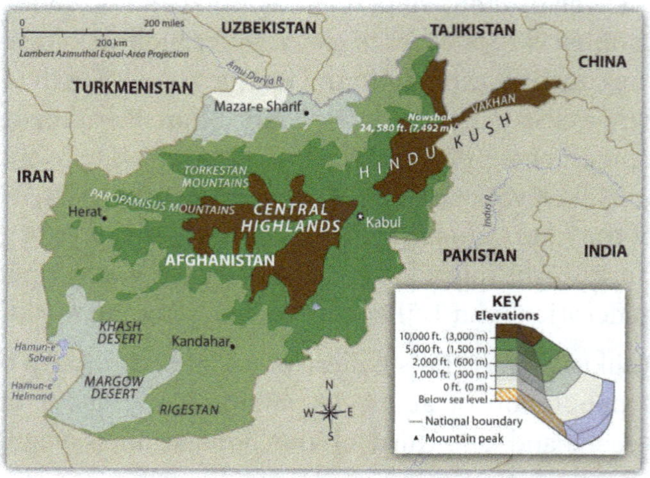

Figure 8.67: Afghanistan Kabul is the capital, and the southern city of Kandahar is the second-largest city. Map courtesy of CIA–UT Austin Map Library – public domain.

The Soviet Invasion and the Taliban

In 1979, the Soviet Union took advantage of ongoing ethnic warfare in Afghanistan to inject itself into the country. The Soviets pushed in from the north and occupied much of Afghanistan until they completely withdrew in 1989. During the Soviet occupation, the United States supported anti-Communist resistance groups such as the Mujahideen with money, arms, and surface-to-air missiles. The missiles were instrumental in taking out Soviet aircraft and MiG fighters, which caused a critical shift in the balance of power in the war. One of the major connections between the for the Central Intelligence Agency (CIA) and Mujahideen was a Saudi national named **Osama bin Laden**. Support from the CIA through bin Laden to the Mujahideen was instrumental in defeating the Soviets.

The power vacuum left by the retreating Soviets allowed conflicts to reemerge between the many ethnic factions in Afghanistan. Dozens of languages are spoken in Afghanistan; the top two are Pashtu and Afghan Persian-Dari. There are also a dozen major ethnic groups; the top two are Pashtun and Tajik. The groups regularly fight among themselves, but they have also been known to form alliances. Rural areas are usually led by clan leaders who are not part of any official arm of a national government. Afghanistan is a place where forming any national unity or identity is not easy. The national government in the capital city of Kabul has little influence in the country's rural regions.

The Soviet invasion brought the internally warring factions together for a short period to focus on the Soviet threat. Chaos and anarchy thrived after the Soviet forces withdrew, but the Islamic fundamentalist group known as the Taliban came forward to fill the power vacuum. One objective of the Taliban was to use Islam as a unifying force to bring the country together. The problem with that concept was that there was much diversity in how Islam was practiced by the numerous local groups. Many of the factions in Afghanistan opposed the Taliban; one such group being the **Northern Alliance**, which was an association of groups located in the northern portion of the country. The civil war between the Taliban and those that opposed them resulted in the deaths of more than fifty thousand people by 1996 when the Taliban emerged to take power in Kabul. The Taliban is a Sunni Muslim group that adheres to strict Islamic laws under the Wahhabi branch of the faith similar to that of Saudi Arabia. Under Taliban rule, women were removed from positions in hospitals, schools, and work environments and had to wear **burkas** (also spelled **burqas**) and be covered from head to toe, including a veil over their faces. Violators were either beaten or shot. The Taliban brought a sense of militant order to Kabul and the regions under their control. Various factions such as the **Northern Alliance** did not share the Taliban's strict Islamic views and continued to oppose their position in power.

Al-Qaeda and the US Invasion

After the war against the Soviet Union was over, the US role in Afghanistan diminished. The groups that the United States had supported continued to vie for power in local conflicts. Osama bin Laden remained in Afghanistan and established training camps for his version of an anti-Western resistance group called al-Qaeda. Just as he had opposed the Soviet Union, he now opposed the United States, even though the United States had supported him against the Soviets. The Saudi government allowed the United States to establish military bases in Saudi Arabia during Operation Desert Storm in the Persian Gulf War, and this was one reason for bin Laden's opposition; he believed that non-Muslims should not be on the same ground as the Muslim holy sites of Mecca and Medina.

The 9-11 attack in New York City was traced back to al-Qaeda and bin Laden, who was residing in Afghanistan at the time. In a military action dubbed Operation Enduring Freedom, the United States invaded Afghanistan in 2001, removed the Taliban from power, and dismantled the al-Qaeda training camps. Although bin Laden escaped, the terror of the Taliban was temporarily reduced. Women were allowed to return to the workplace, and the rebuilding of the country became a priority. The country was devastated by war and is divided by the human geography because of the various ethnic and traditional groups. Afghanistan is one of the most impoverished places on Earth. The armed conflicts in Afghanistan did not end with the US invasion. After regrouping, the Taliban rallied its supporters on the Pakistani side of the border and returned to the fighting front in Afghanistan against North Atlantic Treaty Organization (NATO) and US forces.

Figure 8.68: Propaganda Poster in Afghanistan with Image of Osama bin Laden Osama bin Laden was killed by US Navy Seals in 2011. US Department of Defense – public domain.

Fighting between Western forces and the Taliban in Afghanistan continued to provide the exiled bin Laden a platform to promote his al-Qaeda terrorist activities from his hiding place. Efforts to locate and marginalize bin Laden continued through to the US presidency of Barack Obama. In May of 2011, on orders from President Obama, a team of US Navy Seals were sent into the city of Abottabad, Pakistan, to a private compound where intelligence indicated that bin Laden was hiding. In the confrontation, the US Navy Seal team killed bin Laden. The entire operation was conducted without the awareness of the Pakistani government. This event may have impacted al-Qaeda but has not likely diminished the fighting in Afghanistan.

The country is the world's largest producer of opium, a product extracted from a poppy plant seedpod that can also be refined into heroin. The expanding poppy cultivation as well as a growing drug trade may account for one-third of the country's income. More than 80 percent of the heroin consumed in Europe is grown in Afghanistan. The drug trade has only multiplied the problems in this devastated country. Prudent and effective methods for the government to address the drug trade are matters for debate and negotiation. Most of the country is ruled by warlords and clan leaders who have few resources other than tradition and custom. Afghanistan's infrastructure has been destroyed through warfare, and its government is dependent on foreign aid; without it, this country cannot recover to integrate itself with the global economy. Central Asia has enormous oil and natural gas reserves, and the core economic regions of the world will continue their work to extract these resources for economic gain.

Operation Enduring Freedom

The US Department of Defense issued an official statement on Afghanistan in 2008 (Source DoD 2008): In response to the events of September 11, 2001, the U.S. and its allies launched an invasion of Afghanistan to overthrow the Taliban regime and destroy the al-Qaeda terrorist network it supported. In the years since, the International Security Assistance Force, under NATO leadership, has taken charge of extensive provincial reconstruction and stabilization efforts, helping set the economic, political and security conditions for the growth of an effective, democratic national government in Afghanistan. As the lead member of the international coalition, the U.S. contributes troops to both the ISAF [International Security Assistance Force] mission and *Operation Enduring Freedom*, tasked with pursuing al-Qaeda throughout Afghanistan's inhospitable border region with Pakistan.

The Western military troops aligned themselves with Afghan groups such as the Northern Alliance to defeat the Taliban and remove the al-Qaeda presence.

Figure 8.70: Voting in Afghanistan After voting in Afghanistan, individuals dipped their fingers in ink to indicate they had voted and were not allowed to vote more than once. This photo indicates that women were also allowed to vote in Afghanistan in the 2005 elections for provincial councils and parliamentary positions. US Department of Defense – public domain.

Democratic elections were held for the office of president in Afghanistan beginning in October of 2004. Hamid Karzai was the country's first elected president in the twentieth century. He was reelected as president in 2009 under the cloud of claims of election fraud. The right to vote was restored to women in the 2004 election. To combat voter fraud, people would dip their fingers in ink to indicate they had voted and to ensure they did not vote more than once. Voting has not been a smooth process because democratic rule is new to Afghanistan. Afghanistan's future is insecure. Most of Afghanistan is still ruled by warlords and clan leaders. The Taliban has sustained its support in Afghanistan from bases on the Pakistani side of the border, and United Nations (UN) and NATO troops continue to confront the Taliban and work toward stability. Kandahar, the second-largest city in Afghanistan, is located in the south, an area where support for the Taliban is stronger than it is in Kabul in the north.

Resources and Globalization

In 2010, a US government report indicated that vast amounts of mineral wealth were discovered in Afghanistan by American geologists and Pentagon officials. Enormous deposits of iron, copper, gold, cobalt, and rare industrial minerals such as lithium are reported to be present in Afghanistan. Total reserves are unknown or have not been released but if extracted would result in trillions of dollars of economic gain for the country. Lithium is highly sought after and is used in the manufacturing of batteries, computers, and electronic devices. The report indicated that Afghanistan could become the world's premier mining country.

Figure 8.71: Vastness of Afghanistan In this photo is the Lataband Road between Kabul and Surobi. The dry, treeless expanse of this region is home to Afghanistan's vast mineral wealth. Wikimedia Commons – CC BY-SA 4.0.

Discovery of vast resources helps place the war in Afghanistan in perspective with respect to global competition over the control of resources. It has been reported that China has already offered millions of dollars in incentive money to Afghan government officials to allow its country to mine copper. Bribery and corruption in the Afghan government is a serious impediment to a stable political environment, but criminal activities are projected to persist and swell with the potential for additional mining wealth. Afghanistan does not have a long-standing tradition of mining. Agriculture has been the main focus of economic activity for the rural communities. A newfound potential for mineral wealth will change the future of Afghanistan. It will be interesting to watch how Afghanistan adapts to and benefits from the discovery of previously unknown resources.

Key Takeaways

- Central Asia is a landlocked region that receives little rainfall. Two large desert regions are located at the region's core. Vast grasslands called steppes dominate the northern sector. High mountains to the east provide a border between Central Asia and China.
- Central Asia (a.k.a. Turkestan) has been dominated by the Soviet Union during the twentieth century. The transition to independence

has challenged the region to adjust to changes in political and economic systems.

- The demise of the Aral Sea is being caused by the diversion of water from the two rivers flowing into the sea. A once-thriving fishing industry has been destroyed and environmental damage has been catastrophic.
- Armed conflicts in Afghanistan continue between Western military forces and the Taliban over the control of the country. Clan leaders are a main component of the political fabric of the country. The diversity and fragmentation of the country make it almost impossible to govern.
- Afghanistan is one of the poorest countries in the world based on data on standards of living. However, the country has one of the largest deposits of valuable minerals and ores in the world waiting to be extracted. The deposits are the target of multinational corporations around the globe.
- The following is a summary of trends in Central Asia:
    - Shift from Soviet Union to independent states
    - Rise in authoritarian governments and corruption
    - Increase of cultural influence of Islamic institutions
    - Global economic focus on extractive activities
    - Increase in military activity over control of resources
    - Decline in agriculture as an economic base
    - Continued concern for environmental problems

Discussion and Study Questions

1. Why is Central Asia often referred to as Turkestan? What does the suffix "stan" indicate?
2. Which country in Central Asia has a forward capital and why?
3. How is the Tengiz basin important to the global economy? Where is it located?
4. How have problems with the Aral Sea affected the people of the region?
5. What is attracting multinational corporations to Central Asia? How do the corporations impact politics?
6. What happened in Tajikistan after it received independence in 1991? How did this affect the country?
7. List at least five general trends that have been occurring in Central Asia since the collapse of the Soviet Union.
8. Why did the CIA get involved in the 1980s war in Afghanistan? Who acted as a CIA contact?

9. What role did the Taliban once have in Afghanistan's government? Why were not they allies with Iran?
10. What are the main methods of gaining wealth in Afghanistan today? How might this change in the future?

Geography Exercise

Identify the following key places on a map:

- Amu Darya River
- Aral Sea
- Aydar Lake
- Caspian Sea
- Hindu Kush
- Kandahar
- Kara Kum Canal
- Kara Kum Desert
- Karakorum Ranges
- Kyzyl Kum Desert
- Little Aral Sea
- Pamir Knot
- Pamirs
- Samarkand
- Syr Darya River
- Tengiz basin
- Tian Shan

8.8. CONCISE SUMMARY

North Africa, Southwest Asia, and Central Asia (Turkestan) comprise the region of North Africa and Southwest Asia, characterized by the predominance of Islam, desert type B climates, and the exportation of oil and natural gas. Vast, extensive deserts constitute the majority of the geographical expanses in all three sections of the realm. Water is crucial, as the majority of the people has historically relied on agriculture for their livelihood. Division, warfare, and conflict have been persistent phenomena. The Arab Spring of 2011 encompassed a series of widespread citizen protests across numerous Arab nations, wherein the populace sought political and economic reform.

The three principal monotheistic religions—Judaism, Christianity, and Islam—originated in the Middle East and can be linked to a shared patriarch, Abraham. The majority of the population in the realm adheres to Islam. The Five Pillars of Islam serve as a framework for Muslim conduct. Mecca, the most sacred city for Muslims globally, is the site where the prophet Muhammad initiated the religion. The primary sects of Islam are Sunni, comprising 84 percent, and Shia, accounting for 15 percent. Additional minor divisions exist, such as Sufi. North Africa encompasses the nations adjacent to the Mediterranean Sea and Sudan along the Red Sea. The majority of individuals in North Africa adhere to Islam. The densely populated regions are situated along the coastline or adjacent to the Nile River.

The Maghreb encompasses the area containing the Atlas Mountains to the west. The Atlas Mountains retain moisture, facilitating precipitation for agricultural activities in the mountain valleys. North African nations have had political uprisings in which citizens have called for democratic reforms and enhanced civil rights. Egypt is the most populous Arab nation and serves as the foundation of the Arab world. The African Transition Zone is the area where the arid conditions of the Sahara Desert converge with the tropical conditions of Equatorial Africa. Islam predominates in the north, while Christianity and indigenous religions prevail in the south. Sudan is a location where the two factions have engaged in conflict and division. The tribal Christian population of southern Sudan seceded from the Arab Muslim north to establish the sovereign nation of South Sudan. Ethnic cleansing has occurred in the western portion of Sudan, specifically in Darfur.In 1945, Palestine was bifurcated into an Arab state and a Jewish state. The Jewish state was established as the nation of Israel in 1948. Since then, a persistent conflict has existed between the predominantly Arab Palestinians and the Jewish State of Israel. The West Bank and the Gaza Strip are Palestinian territories striving for equal state recognition. The resolution of this division is contentious; both parties present mutually exclusive assertions. The adjacent Arab nations of Jordan, Syria, and Lebanon possess distinct political circumstances.

The Arabian Peninsula comprises several states governed by a singular monarch, sultan, emir, or sheikh, with the exception of Yemen, which has a democratically elected president. Saudi Arabia is the location of the two principal cities of Islam, Mecca and Medina. Saudi Arabia is the largest nation on the peninsula and possesses around 25 percent of the world's known oil reserves. Oil and natural gas constitute the primary export commodities and the principal source of revenue for the region. Yemen is predominantly an agricultural nation characterized by significant population growth and a rural majority.The majority of the population in Turkey is of Turkish ethnicity, while the predominant ethnicity in Iran is Persian. Turkey is the remnant of the formerly vast Ottoman Empire. Iran is a nation characterized by mountains and deserts that was formerly the epicenter of the Persian Empire. Turkey comprises a segment of its territory in Europe, while the majority encompasses the vast Anatolian Plateau in Asia. Turkey is a democratic republic with elected public officials. Iran is a theocratic regime whose Islamic authorities dictate governmental inclusion. Each country has a substantial population of approximately seventy million.

Iraq was at the center of the Fertile Crescent of Mesopotamia, which spawned early human civilizations. Iraq was invaded by the United States in 2003 and has been working to establish a democratic government. The country is divided between a Shia majority and a Sunni minority. It is an Arab country with a large minority group of Kurds in the north. Iraq has enormous oil reserves that have not yet been developed for exploitation, which has attracted the attention of the world's energy-dependent core regions.

Central Asia (Turkestan) is a landlocked region with mainly type B climates with cooler temperatures in the higher elevations of the eastern mountains. The prominently Muslim region has been transitioning from a Soviet-controlled domain to independent states. Some countries are making the transition with less difficulty than others. The Aral Sea has been one of the major environmental disasters in modern history. Globalization activities have targeted the region for its immense amount of natural resources—mainly oil, natural gas, and minerals.

Afghanistan has been conquered by many empires and was a buffer state between British and Russian colonial efforts. The country has a dry, rugged terrain and a devastated economy. The people are poor with little infrastructure. The many tribal groups have made it difficult for a centralized government to be effective. The United States invaded Afghanistan in 2001 to remove al-Qaeda training camps run by Osama bin Laden. North Atlantic Treaty Organization (NATO) forces have taken over the mission in Afghanistan to defeat the Taliban extremists and support stability for the country. Afghanistan has enormous amounts of mineral resources that are in high demand by the world's core industrial regions.

9 Chapter

SOUTH ASIA

9.1. INTRODUCTION: AN OVERVIEW

South Asia, a vibrant and diverse region, is home to over a billion people and a rich tapestry of cultures, languages, and histories. Encompassing countries like India, Pakistan, Bangladesh, Nepal, Bhutan, Sri Lanka, and the Maldives, it showcases an extraordinary blend of traditions, from ancient civilizations to modern developments. The region boasts stunning landscapes, from the towering Himalayas to lush coastal plains, and is known for its culinary diversity, ranging from spicy curries to fragrant biryanis. Despite facing challenges such as political tensions, economic disparities, and environmental issues, South Asia continues to evolve, with its youth driving innovation and cultural exchange in an increasingly interconnected world.

Figure 9.1 Main Features of South Asia

The landmass of South Asia was formed by the Indian Plate colliding with the Eurasian Plate. This action started about seventy million years ago and gave rise to the highest mountain ranges in the world. Most of the South Asian landmass is formed from the land in the original Indian Plate. Pressure from tectonic action against the plates causes the Himalayas to rise in elevation by as much as one to five millimeters per year. Destructive earthquakes and tremors are frequent in this seismically active realm. The great size of the Himalayas has intensely influenced the beliefs and traditions of the people in the realm. Some of the mountains are considered sacred to certain religions that exist here.

The Physical Geography

Figure 9.2: Trekking Trail on the Way to Mt. Everest in the Himalayas of Northern Nepal Mt. Everest is the world's highest peak at 29,035 feet. The Himalayas are the highest mountain chain in the world and create a natural border between South Asia and China. Steve Hicks – The Himalayas sumed up in one picture! – CC BY 2.0.

The Himalayan Mountains dominate the physical landscape in the northern region of South Asia. **Mt. Everest** is the tallest peak in the world, at 29,035 feet. Three key rivers cross South Asia, all originating from the Himalayas. The **Indus River**, which has been a center of human civilization for thousands of years, starts in Tibet and flows through the center of Pakistan. The **Ganges River** flows through northern India, creating a core region of the country. The **Brahmaputra River** flows through Tibet and then enters India from the east, where it meets up with the Ganges in Bangladesh to flow into the Bay of Bengal. While the northern part of this region includes some of the highest elevations in the world, the Maldives in the south has some of the lowest elevations, some barely above sea level. The coastal regions in southern Bangladesh also have low elevations. When the seasonal reversal of winds called the monsoon arrives every year, there is heavy flooding and its effect on the infrastructure of the region is disastrous. The extensive **Thar Desert** in western India and parts of Pakistan, on the other hand, does not receive monsoon rains. In fact, much of southwest Pakistan—a region called **Baluchistan**—is dry, with desert conditions.

The mountains on the border between Pakistan and Afghanistan extend through Kashmir and then meet up with the high ranges of the Himalayas. The Himalayas create a natural barrier between India and China, with the kingdoms of Nepal and Bhutan acting as buffer states with Tibet. Farther south along the east and west coasts of India are shorter mountain ranges called ghats. The **Western Ghats** reach as high as eight thousand feet, but average around three thousand feet. These ghats are home to an extensive range of biodiversity. The **Eastern Ghats** are not as high as the Western Ghats, but have similar physical qualities. The ghats provide a habitat for a wide range of animals and are also home to large coffee and tea estates. The **Deccan Plateau** lies between the Eastern and Western Ghats. The **Central Indian Plateau** and the **Chota-Nagpur Plateau** are located in the central parts of India, north of the two Ghat ranges. The monsoon rains ensure that an average of about fifty-two inches of rain per year falls on the Chota-Nagpur Plateau, which has a tiger reserve and is also a refuge for Asian elephants.

The Monsoon

A monsoon is a seasonal reversal of winds that is associated with heavy rains. The summer monsoon rains—usually falling between June and September—feed the rivers and streams of South Asia and provide the water needed for agricultural production. In the summer, the continent heats up, with the Thar Desert fueling the system. The rising hot air creates a vacuum that pulls in warm moist air from the Bay of Bengal and the Indian Ocean. This action shifts moisture-laden clouds over the land, where the water is precipitated out in the form of rain.

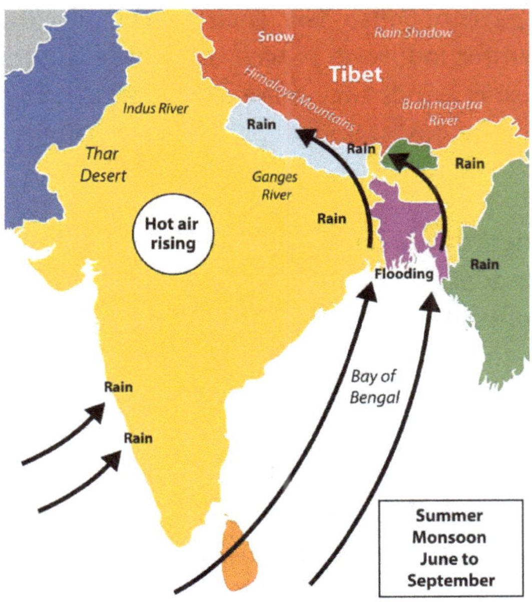

Figure 9.3: The Monsoon System in South Asia. Wikimedia Commons = public domain.

The monsoon rains bring moisture to South Asia right up to the Himalayas. As moisture-laden clouds rise in elevation in the mountains, the water vapor condenses in the form of rain or snow and feeds the streams and basins that flow into the major rivers, such as the Brahmaputra, Ganges, and Indus. The Western Ghats creates a similar system in the south along the west coast of India. Parts of Bangladesh and eastern India receive as much as six feet of rain during the monsoon season, and some areas experience severe flooding. The worst-hit places are along the coast of the Bay of Bengal, such as in Bangladesh. There is less danger of flooding in western India and Pakistan, because by the time the rain clouds have moved across India they have lost their moisture. Desert conditions are evident in the west, near the Pakistan border in the great Thar Desert. On average, fewer than ten inches of rain fall per year in this massive desert. On the northern rim of the region, the height of the Himalayas restricts the warm moist monsoon air from moving across the mountain range. The Himalayas act as a precipitation barrier and create a strong rain shadow effect for Tibet and Western China. The monsoon is responsible for much of the rainfall in South Asia.

By October, the system has run its course and the monsoon season is generally over. In the winter, the cold, dry air above the Asian continent blows to the south, and the winter monsoon is characterized by cool, dry winds coming from the north. South Asia experiences a dry season during the winter months. A similar pattern of rainy summer season and dry winter season is found in other parts of the world, such as southern China and some of Southeast Asia. A final note about the monsoons: small parts of South Asia, such as Sri Lanka and southeastern India, experience a rainy winter monsoon as well as a rainy summer monsoon. In their case, the winter monsoon winds that come down from the north have a chance to pick up moisture from the Bay of Bengal before depositing it on their shores.

Early Civilizations

The Indian subcontinent has a long history of human occupation, and is an area where cities independently developed and civilization emerged. The earliest civilization on the subcontinent was the **Indus Valley Civilization**, in existence from about 3300 BCE to 1500 BCE. This Bronze Age civilization started as a series of small villages that became linked in a wider regional network. Urban centers developed into various religious and trade networks that spanned as far as Central Asia, Southwest Asia, and, perhaps, Egypt. The civilization is known for its planned structures. The cities and villages of the urban phases were planned with major streets going north/south and east/west. It had a system of drains that channeled waste water outside the city. Additionally, this civilization had a homogeneous material culture. Its artifacts of pottery and metallurgy all had a very similar style that was spread over a vast land area, a fact that aided in the recognition of the expanse of the culture.

Invasions by outsiders have the potential effect of bringing with them an influx of new ideas, concepts, and technology. Likewise, the Indus Valley Civilization no doubt had an impact on the region that it encompassed. Little is known of the historical events of earlier times. Some of the evidence we rely on today to discern historical events is gleaned from language, religion, and ethnicity. Significant to South Asia is the presence of Indo-European languages. It is presumed that these languages were brought to the region by immigrants from the west, where these languages were dominant. Aryans from Persia and other cultures might have diffused languages such as Hindi to South Asia, which later may have led to Hindi, for example, becoming the lingua franca of the region.

The northern plains of South Asia, which extend through the Ganges River valley over to the Indus River valley of present-day Pakistan, were fertile grounds for a number of empires that controlled the region throughout history. After the decline of the Indus Valley Civilization, various phases of Iron Age traditions emerged. Most of this Iron Age culture is defined by the presence of iron metallurgy and distinctive characteristics of ceramics.

The **Mauryan Empire** existed between 322 and 185 BCE and was one of the most extensive and powerful political and military empires in ancient India. This empire was founded by Chandragupta Maurya in 322 BCE, who began to extend his regime westward, easily conquering areas that had been disrupted by the expansion of Alexander the Great's armies. The Mauryan Empire was prosperous and greatly expanded the region's trade, agriculture, and economic activities. This empire created a single and efficient system of finance, administration, and security. One of the greatest emperors in the Mauryan dynasty was Ashoka the Great, who ruled over a long period of peace and prosperity. Ashoka embraced Buddhism and focused on peace for much of his rule. He created hospitals and schools and renovated major road systems throughout the empire. His advancement of Buddhist ideals is credited with being the reason most of the population on the island of Sri Lanka is Buddhist to this day.

Islam became a powerful force in South Asia upon its diffusion to the subcontinent. Muslim dynasties or kingdoms that ruled India between 1206 and 1526 are referred to collectively as the **Delhi Sultanate**. The Delhi Sultanate ended in 1526 when it was absorbed into the expanding **Mughal Empire**. The Islamic Mughal Empire ruled over much of northern and central India from the 1500s to about the middle of the nineteenth century. After 1725, it began to decline rapidly because of a combination of factors, with European colonialism adding the finishing touch. The Mughal Empire had been religiously tolerant but Muslim oriented. The classic period of this empire began in 1556 and ended in 1707. Many of the monuments we associate with India, including the Taj Mahal, the Red Fort in Lahore, and the Agra Fort, were built during the classical period.

Colonialism in South Asia

The force of colonialism was felt around the world, including in South Asia. South Asia provides an excellent example of colonialism's role in establishing most of the current political borders in the world. From the sixteenth century onward, ships from colonial Europe began to arrive in South Asia to conduct trade. The British East India Company was chartered in 1600 to trade in Asia and India. They traded in spices, silk, cotton, and other goods. Later, to take advantage of conflicts and bitter rivalries between kingdoms, European powers began to establish colonies. Britain controlled South Asia from 1857 to 1947.

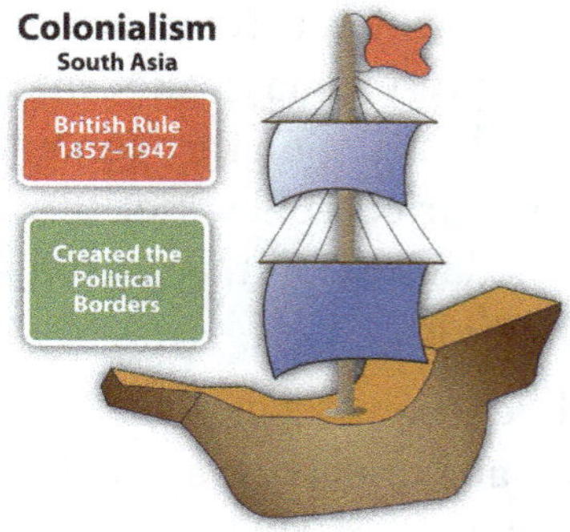

Figure 9.4: British colonialism in South Asia began in 1857 and lasted until 1947.

Goa is the smallest state in modern-day India. In the sixteenth century, it was first encountered by Portuguese traders, who annexed it shortly after arriving. Goa was a colony of Portugal for the next 450 years. By the mid-1800s, most of the population of the tiny area had been forcibly converted to Christianity. Many of the Hindu traditions, however, survived in the region. Hindu holidays are celebrated among the expatriate community in India. Christian holidays are also celebrated, especially Christmas and Easter. The cathedrals and secular architecture in many of the historic buildings of Goa are European in style, reflecting its Portuguese origins. This architecture is locally termed "Indo-Portuguese." Goa was one of the longest-held colonial possessions in the world. It was finally annexed to India in 1961.

The British no longer controlled South Asia after 1947. Local resistance and the devastating effects of World War II meant the British Empire could not be controlled as it once was. Great Britain pulled away from empire building to focus on its own redevelopment. Upon the British withdrawal from India, Britain realized the immense cultural differences between the Muslims and Hindus and created political boundaries based on those differences. **West Pakistan** was carved out of western India; **East Pakistan** was carved from eastern India. However, the new borders separating Hindu and Muslim majorities ran through population groups, and some of the population now found itself to be on the wrong side of the border. The West Pakistan-India partition grew into a tragic civil war, as Hindus and Muslims struggled to migrate to their country of choice. More than one million people died in the civil war, a war that is still referred to in today's political dialogue between Pakistan and India. The **Sikhs**, who are indigenous to the **Punjab** region in the middle, also suffered greatly. Some people decided not to migrate, which explains why India has the largest Muslim population of any non-Muslim state.

Another civil war would erupt in 1973 between West Pakistan and East Pakistan. When the states were first created in 1947, they operated under the same government despite having no common border and being over nine hundred miles apart and populated by people with no ethnic similarities. The civil war lasted about three months and resulted in the creation of the sovereign countries of Pakistan and Bangladesh (formerly East Pakistan). The name Bangladesh is based on the Bengali ethnicity of most of the people who live there. Both Pakistan and Bangladesh are among the top ten most populous countries in the world.

Language is probably one of the more pervasive ways that Europeans affected South Asia. In modern-day India and Pakistan, English is the language of choice in secondary education (English-medium schools). It is often the language used by the government and military. Unlike many other Asian countries, much of the signage and advertising in Pakistan and India is in English, even in rural areas. Educated people switch back and forth, using English words or entire English sentences during conversation in their native tongue. Some scholars have termed this Hinglish or Urglish as the base languages of northern India and Pakistan are Hindi and Urdu, respectively.

The British game of cricket is an important cultural and national sport within this Asian subcontinent. The constant conflict between the nations of India and Pakistan are reflected in the intense rivalry between their national cricket teams. The Cricket World Cup is held every four years and is awarded by the International Cricket Council. South Asian countries have won the Cricket World Cup three times: India (1983), Pakistan (1992), and Sri Lanka (1996).

Population in South Asia

South Asia has three of the ten most populous countries in the world. India is the second largest in the world, and Pakistan and Bangladesh are numbers five and six, respectively. Large populations are a product of large family sizes and a high fertility rate. The rural population of South Asia has traditionally had large families. Religious traditions do not necessarily support anything other than a high fertility rate. On the other hand, the least densely populated country in South Asia is the Kingdom of Bhutan. Bhutan has a population density of only fifty people per square mile. Bhutan is mountainous with little arable land. More than a third of the people in Bhutan live in an urban setting. Population overgrowth for the realm is a serious concern. An increase in population requires additional natural resources, energy, and food production, all of which are in short supply in many areas.

Table 9.1 Demographics of South Asia and the World's Most Populous Countries

Rank	Country name	Population in millions	Total population density	Physiologic density	Fertility rate	Population growth rate (%)	Doubling time in years†	Perc-entage urban	Gross domestic product per capita ($)
1	China	1,336	361	2,405	1.54	0.49	143	47	7,600
2	**India***	1,189	937	1,912	2.62	1.34	52	30	3,500
3	United States	313	84	468	2.06	0.96	73	82	47,200
4	Indonesia	245	331	3,013	2.25	1.07	65	44	4,200
5	Brazil	203	62	884	2.18	1.13	62	87	10,800
6	**Pakistan***	187	604	2,414	3.17	1.57	45	36	2,500
7	**Bangladesh***	158	2,852	5,186	2.60	1.57	45	28	1,700
8	Nigeria	155	435	1,319	4.73	1.94	36	50	2,500
9	Russia	138	21	301	1.42	-0.47		73	15,900
10	Japan	126	867	7,225	1.21	-0.28		67	34,000
11	Mexico	113	149	1,149	2.29	1.10	64	78	13,900
41	**Nepal***	29	525	3,379	2.47	1.59	44	19	1,200
57	**Sri Lanka***	21	862	6,001	2.2	0.93	75	14	5,000
165	**Bhutan***	0.700	50	1,697	2.2	1.2	58	35	5,500
176	**Maldives***	0.400	3,438	26,194	1.81	-0.15		40	6,900

* Countries noted with an asterisk are part of South Asia

† Empty cell indicates a negative population doubling time.

Source: CIA World Factbook, June 2011, accessed September 13, 2011, https://www.cia.gov/library/publications/the-world-factbook/geos/xx.html.

South Asia's growing population has placed exceedingly high demands on agricultural production. The amount of area available for food production divided by the population may be a more helpful indicator of population distribution than total population density. For example, large portions of Pakistan are deserts and mountains that do not provide arable land for food production. India has the Thar Desert and the northern mountains. Nepal has the Himalayas. The small country of the Maldives, with its many islands, has almost no arable land. The number of people per square mile of arable land, which is called the physiologic density, can be an important indicator of a country's status. Total population densities are high in South Asia, but the physiologic densities are even more astounding. In Bangladesh, for example, more than five thousand people depend on every square mile of arable land. In Sri Lanka the physiologic density reaches to more than 6,000 people per square mile, and in Pakistan it is more than 2,400. The data are averages, which indicate that the population density in the fertile river valleys and the agricultural lowlands might be even higher.

Figure 9.5: Crowded Street in New Delhi, India Urban areas of South Asia are expanding rapidly. John Halsam – IMG_3555 – CC BY 2.0.

The population of South Asia is relatively young. In Pakistan about 35 percent of the population is under the age of fifteen, while about 30 percent of India's almost 1.2 billion people are under the age of fifteen. Many of these young people live in rural areas, as most of the people of South Asia work in agriculture and live a subsistence lifestyle. As the population increases, the cities are swelling to accompany the growth in the urban population and the large influx of migrants arriving from rural areas. Rural-to-urban shift is extremely high in South Asia and will continue to fuel the expansion of the urban centers into some of the largest cities on the planet. The rural-to-urban shift that is occurring in South Asia also coincides with an increase in the region's interaction with the global economy.

The South Asian countries are transitioning through the five stages of the index of economic development. The more rural agricultural regions are in the lower stages of the index. The realm experienced rapid population growth during the latter half of the twentieth century. As death rates declined and family size remained high, the population swiftly increased. India, for example, grew from fewer than four hundred million in 1950 to more than one billion at the turn of the century. The more urbanized areas are transitioning into stage 3 of the index and experiencing significant rural-to-urban shift. Large cities such as Mumbai (Bombay) have sectors that are in the latter stages of the index because of their urbanized work force and higher incomes. Family size is decreasing in the more urbanized areas and in the realm as a whole, and demographers predict that eventually the population will stabilize.

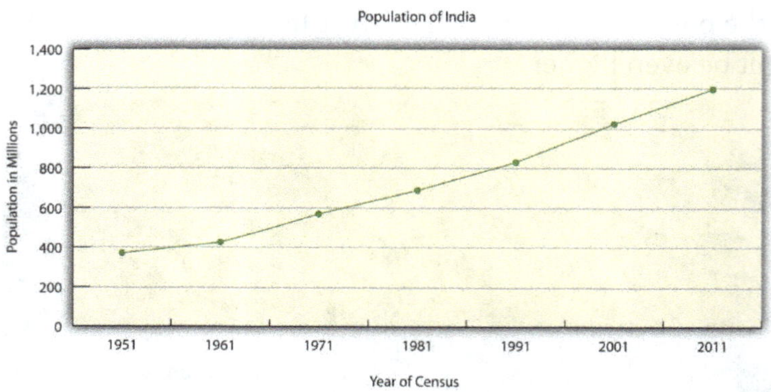

Figure 9.6: Population Growth in India. CIA World Factbook – public domain.

At the current rates of population growth, the population of South Asia will double in about fifty years. Doubling the population of Bangladesh would be the equivalent of having the entire 2011 population of the United States (more than 313 million people) all living within the borders of the US state of Wisconsin. The general rule of calculating doubling time for a population is to take the number seventy and divide it by the population growth rate. For Bangladesh the doubling time would be $70 \div 1.57 = 45$ years. The doubling time for a population can help determine the economic prospects of a country or region. South Asia is coming under an increased burden of population growth. If India continues at its current rate of population increase, it will double its population in fifty-two years, to approximately 2.4 billion. Because the region's rate of growth has been gradually in decline, this doubling time is unlikely. However, without continued attention to how the societies address family planning and birth control, South Asia will likely face serious resource shortages in the future.

Key Takeaways

- All the South Asian countries border India by either a physical or a marine boundary. The Himalayas form a natural boundary between South Asia and East Asia (China). The realm is surrounded by deserts, the Indian Ocean, and the high Himalayan ranges.
- The summer monsoon arrives in South Asia in late May or early June and subsides by early October. The rains that accompany the monsoon account for most of the rainfall for South Asia. Water is a primary resource, and the larger river systems are home to large populations.
- The Indus River Valley was a location of early human civilization. The large empires of the realm gave way to European colonialism. The British dominated the realm for ninety years from 1857 to 1947 and established the main boundaries of the realm.
- Population growth is a major concern for South Asia. The already enormous populations of South Asia continue to increase, challenging the economic systems and depleting natural resources at an unsustainable rate.

Discussion and Study Questions

1. Why are the Himalayan Mountains continuing to increase in elevation? Which of the countries of South Asia border the Himalayas?
2. What are the three major rivers of South Asia? Where do they start and what bodies of water do they flow into? Why have these river basins been such an important part of the early civilizations of the realm and why are they core population areas today?
3. Why does the monsoon usually arrive in late May or early June? What is the main precipitation pattern that accompanies the monsoon? Why is the monsoon a major source of support for South Asia's large population?
4. What changes did British colonialism bring to South Asia? When did the British control South Asia? Why do you think the British lost control when they did?
5. Why is the high population growth rate a serious concern for South Asian countries? What can these countries do to address the high population growth rate?
6. How can Pakistan have a higher fertility rate than Bangladesh but still have the same growth rate and doubling time?
7. Why would the country of the Maldives be concerned about climate change?
8. How would you assess the status of each country with regard to the index of economic development?

9. What are the three dominant religions of the realm? How did religion play a role in establishing the realms' borders? What happened to East Pakistan?
10. How can the principle of doubling time be used to assess a country's future potential? What is the general formula to calculate a population's doubling time?

Geography Exercise

Identify the following key places on a map:

- Arabian Sea
- Baluchistan
- Bay of Bengal
- Brahmaputra River
- Central Indian Plateau
- Chota-Nagpur Plateau
- Deccan Plateau
- Eastern Ghats
- Ganges River
- Himalayas
- Indian Ocean
- Indus River
- Kashmir
- Mt. Everest
- Punjab
- Thar Desert
- Western Ghats

9.2. THE PERIPHERAL STATES OF SOUTH ASIA

The Indian subcontinent is a substantial landmass extending into the Indian Ocean, situated on the southern flank of Asia, bordered by Afghanistan to the west, Myanmar to the east, and positioned south of China. The Indian perimeter encompasses the southern nations of the Maldives and Sri Lanka, as well as the northern territories of Punjab, Kashmir, Nepal, and Bhutan. This landmass possesses an extensive tectonic history, resulting from the collision between the Indian Tectonic Plate and the Eurasian Plate. The tectonic collision has resulted in the formation of the highest mountain ranges in the world, specifically the Himalayas, located in the northern and northwestern regions of the Indian subcontinent.

The Karakoram Mountains are situated in northern Pakistan and Kashmir. Alongside the other Himalayan mountain ranges, they create an arc that extends across the complete northern boundary of South Asia. Nepal and Bhutan are situated in the Himalayas. The Himalayan mountains contain some of the tallest peaks in the area, notably Mount Everest, situated on the boundary of Nepal and China, and K2, located in Pakistan. The Hindu Kush mountain ranges are located in western Pakistan and western Afghanistan. The Indus River originates in the northern region of the Karakoram mountains and forms an extensive, fertile floodplain. The northern region of the Indus River System comprises four primary tributaries.

Collectively, these rivers form the five rivers of the Punjab area in Pakistan and India; "Punjab" translates to "land of the five rivers" in Punjabi.

The Punjab

The Punjab is a fertile agricultural region with a high population density located on the border between India and Pakistan. Areas of the Punjab lie in both India and Pakistan. Where there is ample fresh water and bountiful food production, there is usually high population density. The Punjab is the most densely populated region in Pakistan. India has a separate state called the Punjab. Its river valleys are excellent areas for agricultural production and contribute heavily to the provisions needed to feed the enormous populations of the two countries.

The Punjabi people are found in the Punjab State of India and the Punjab Province of Pakistan. This large cultural area was separated into two countries during Partition at the time of Indian independence and the creation of Pakistan in 1947. Most of the people in the Punjab speak Punjabi, an Indo-European language. In a larger context, people with Punjabi background are considered one of the main ethnic groups in South Asia. Punjabis account for about 45 percent of the population of Pakistan.

In Pakistan, the Punjabis are grouped in clans and groups that correspond with traditional occupations. Traditionally, Punjabis are farmers and warriors, and in modern times are associated with agricultural professions and military life. Punjabis in Pakistan are predominantly Muslim, although a Christian minority exists. Indian Punjabis belong to traditional groups, including many of the same groups as in Pakistan, but many more. Most of the Muslim populations of the Punjab migrated to Pakistan in 1947 and, thus, most of the Indian Punjabis are Sikh with a Christian minority.

Figure 9.7: Sikhs protest against India's opposition to their proposed homeland (nation-state) in the Punjab called Khalistan. People gravitate toward nation-state systems, but globalization supports integration across political boundaries. Elliot Brown – Remembering the Sikh Genocide – CC BY-SA 2.0.

The Punjab region of Pakistan and India is the homeland of the Sikhs, people who follow a religion that is different from Islam or Hinduism. Sikhism was founded by Guru Nanak Dev (1469–1538). All distinctions of caste, creed, race, or gender are rejected in this religion. In Sikhism there is no priestly class. Every person is equally and fully responsible for leading a moral life, which eventually leads to universal salvation. Heaven and hell are not physical places, and God is the cosmic universal spirit. Historically, traditional Sikh men wear turbans on their heads and never cut their hair or beard. Sikhism is a universal religion. A prominent Sikh landmark and spiritual center is their Golden Temple, located in the city of Amritsar in the Indian state of the Punjab.

Sikhism is a system of religious philosophy and expression, known as the Gurmat or the counsel of the gurus, or the Sikh Dharma (or way of life). Sikhism comes from the Hindi and Punjabi word *sikhna*, which means "to learn." The principal belief in Sikhism is faith in the universal God. Sikhism promotes the pursuit of salvation through discipline and personal meditation on the name and message of God. However, it must be mentioned that Sikhs have a nonanthropomorphic concept of God; that is, Sikhs do not envision God as having any form or shape or mind similar to that of humans. Sikhism has become the fifth-most widely adhered to religion on Earth.

During British colonial occupation of South Asia, Sikhs were elevated to positions of power to help the British rule over Muslim and Hindu populations. The Sikhs are often overshadowed by the large Hindu and Muslim populations in the realm. Many of the Sikhs would like to have their own nation-state, free from Muslim or Hindu domination, and would like to see the Punjab region become the new homeland for this nation-state, called Khalistan. They have held rallies and demonstrations to promote the creation of Khalistan. The Indian government has, however, cracked down on militant movements that support the Khalistan concept. The Khalistan movement was more popular in the 1970s and 1980s and has been scaled down in recent decades. Proponents still attempt to attract young people and foreign donations to its cause. Khalistan does not have the support of the Pakistani and Indian governments and is not likely to become a reality any time soon. Khalistan is an example of the devolutionary push for a nation-state political unit for a particular group of people with similar aspirations or heritage.

The Kingdom of Kashmir

Located in the high mountains of the north is the former Kingdom of Kashmir, a separate kingdom before the British divided South Asia. In 1947, when the British drew the boundary between India and Pakistan, the leader of Kashmir, the maharajah, chose not to be a part of either country but to remain independent. About 75 percent of the population in Kashmir was Muslim; the rest, including the maharajah, were mainly Hindu. This arrangement worked for a time, until the Muslim majority was encouraged by their fellow Muslims in Pakistan to join Pakistan. After a Muslim uprising, the maharajah asked the Indian military for assistance. India was more than pleased to oblige and saw it as an opportunity to oppose Pakistan one more time. Today Kashmir is divided, with Pakistan controlling the northern region, India controlling the southern region, and China controlling a portion of the eastern region. A cease-fire has been implemented, but outbreaks of fighting have occurred. The future of Kashmir is unclear. None of the countries involved wants to start a large-scale war, because they all have nuclear weapons.

Figure 9.8: The Issues with Kashmir Pakistan controls the northern areas, India controls Jammu and Kashmir, and China controls the eastern portion, labeled Aksai Chin on this map. All three countries have nuclear weapons, and it seems apparent that none of the countries wants to start a nuclear war. Updated from map courtesy of University of Texas Libraries.

The conflict in Kashmir is about strategic location and control of water rather than labor and resources. It is unknown whether there are abundant minerals in the mountains in Kashmir to be mined, but regardless, there is little mining activity going on, not enough to cause conflict. One of the main physical geography features of importance is water. The Indus River flows through Kashmir from Tibet and into Pakistan. The control of this river system is critical to the survival of people living in northern Pakistan. If India were to place a dam on the river and divert the water to their side of the border, to the dry regions of the south, Pakistan could suffer a water shortage in the northern part of the country. Another aspect of the Kashmir conflict goes back to the division of Pakistan and India, which pitted Muslims against Hindus along the border region. The religious differences have come to the surface again in the conflict over the control of Kashmir. Extremist movements within Kashmir by the Muslim population have fueled the division between those who support Pakistan and those who support Hindu-dominated India.

The Kingdom of Bhutan

Landlocked and mountainous, the small Kingdom of Bhutan is remotely located next to the high Himalayas between China and India. The mountain peaks reach more than twenty-three thousand feet. Bhutan is about half the size in physical area of the US state of Kentucky and has fewer than one million people. The southern plains are warm, with subtropical weather, but the higher altitudes of the snow-capped mountains have polar-type climates. The local people call their country the "Land of the Thunder Dragon" because of the harsh storms they experience. Bhutan has large areas of natural habitat that have not been disturbed by human activity. The natural environment and the unique heritage and culture of the people make Bhutan an attractive destination for world travelers.

Bhutan is a small country without much industry or high-tech corporate involvement. Forestry and agriculture are the main economic activities, which account for approximately 60 percent of the country's population. Grazing livestock and subsistence agriculture are the primary types of farming. Increasing the country's modest infrastructure is hampered by its high mountains and remote location. India is Bhutan's main trading partner and has played an important role in the country's development and economic situation.

Figure 9.9: Bhutan's Famous Taktshang Buddhist Monastery, Commonly Known as the Tiger's Nest The unique landscapes and cultural experience that Bhutan offers to travelers have promoted tourism as an increasing economic activity. Avinash singh – taktsand – CC BY 2.0.

Modern transportation and communication technologies are being introduced in Bhutan and are changing how the country is connected to the rest of the global economy. Satellite and cable television, mobile phone networks, the Internet, and major airline service are opening the doors of opportunity and interaction between the people of Bhutan and the rest of the world. Introduction of technological services has prompted Bhutan's government to take steps to protect its environment and unique heritage. Tourism has become a major focus of the changes. The country has stepped up its efforts to develop tourism but has targeted a specific type of traveler. Bhutan is an expensive place to visit, which has been the biggest deterrent for travelers. Visitors from places other than India and Bangladesh must agree to strict requirements set by the suppliers of Bhutanese tourism, including large daily fees just to be in the country. Tourism is increasing in Bhutan but remains highly selective in its requirements and regulations. These measures are to ensure that the environmental health of the country remains intact and that there is minimal cultural impact from outsiders.

Buddhism is the state religion and is followed by about 75 percent of the population. Hinduism is the second-largest religion and is followed by the other 25 percent of the population. One of the principles of the government in regulating development projects has been the concept of gross national happiness (GNH), which is used as a guide to determine the impact of a project on the culture and people of Bhutan. The stern measures regarding development have protected the country from serious environmental degradation and have helped to sustain the lifestyles of the Bhutanese people. Some measures may appear harsh to outsiders, but the country is implementing these measures to promote the health and well-being of its people. For example, tobacco products are banned from being sold in the country. Democratic elections are becoming standard after centuries of rule by a monarchy. The intent of the transition is to provide the people with more direct control of their government and country.

Interesting points about the culture of Bhutan include the issue of marriage. Marriages based on love are becoming more common in the cities, while arranged marriages remain a tradition in many of the smaller villages. Under the current legal system, women have the right to inheritance. Homes and personal possessions are passed down through a family's female children. Traditionally, male children do not inherit. Men are expected to earn their own livelihood and if they get married will most often live in the wife's house.

Archery is the national sport of Bhutan. Most villages regularly hold archery competitions, which usually include festivities of serving food and conducting community events.

The Kingdom of Nepal

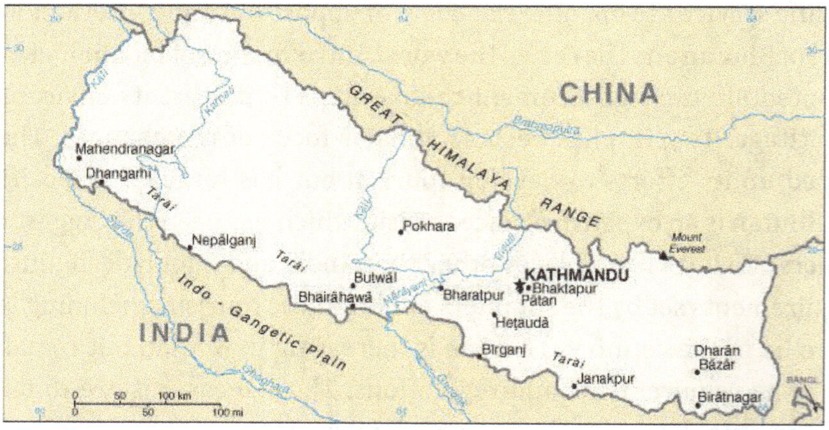

Figure 9.10: The Himalayan Mountains and the Tarai Lowlands of Nepal. CIA World Factbook – public domain.

Bordering the highest mountain range in the world, the Himalayas, the country of Nepal is isolated from any seacoast and buffered from the outside world by India and China. Nepal is about the same size in physical area as Bangladesh, and is home to almost thirty million people. More than 80 percent of its people work the land in a region that is suffering from severe deforestation and soil erosion. Trees are cut down to build houses, to cook food, and to keep warm. Without trees to hold the soil, the monsoon rains wash soil from the mountain fields into the valleys. The combination of the fast-growing population with the loss of food-growing capacity means it is only a matter of time before a major crisis occurs in Nepal. Nepal's best farmland is in the **Tarai lowlands** of southern Nepal, while the north is quite mountainous. The towering elevation of the Himalayas restricts human habitation in the north. High population growth has also been outstripping the country's economic growth rate in recent years.

Nepal has an abundance of tourist attractions, **Mt. Everest** being its best known. In addition, there are hundreds of ancient temples and monasteries. Swift flowing streams and high-mountain terrain support a modest trekking industry. Visitors to Nepal have an opportunity to glimpse a rich culture that few outsiders can witness. The downside is that tourism demands an investment in infrastructure and services. Such investments direct funds away from schools, medical clinics, and public services needed by the Nepalese people. Income from tourism is needed and always welcome, but the trade-off with investments is a difficult choice to make. Tourism in Nepal is not as restrictive as that of Bhutan, and the unique physical and cultural landscapes will continue to draw travelers from throughout the world.

Figure 9.11: The Deforested Landscape of Nepal Deforestation is a serious problem. The trees are cut down for firewood and building materials. The winding road connects Kathmandu with Tibet. Wikimedia Commons – public domain.

Hinduism is the main religion in Nepal, but a blend of Buddhism is more prevalent in the north. The guardian deity of Nepal is Shiva. **Pashupatinath Temple**, the world's most significant Shiva worship site, is located in the capital city of Kathmandu. This Shiva temple is not only a UNESCO World Heritage Site but a major destination for Hindu pilgrims from around the world. Buddhist and Hindu beliefs often mix in Nepal. There are certain situations where the same deities and temples can be honored or worshipped by members of both religions.

In a different part of Nepal, Lumbini—near the city of Bhairahawa (Siddharthanagar), on the border with India—is another UNESCO World Heritage Site, this one focusing on the birthplace of the Siddhartha Gautama, the founder of Buddhism. Prince Siddhartha Gautama was born about 563 BCE, near Lumbini. The city has recognized the prince with a number of monasteries and temples built in his honor. An entire development zone is restricted to nothing but monasteries and temples. No other commercial or public buildings—such as hotels, shops, or businesses—can be constructed in the zone. The different branches of the Buddhist faith each have their own specific designated sections of the zone. Lumbini is a major pilgrimage site for Buddhist believers from around the world.

In the late 1700s, local states of Nepal were consolidated into the one kingdom and ruled by monarchy. The kingdom was ruled by royal families until the mid-1900s. Nepal has been free of British influence since 1947, but has had trouble establishing a stable central government. The royal family in charge of the kingdom was replaced in 1951, and further democratic reforms were made in 1990. Communist partisans from China have been active in insurgent activities. Frequent protests and civil unrest have caused political instability, which has discouraged tourism and has depressed the economy even further. The Maoist Communist movement and other opposition political parties held mass protests, culminating in a peace accord. Ensuing elections created the establishment of a federal democratic republic. The first president of Nepal was sworn into office in 2008. There is still much tension in the country between those loyal to the royal family and those wanting the royal family to be dissolved. Without a stable government, economic and political progress in Nepal will be a serious challenge.

Sri Lanka

Sri Lanka is a beautiful island about half the size of Nepal. The island has a warm, tropical type A climate, with forested hills and mountains in the center. Rivers flow from the center outward to water the farm fields of rice and other crops. The best farmland in located in the Sinhalese-controlled areas of the southwestern portions of the island. Cinnamon is native to the island and has been cultivated since colonial times as an important export. Coconuts, coffee, and tea are also important export products. The island is home to various national parks, four biosphere reserves, and several wild elephant herds. Sri Lanka has the potential to become a major tourist destination with high incomes and a hub for international trade. Factors working against Sri Lanka reaching its potential are not based on its physical geography or location, but rather they are linked back to colonialism and cultural or ethnic divisions between the Sinhalese majority and the Tamil minority.

The people of South Asia follow various religions. Pakistan and Bangladesh are Muslim. India has a Hindu majority. About 90 percent of the people of Nepal are considered Hindu but many follow a unique blend of Hindu and Buddhist beliefs.

Figure 9.12: Picking Tea in Sri Lanka Tea, coffee, and cinnamon are export products of Sri Lanka. Tallis Keeton – My mother's Sri Lanka 7 – CC BY 2.0.

The small, mystical, mountainous kingdom of Bhutan is Buddhist. Sri Lanka has its own unique circumstances and is a mixed country with a strong Buddhist majority and an active Hindu minority. The conflict between the majority and the minority ethnic groups fueled a low-level civil war on the island for decades. Differences in religion, ethnicity, and politics have brought the country to halt on various occasions.

Sinhalese people from somewhere in northern India moved to the island of Sri Lanka about 2,500 years ago. The Sinhalese brought with them Buddhism and the Sinhala language, which belongs to the Indo-European language family. They established themselves on the island for centuries. Sri Lanka was first colonized by Portugal, then Holland. When the British colonized South Asia, they took control of Sri Lanka. It was called **Ceylon** at that time and changed its name to Sri Lanka in 1972. The higher elevations of the center of the island were excellent for tea production; British colonizers established tea plantations there. To work the plantations Britain brought thousands of additional Tamil laborers from southern India across the **Polk Strait** to Ceylon. Most of the Tamil speak a **Dravidian** language and follow the Hindu religion.

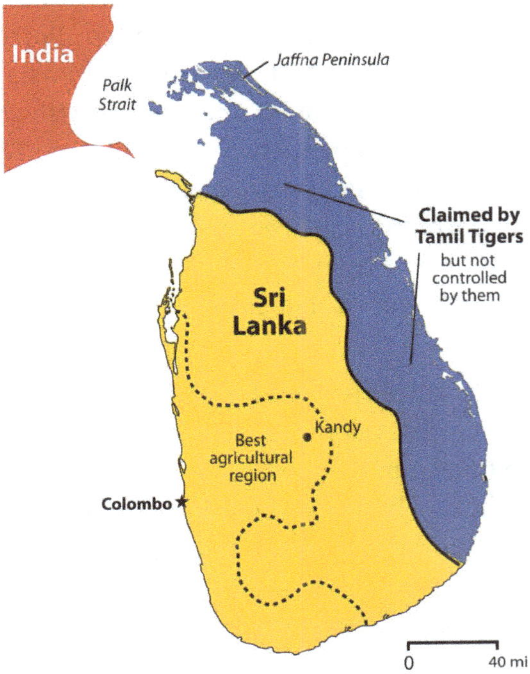

Figure 9.13: Claims of Tamil in Sri Lanka

The Tamil Tigers laid claim to a large portion of the eastern part of the island but did not control it. The only Tamil-controlled areas were in the far north, including the Jaffna Peninsula and a small area around it.

When the British were forced out of South Asia and left Ceylon, the Tamils remained on the island. The Tamils now make up only 10 percent of the population and live mainly in the northeastern region of the island. They have been pressuring the Sinhalese majority to split the island politically and grant them independence. An insurgent civil war was waged for decades between the Tamil guerillas—called the **Tamil Tigers**—and the Sinhalese government. About sixty to eighty thousand people died in this conflict. Originally only controlling the **Jaffna Peninsula**, the Tamil Tigers later made claims on a large portion of the northeastern part of the island. The Tamil Tigers created a government in the north called *Eelam* and wanted to legitimize it. The Sri Lankan president announced an end to the civil war in 2009, and the Tigers admitted defeat at that time. This civil war devastated Sri Lanka's tourism industry and discouraged foreign investments, further reducing economic opportunities for the island.

The Maldives

Just north of the Equator in the Indian Ocean to the southwest of India lie the Maldives, a group of low-lying islands that consists of twenty-six atolls encompassing a territory of only about 115 square miles. Within the atolls are approximately 1,200 small islands, of which about 200 are inhabited. Portugal controlled the Maldives from 1558 during their colonial expansion into Asia. Holland took over from the Portuguese in 1654. The Maldives became a British protectorate in 1887, which lasted until 1965, when independence was achieved. Three years later the country became a republic. The Maldives is a country with many extremes. It is Asia's smallest nation in both physical area and population. The island nation has the smallest physical area of any country with a majority Muslim population. The average elevation—four feet, eleven inches above sea level—is the lowest in the world for any country.

Fishing and tourism are the chief methods for Maldivians to earn a living. Tourism has increased in recent years. The many islands and atolls are attractive destinations for world travelers. The first tourist resort opened in 1972. Since that time, dozens of world-class resort facilities have opened for business across the archipelago. Tourism is the country's number one means of gaining wealth. The coral reefs that make up the island chain are excellent for diving and water sports. The tropical climate and miles of sandy beaches provide for an attractive tourism agenda.

The Maldives is an example of an entire country that could be in danger of flooding because of climate change if polar ice melts and sea levels rise. Concerns over the future of the islands gave reason for the president of the country to announce a plan in 2008 to purchase land in other countries in case sea levels rise to a point where the Maldives are no longer habitable. The purchase of land from tourism receipts would provide a place for the Maldivians to move in case they had to evacuate the islands. The administration of the Maldives has worked hard to lobby the international community to address the increase in greenhouse gas emissions and the possibility of an increase in sea level caused by the global warming aspect of climate change.

Key Takeaways

- The Punjab is a highly productive agricultural region located partly in Pakistan and partly in India. The region is home to the Sikh population, which has proposed having its own nation-state.
- Kashmir is divided between Pakistan, China, and India. The religious differences and the control of a valuable water source are at the core of this conflict.
- Nepal borders the Himalayan Mountains and has an economy based on agriculture and tourism. High population growth has been stripping the land of trees, causing serious deforestation issues and soil erosion.
- The Kingdom of Bhutan has placed major restrictions on tourism to protect its environment and limit outside influences on its culture.
- The beautiful tropical island of Sri Lanka experienced a low-level civil war for decades between the Sinhalese Buddhist majority and the Tamil Hindu minority. The island has an excellent location and potential for economic development.
- The archipelago of the Maldives is a small country that depends on tourism for its economic survival. The low elevation of its land area makes it subject to flooding due to the effects of climate change.

Discussion and Study Questions

1. Why is the Punjab region vital to both Pakistan and India?
2. What has been one political goal of the Sikhs living in the Punjab?
3. What three countries control parts of the Kingdom of Kashmir?
4. Why is the region of Kashmir vital to South Asia's viability?
5. Why is Bhutan so selective about admitting tourists?
6. Explain Bhutan's guiding philosophy regarding development.
7. Why is Nepal experiencing environmental degradation?
8. What was the civil war about in Sri Lanka? What did each side want?
9. What is it about each country listed that is attractive to the tourism industry?
10. What plan was announced to address the Maldives's environmental concerns?

Geography Exercise

Identify the following key places on a map:

- Jaffna Peninsula
- Lumbini
- Mt. Everest
- Polk Strait
- Tarai lowlands

9.3. PAKISTAN AND BANGLADESH

Pakistan and Bangladesh are distinct and sovereign nations geographically separated by India. Historically, this was not consistently true: from 1947 to 1971, they were governed by the same administration. The two nations has certain characteristics in common. Both possess Muslim majorities and exhibit significant population densities. These nations rank among the ten most populous in the world. Their demographics are predominantly youthful and rural, with agriculture as the principal economic activity in each nation. The migration from rural to urban areas is a significant trend influencing urban development. Infrastructure is deficient in numerous regions of each nation. The same elements suggest that both Pakistan and Bangladesh would encounter similar challenges in catering to their substantial populations and safeguarding their natural surroundings.

The Muslim League facilitated the establishment of a unified Pakistan, a largely Muslim nation for South Asian Muslims. Pakistan was established from the former Indian regions of Sindh, North West Frontier Provinces, West Punjab, Baluchistan, and East Bengal. Pakistan was established with two distinct geographical regions, characterized by religious majority. East Bengal, located in eastern India, was referred to as East Pakistan, and the rest, divided by over one thousand miles, was designated as West Pakistan. The two physical entities were politically unified.

In 1947, East and West Pakistan, governed by a single administration, attained independence from their colonial ruler, Britain. In 1956, Pakistan (both East and West) enacted its constitution and established itself as an Islamic republic. In 1970, a significant cyclone struck the coast of East Pakistan, and the central government in West Pakistan responded inadequately to the destruction. The Bengali populace expressed indignation at the government's disregard for their needs in relation to the storm and other issues. The Indo-Pakistan War altered the circumstances. During this conflict, East Pakistan, supported by the Indian military, confronted West Pakistan and proclaimed independence, resulting in the establishment of Bangladesh in 1972. West Pakistan evolved into the present-day nation of Pakistan.

Pakistan

The physical area of Pakistan is equivalent to the US states of Texas and Louisiana combined. Much of Pakistan's land area comprises either deserts or mountains. The high Himalayan ranges border Pakistan to the north. The lack of rainfall in the western part of the country restricts agricultural production in the mountain valleys and near the river basins. The Indus River flows roughly northeast/ southwest along the eastern side of Pakistan, flowing into the Arabian Sea. River sediments are deposited in large areas found between river channels and oxbow lakes formed from the constantly changing river channels. These "lands between the rivers" are called "doabs" and represent some of the most fertile land in the Indian subcontinent. The Indus River flows from the northern part of the Karakoram mountains and creates a large, fertile flood plain that comprises much of eastern Pakistan. Pakistan has traditionally been a land of farming. The Indus River Valley and the Punjab are the dominant core areas where most of the people live and where population densities are remarkably high.

Figure 9.14: The two core areas of Pakistan are the Punjab and the Indus River Valley. CIA World Factbook – public domain.

Approximately 64 percent of the population lives in rural areas and makes a living in agriculture. Most of the people are economically quite poor by world standards. In spite of the rural nature of the population, the average family size has decreased from seven to four in recent decades. Nevertheless, the population has exploded from about 34 million in 1951 to about 187 million as of 2011. About half of the population is under the age of twenty; 35 percent is under the age of fifteen. A lack of adequate medical care, an absence of family planning, and the low status of women have created an ever-increasing population, which will have dire consequences for the future of Pakistan. Service and infrastructure to address the needs of this youthful population are not available to the necessary degree. Schools and educational opportunities for children are rarely funded at the needed levels. As of 2010, only about 50 percent of Pakistan's population was literate.

Figure 9.15: The Provinces and Territories of Pakistan. Wikimedia Commons – public domain.

The capital of Pakistan when it was under British colonialism was Karachi, a port city located on the Arabian Sea. To establish a presence in the north, near Kashmir, the capital was moved to Islamabad in 1960. This example of a *forward capital* was an expression of geopolitical assertiveness by Pakistan against India. The lingua franca of the country for the business sector and the social elite continues to be English, even though Urdu is considered the national language of Pakistan and is used as a lingua franca in many areas. More than sixty languages are spoken in the country. There are as many ethnic groups in Pakistan as there are languages. The three most prominent ethnic groups are Punjabis, Pashtuns, and Sindhis.

Regions of Pakistan

The three main physical geographic regions of Pakistan are the Indus River Basin, the Baluchistan Plateau, and the northern highlands. These physical regions are generally associated with the country's main political provinces. The four main provinces include the **Punjab**, **Baluchistan** (Balochistan), **Sindh** (Sind), and **Khyber Pakhtunkhwa** (North West Frontier). To the north is the disputed region of Kashmir known as the **Northern Areas.** Each of these regions represents a different aspect of the country. The North West Frontier has a series of **Tribal Areas** bordering Afghanistan that have been traditionally under their own local control. Agents under Tribal Agencies have attempted to administer some type of structure and responsibility for the areas, with little success.

The Punjab

Figure 9.16: Donkey Cart on Busy Street in Lahore, Pakistan, in the Punjab This is an example of traditional transportation mixing with modern technology. Lahore is a large city with a wide range of methods of conducting business. Guilhem Vellut - A guy and his donkey - CC BY 2.0.

As explained previously, the Punjab is a core area of Pakistan, and has about 60 percent of Pakistan's population. The five rivers of the Punjab border India and provide the fresh water necessary to grow food to support a large population. Irrigation canals create a water management network that provides water throughout the region. The southern portion of the Punjab includes the arid conditions of the **Thar Desert.** The northern sector includes the foothills of the mountains and has cooler temperatures in the higher elevations. The Punjab is anchored by the cities of **Lahore**, Faisalabad, and Multan. Lahore is the cultural center of Pakistan and is home to the University of the Punjab and many magnificent mosques and palaces built during its early history. In the 1980s, many Punjabis migrated to Europe, the Middle East, and North America seeking opportunities and employment. This diaspora of people from the Punjab provided cultural and business ties with Pakistan. For example, trade connections between the Punjab and the United States are increasing. The Punjab is the most industrialized of all the provinces. Manufacturing has increased with industries producing everything from vehicles to electrical appliances to textiles. The industrialization of the Punjab is an indication of its skilled work force and the highest literacy rate in Pakistan, at about 80 percent.

Baluchistan

Figure 9.17 Man with His Camel in the Desert Region of Baluchistan in Western Pakistan Kashif Muhammad Farooq – CAMEL @ GADANI – CC BY 2.0.

Baluchistan (Balochistan) encompasses a large portion of southwest Pakistan to the west of the Indus River. The region connects the Middle East and Iran with the rest of Asia. The landscape consists of barren terrain, sandy deserts, and rocky surfaces. Baluchistan covers about 44 percent of the entire country and is the largest political unit. The sparse population ekes a living out of the few mountain valleys where water can be found. Local politics provides the basic structure for society in this region. Within the Baluchistan province of Pakistan are several coastal and interior rivers; the interior rivers flow from the Hindu Kush Mountains of Afghanistan, while most of the rivers along the coastal deserts from west of Karachi to the Iranian border are seasonal in nature and provide one of the few sources of fresh water in those coastal regions. Much of the coastal region is arid desert with sand dunes and large volcanic mountainous features.

The Sindh

The Sindh (Sind) region of the southeast is anchored by **Karachi**, Pakistan's largest city and major port. The Indus River is the border on the west and the Punjab region lies to the north. To the east of the Sindh is the border with India and the great Thar Desert. The Sindh is a region that misses out on the rains from the summer monsoon and the retreating monsoon season, when the winds sweep in from the north over South Asia. The city of **Hyderabad, Pakistan**, is located along the Indus River, which is a key food-growing area. Food crops consist of wheat and other small grains, with cotton as a major cash crop that helps support the textile industry of the region.

Hyderabad, Pakistan, is not to be confused with a large city with the same name in India.

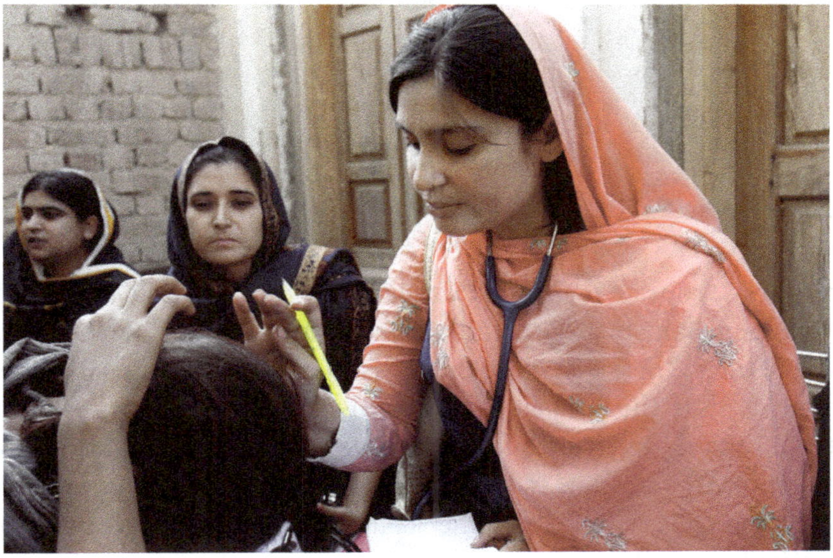

Figure 9.18: Female Doctor Examining Patient from a Mobile Medical Clinic in the Sindh Region of Pakistan. DFID – UK Department for International Development – A female doctor with the International Medical Corps – CC BY 2.0.

Rural-to-urban shift has pushed large numbers of Sindh residents into the city of Karachi to look for opportunities and employment. In previous sections, slums and shantytowns have been described and explained for cities such as Mexico City and São Paulo; Karachi has similar development patterns. The central business district has a thriving business sector that anchors the southern part of the country. The city has a large port facility on the Arabian Sea. As a city of twelve to fifteen million people or more, there are always problems with a lack of public services, law enforcement, or adequate infrastructure. Urban centers usually have a strong informal economy that provides a means for many of the citizens to get by but is outside the control of the city or national government. The Sindh is the second-most populous region of Pakistan, after the Punjab.

Khyber Pakhtunkhwa (The North West Frontier)

Figure 9.19 Man Firing AK-47 in the North West Frontier of Pakistan Kai Hendry – dcp_9062 – CC BY 2.0.

The North West Frontier is a broad expanse of territory that extends from the northern edge of Baluchistan to the Northern Areas of the former Kingdom of Kashmir. Sandwiched between the tribal areas along the Afghanistan border and the well-watered lands of the Punjab, the Khyber Pakhtunkhwa Province is dominated by remote mountain ranges with fertile valleys. The famous **Khyber Pass**, a major chokepoint into Afghanistan, is located here. The frontier is a breeding ground for anti-Western culture and anti-American sentiments, mainly fueled by the US military activity in Afghanistan. The Taliban movement that once controlled the government of Afghanistan has been active and generally more organized in this region than in Afghanistan. A push for more fundamentalist Islamic law has been a major initiative of the local leaders. Support for education and modernization is minimal. The government of Pakistan has also stepped up its military actions in the region to counter the activities of the militant Islamic extremists.

The Tribal Areas

The North West Frontier borders the Tribal Areas, where clans and local leaders are standard parts of the sociopolitical structure. These remote areas have seldom been fully controlled by either the colonial governments (the British) or the current government of Pakistan. There are about seven main areas that fall under this description. Accountability for the areas has been difficult and even when the national government stepped in to exercise authority, there was serious resistance that halted any real established interaction. These remote areas are where groups such as al-Qaeda and the Taliban often find safe haven. South and North Waziristan are two of the main areas that have been controlled by Tribal Agencies and not directly by the Pakistani government.

Northern Areas with Disputed Kashmir

Pakistan's Northern Areas include the territories that were once part of the Kingdom of Kashmir, the boundaries of which are disputed with India. The region is, in other words, interconnected with the issues related to Kashmir that involve Pakistan, India, and China. There are two main political entities: the large northern section bordering Afghanistan is called **Gilgit-Baltistan**, and the narrow section near Islamabad is called **Azad Kashmir** (Azad Jammu and Kashmir). The Northern Areas are highlands, bordered to the north by the towering Karakoram and Pamir mountain ranges. K2, the world's second highest mountain, which reaches 28,250 feet, is located here. The Northern Areas are sparsely populated except for the Indus River valley. The conflicts over these territories fuel nationalistic forces in both Pakistan and India. The conflicts are as much between Islam and Hinduism as they are between political factions. The early war between India and Pakistan over the border that the British placed between them in 1947 almost seems to be reenacted in the more recent conflicts over the region of Kashmir.

Figure 9.20: The Highlands of the Northern Areas in Pakistan. Junaidrao – This is an absolutely Amazing View – CC BY-NC-ND 2.0.

Religion and Politics in Pakistan

Today most of the people living in Pakistan are Muslim. About 85 percent of the Muslim population in Pakistan is Sunni and about 15 percent of the Muslim population is Shia, which is consistent with the percentages of the two Islamic divisions worldwide. Islam is considered the state religion of Pakistan. The state is a federal republic with a parliamentarian style of government. As an *Islamic state* following the Sharia laws of the Koran, it has been a challenge for Pakistan to try to balance instituting democratic reforms while staying true to fundamental Islamic teachings. Pakistan has held elections for government leaders, and the status of women has improved. Women have held many governmental and political positions, including prime minister. The military has been a foundation of power for those in charge. As a result of weak economic conditions throughout the country, it has been the military that has received primary attention and is the strongest institution within the government. Pakistan has demonstrated its nuclear weapons capability in recent years, which established it as a major player in regional affairs.

Pakistan has suffered from inadequate funding for public schools. As a rule, the wealthy urban elites have been the only families who could afford to send their children to college. With half the population consisting of young people, there are few opportunities to look forward to in Pakistan. Education has been supported in the form of Islamic religious schools called madrassas, which teach children the Koran and Islamic law. Much of the funding for religious schools comes from outside sources such as Saudi Arabia. The result is a religious education that does not provide the skills needed for the modern world. Pakistan has worked to build schools, colleges, and universities to educate its people. The situation is that population growth has been outpacing what little budget was allocated for educational purposes.

The government of Pakistan has struggled to meet the challenge posed by the democratic structure of its constitution. The combination of a federal republic and an Islamic state creates a unique and at times difficult balance in administrative politics. The legislative body of Pakistan consists of a National Assembly and a Senate. The leader of the National Assembly is the prime minister. The elected president not only is in charge of the military but is also head of state. The military establishment and the Inter-Services Intelligence (ISI) agency hold major political power in Pakistan. The political leadership has often vacillated between military and civilian rule. Transitions between the two types of leadership have been conducted through civil unrest or political demonstrations in the streets.

Benazir Bhutto: The First Female Prime Minister of Pakistan

Women's roles in Pakistani leadership have been complex. To understand the impact of electing the first woman prime minister in Pakistan, one has to go back to 1972. At that time, East Pakistan gained its independence and changed its name to Bangladesh. At roughly the same time, Pakistan elected President Zulfikar Ali Bhutto to power. After ruling for five years, Zulfikar Ali Bhutto was ousted from power and sentenced to death by General Zia, who became the next president. General Zia was the first military general to also be president. He allowed the Islamic Sharia law to be introduced into the legal system, which bolstered the influence of Islam on the military and government services. In 1988, General Zia was killed in an unexplained plane crash. The daughter of the first president—Zulfikar Ali Bhutto—was then elected as prime minister. The thirty-five-year-old Benazir Bhutto was the first female prime minister of Pakistan.

Benazir Bhutto's tenure as prime minister was short-lived. After a year and a half, the president removed her from office on accusations of corruption. She ran again in 1993 and was reelected to the prime minister position. Charges of corruption continued and she was removed as prime minister a second time in 1996. Political corruption, or the accusations of such a charge, is not uncommon in Pakistan or other countries with volatile political situations. To keep from being prosecuted by her opposition, Benazir Bhutto left Pakistan in 1998 and lived in Dubai. She did not return to Pakistan until 2007, under an agreement reached with the military general who was president at the time, General Pervez Musharraf. Benazir Bhutto was given amnesty for any and all claims against her.

Figure 9.21: Benazir Bhutto visits the United States in 1989 while she was the prime minister of Pakistan. Wikimedia Commons – public domain.

The elimination of the corruption charges against her allowed Benazir Bhutto to become a candidate for the office of president. She organized an effective campaign. Her campaign for the presidency energized the political landscape of the country. In December of 2007, her campaign was cut short. Benazir Bhutto was assassinated after departing a political rally. She was assassinated a few weeks before the polls were to open for the election. Benazir Bhutto was the leading opposition candidate. Her death rallied support against General Musharraf with continued calls for his removal from office. General Musharraf resigned from the presidency in 2008. The winner of the presidential elections that followed was none other than Benazir Bhutto's husband, Asif Ali Zardari. President Zardari blamed the Taliban for the assassination of his wife.

Environmental Issues in Pakistan

Pakistan is confronted with severe environmental issues. One of the concerns is a fault zone that runs directly through the region. Pakistan's regions are divided by the Eurasian tectonic plate and the Indian tectonic plate. Shifting tectonic plates cause earthquakes; there have been several major ones in Pakistan's recent history. For instance, an earthquake of 7.6 magnitude in 2005 in the northern region of the North West Frontier claimed about eighty-nine thousand lives in Pakistan and a few thousand others in neighboring India and Afghanistan. About 150 aftershocks were felt the next day, many of which were over the 6.0 level on the Richter scale. This particular event in 2005 was listed as the fourteenth most devastating earthquake on record at the time. It left over three million people without a home in Pakistan alone and more people were left homeless in neighboring areas. Hardest hit outside of Pakistan was the portion of Kashmir that is controlled by India. Many countries, including the United States, stepped up and supported the aid effort to reach people in the devastated region. Major earthquakes that cause devastation for the large populations here are common along this tectonic plate boundary.

Figure 9.22: Helping Pakistanis A US Army soldier and Pakistani troops help Pakistani residents as they disembark from a US Army helicopter in Khwazahkela, Pakistan, as part of relief efforts to help flood victims on August 5, 2010. Heavy rains forced thousands of residents to flee rising flood waters. US forces partnered with the Pakistani military to coordinate evacuation and relief efforts. The U.S. Army – Helping Pakistanis – CC BY 2.0.

Another environmental issue in Pakistan is water pollution. Raw sewage discharges into the rivers and streams and contaminates the drinking water for many Pakistanis. Most of the population lives in rural areas and relies on natural untreated water for their consumption. The water sources are heavily polluted, triggering disease and health problems. The urban areas lack public water works to handle fresh water supplies or to dispose of sewage properly. Industrial wastes and agricultural runoff also pollute and damage water supplies. Floods and natural runoff can carry pollutants from the land or urban areas into the rivers and streams that are used by human communities. The net effect of all these factors is that a majority of Pakistan's population lacks safe drinking water.

Deforestation is another environmental problem in Pakistan, because the demand for wood for cooking fuel and building is on the rise. Only about 2.5 percent of Pakistan is forested. In the last two decades, Pakistan has lost about one-fourth of its forest cover. The removal of forests causes widespread soil erosion during heavy rains and decreases natural habitat for organisms and wild animals. Efforts to protect the biodiversity of the country have been minimal and are complicated by the increase in population, which is expected to double in about forty-five years if population growth remains on its current trajectory. The reality is that deforestation is likely to continue in Pakistan, with little hope of a solution anytime soon.

Bangladesh

Bangladesh is a low-lying country that is associated with the types of marshy environments found in tropical areas and river deltas. The region is extremely prone to flooding, particularly during the monsoon season because of the high amount of rainfall. One of the most important rivers of Bangladesh flows southward from the Himalayas through India and into Bangladesh. While in India, this river is known as the **Brahmaputra River**, but when it enters Bangladesh, it is known as the **Jamuna River**. It provides a major waterway for this region and empties into the **Bay of Bengal**.

Figure 9.23: Bangladesh has about the same geographical area as the US state of Wisconsin. Bangladesh's population estimate in 2011 was 158 million; Wisconsin's was about 5.6 million. CIA World Factbook – public domain.

Contributing to the immense flow of water through the country are the Ganges and the **Meghna** rivers, which join up with the Brahmaputra River near the sea. The **Ganges** flows through northern India and is a major source of fresh water for a large population before it reaches Bangladesh. The Meghna is a collection of tributaries within the boundaries of Bangladesh that flows out of the eastern part of the country. The Meghna is a deep river that can reach depths of almost two thousand feet with an average depth of more than one thousand feet. The hundreds of water channels throughout the relatively flat country provide for transportation routes for boats and ships that move goods and people from place to place. There are few bridges, so land travel is restricted when rainfall is heavy.

Population and Globalization

Imagine a country the size of the US state of Wisconsin. Now imagine half of the entire population of the United States living within its borders. Welcome to Bangladesh. With an estimated population of about 158 million in 2011 and a land area of only 55,556 square miles, it is one of the most densely populated countries on the planet. Most of the population in Bangladesh is rural, agriculturally grounded, and poor. The larger cities, such as the capital of **Dhaka**, have modern conveniences, complete with Internet cafes, shopping districts, and contemporary goods. The rural areas often suffer from a lack of adequate transportation, infrastructure, and public services. Poverty is common; income levels average the equivalent of a few US dollars per day. Remarkably, the culture remains vibrant and active, pursuing livelihoods that seek out every opportunity or advantage available to them.

Figure 9.24: Street Scene in Dhaka, the Capital of Bangladesh. Ben Sutherland – The streets of Dhaka – CC BY 2.0.

There are many ethnic groups in Bangladesh, and many languages are spoken. The official and most widely used language in Bangladesh is Bengali (Bengala), which is an Indo-Aryan language of Sanskrit origin and has its own script. A Presidential Order in 1987 made Bengali the official language for the government of Bangladesh. Bengali is also the main language for the Indian state of West Bengal, which neighbors Bangladesh. English is used as the lingua franca among the middle and upper classes and in higher education. Many minor languages are spoken in Bangladesh and in the region as a whole. Most of the population, about 90 percent, is Muslim, with all but about 3 percent Sunni. There is a sizable minority, about 9 percent, which adheres to Hinduism, Buddhism, Christianity, or animism. The US State Department considers Bangladesh to be a moderate Islamic democratic country.

Bangladesh suffers from a demographic dilemma. Its tropical climate, availability of fresh water, and productive agricultural land have augmented a high population growth rate. The ever-increasing population is a growing concern. The government has stepped up its support for women's health, birth control, and family planning services. This is an Islamic country, but practical approaches and common sense in regard to population control have won out over Islamic fundamentalism, which has not always supported family planning. This poor and highly populated country has experienced political problems. Most of the people work in agriculture, while the remaining population is primarily concentrated in the service sector. Small business enterprises have been encouraged by the issuing of microcredit, or small loans, to assist people in using their skills to earn income.

Globalization is evident in Bangladesh. As a result of the availability of cheap labor, sweat shops have been implemented to manufacture clothing for export to the world markets. The country also receives financial remittances from Bangladeshis working overseas in places such as the oil-rich region of the Middle East, which also is predominantly Muslim and is attracting cheap labor for its economic development projects funded by oil revenues. One example of how Bangladesh has been able to acquire materials such as steel is in the recycling of old ships. Shipping companies that have ships that are no longer viable for modern shipping have brought them to the shores of Bangladesh to be stripped down, taken apart, and the materials recycled.

Environmental Issues

The summer monsoons are both a blessing and a curse in Bangladesh. The blessing of the monsoon rains is that they bring fresh water to grow food. The northeast part of Bangladesh receives the highest amount of rainfall, averaging about eighteen feet per year, while the western part of the country averages only about four feet per year. Most of the rain falls during the monsoon season. Bangladesh can grow abundant food crops of rice and grain in the fertile deltas of the Ganges and Brahmaputra Rivers, rivers that ultimately empty into the Bay of Bengal. About 55 percent of the land area is arable and can be used for farming, but flooding causes serious damage to cropland by eroding soil and washing away seeds or crops. Every year, countless people die because of the flooding, which can cover as much as a third of the country. One of the worst flooding events in Bangladesh's history was experienced in 1998, when river flooding destroyed more than three hundred thousand homes and caused more than one thousand deaths, rendering more than thirty million people homeless.

Most parts of Bangladesh are fewer than forty feet above sea level, and the country is vulnerable to major flooding according to various global warming scenarios. Half of the country could be flooded with a three-foot rise in sea level. Storm surges from cyclones killed as many as one hundred fifty thousand people in 1991. In comparison, about two thousand people died when Hurricane Katrina hit New Orleans in 2006. The high death toll from flooding does not receive its due attention from Western news media. Environmental concerns increase with the long-term projections of climate change. If sea level rises or if storms increase, then the low-lying agricultural lands of Bangladesh would suffer even more flooding and devastation. This would decrease the food supply of a growing population.

Another environmental problem for Bangladesh is deforestation. Wood is traditionally used for cooking and construction. The needs of a larger population have caused widespread deforestation. Brick and cement have become alternative building materials, and cow dung has become a widely used cooking fuel even though it reduces the fertilizer base for agriculture. Even so, these adaptations have not halted the deforestation problem. The main remaining forests are located along the southern borders with India and Burma (Myanmar) and in the northeast sector.

Figure 9.25 Man Working in a Rice Field in Bangladesh Wikimedia Commons – public domain.

Bangladesh is suffer because of widespread water pollution from naturally occurring arsenic that contaminates water wells. The pyrite bedrock underneath much of western Bangladesh has large amounts of arsenic in it. Millions of people drink groundwater contaminated with this arsenic on a daily basis. Arsenic kills people slowly, by building up in their bodies, rotting their fingernails, giving them dark spots and bleeding sores. Arsenic is a slow killer and a carcinogen that increases the risk of skin cancer and tumors inside the body. Villagers in Bangladesh began being affected by these symptoms in the 1970s. In 1993, official tests indicated that up to 95 percent of the wells in one of the villages in the western region were contaminated. The widespread water contamination has also had a social cost. Reports indicate that husbands are sending their disfigured wives back to their families of origin, and some young people are remaining single. Stories are told of people who believe that the health problems are contagious or genetic and can be passed on to children, which causes dilemmas for women who are trying to find a husband.

Women and Banking in Bangladesh

Despite an overall languishing economy, economic success stories in this poor country do exist. The Grameen Bank has been working to empower women in Bangladesh for many years. The bank issues microcredit to people in the form of small loans. These loans do not require collateral. Loans are often issued to impoverished people based on the concept that many of them have abilities that are underutilized and can be transformed into income-earning activities. About 96 percent of these loans are to women, and the average loan is equivalent to about one hundred dollars. Women have proven to be more responsible than men in repaying loans and utilizing the money to earn wealth. The loan recovery rate in Bangladesh is higher than 98 percent (Grameen Support Group, 2011). Microcredit has energized poor women to use their skills to make and market their products to earn a living. More than five million women have taken out such loans, totaling more than five billion dollars. This program has energized local women to succeed. It has been a model for programs in other developing countries.

Key Takeaways

- When Britain's colonialism ended in South Asia in 1947, the Muslim League was instrumental in creating the united Muslim state with both East Pakistan and West Pakistan under one government. East Pakistan broke away and became the independent country of Bangladesh in 1972.
- Both Pakistan and Bangladesh have large populations that are increasing rapidly. Both countries have agriculturally based economies. Rural-to-urban shift is occurring at an ever-increasing rate in both countries. Population growth places a heavy tax on natural resources and social services.
- The political units within Pakistan include four main provinces. Tribal Areas border Afghanistan and are controlled by local leaders. The Northern Areas are disputed with India. Each of the provinces has its own unique physical and human landscapes.

- Earthquakes are common in Pakistan because the country is located on a tectonic plate boundary. Deforestation and water pollution are two other major environmental concerns.
- Bangladesh is a low-lying country with the Brahmaputra River, Ganges River, and the Meghna River flowing into the Bay of Bengal. Flooding is a major environmental concern that has devastated the country on a regular basis.

Discussion and Study Questions

1. What happened to East Pakistan? Why did its name change?
2. Why does Pakistan have a forward capital? Where is it? Where did the capital used to be?
3. Compare the population density of Pakistan, Bangladesh, and the United States.
4. What are the four main provinces of Pakistan? How are they different from each other?
5. How is Bangladesh affected by the summer monsoon? How much rain can it receive annually?
6. What are the main environmental problems in Pakistan? In Bangladesh?
7. What type of government does Pakistan have? What is the law based on?
8. How could both countries address their population growth situation?
9. Who was the first woman prime minister of Pakistan? What happened to her?
10. How has microcredit aided in the economic development of Bangladesh?

Geography Exercise

Identify the following key places on a map:

- Arabian Sea
- Azad Kashmir
- Baluchistan
- Bay of Bengal
- Brahmaputra River
- Ganges River
- Gilgit-Baltistan
- Hyderabad
- Indus River
- Karachi
- Khyber Pakhtunkhwa
- Khyber Pass
- Lahore
- Meghna River
- Northern Areas
- Punjab
- Sindh
- Thar Desert
- Tribal Areas

References

Grameen Support Group, "Grameen Bank—Banking on the Poor," accessed November 14, 2011, http://www.gdrc.org/icm/grameen-supportgrp.html.

9.4. INDIA

India is regarded as the largest democracy globally. Upon examining the historic geography and developmental patterns of India, the intricacies of this Hindu state emerge. The European invaders of South Asia comprised the Dutch, Portuguese, French, and ultimately, the British. Seeking raw supplies, inexpensive labor, and expanding markets, Europeans leveraged their technological advantages to seize control and dominate the regional industrial infrastructure. The East India Company served as the foundation of British operations in South Asia and transformed into the administrative government of the region by 1857. The British administration established an administrative framework to oversee South Asia. The centralized government in India employed numerous Sikhs in administrative roles to oversee the predominantly Muslim and Hindu populace. The English language was established as a lingua franca for the colonies.

Colonialism significantly influenced the formation of contemporary boundaries in South Asia. In addition to unifying the territory under a singular central authority and establishing a common language, India's colonizers enhanced the primary port cities of Bombay, Calcutta, and Madras, today known as Mumbai, Kolkata, and Chennai, respectively. The names of the port cities have been restored to their original Hindi forms. The port cities served as access points for linking goods with marketplaces between India and Europe. Mumbai emerged as the largest metropolis and the economic hub of India. In 1912, the British relocated their colonial capital from Kolkata, the port serving the densely populated Ganges River area, to New Delhi to exploit the interior of India. Chennai served as a port facilitating access to southern India and was crucial to the Dravidian ethnic region.

Britain exploited India by expanding railroad networks from the three principal port cities into the interior, facilitating the transportation of resources from the hinterlands to the ports for export. The Indian Railroad has one of the largest railway networks globally. The issue with colonial railroads was that they did not invariably link cities to one another. The British colonizers established rail connections between the interior and the ports for resource extraction and the export of commercial commodities. Currently, the same port towns serve as pivotal hubs for globalization's import/export activities and continue to be essential industrial centers for South Asia. They are now effectively linked with other cities in India.

Goa is the smallest state in contemporary India. In the sixteenth century, Portuguese traders first discovered it, subsequently annexing it to establish a colony of

Portugal, which persisted for the next 450 years. Goa was one of the longest-held colonial territories globally and was not incorporated into India until 1961. By the mid-1800s, the majority of the populace in this little region had been coerced into Christianity. Despite the preservation of some Hindu customs throughout the colonial era and the celebration of Hindu festivals, Goa is predominantly recognized for its Christian holiday observances, particularly Christmas and Easter. The cathedral and secular architecture of numerous historic structures in Goa exhibit European style, indicative of its Portuguese heritage.

The People of India

Contrasts in India are explicitly evident in the regional differences of its human geography. The north-south contrasts are apparent through the lingua franca and ethnic divisions. The main lingua franca in the north is Hindi. In the Dravidian-dominated south, the main lingua franca is English. The densely populated core region along the Ganges River, anchored on each end by Delhi/New Delhi and Kolkata, has traditionally been called the heartland of India. The south is anchored by the port city of Chennai and the large city of Bangalore. Chennai has been a traditional industrial center. The industrial infrastructure has shifted to more modern facilities in other cities, giving over to a "rustbelt" syndrome for portions of the Chennai region. India is a dynamic country, with shifts and changes constantly occurring. Any attempt to stereotype India into cultural regions would be problematic.

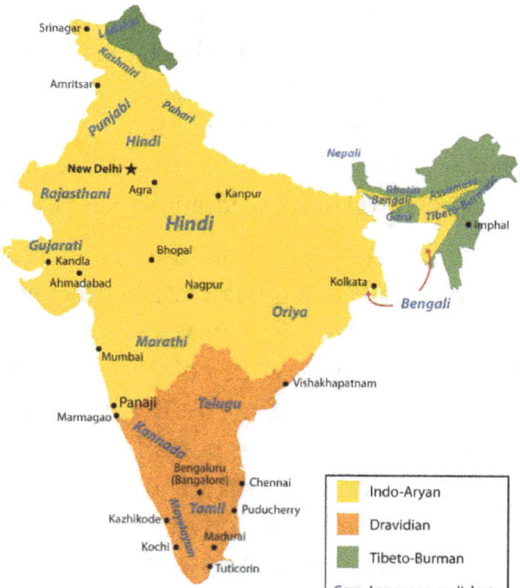

Figure 9.26: The Three Main Language Families in India Hindi is the official language of the government, and both Hindi and English are the lingua franca. Wikimedia Commons – public domain; University of Texas Libraries.

In 2010, India had more than 1.18 billion people, which is about one-sixth of the human population of the earth. An 80 percent majority follow Hindu beliefs. About 13 percent of the population is Muslim. Thirteen may not seem like a high percentage, but in this case it equates to about 140 million people. This is equivalent to all the Muslims who reside in the countries of Iraq, Saudi Arabia, Syria, and Egypt combined. India is sometimes called the third-largest Muslim country in the world, after Indonesia and Pakistan, because of its large Muslim minority. India essentially has two lingua francas: English and Hindi, of which Hindi is the official language of the Indian government. India has twenty-eight states and fourteen recognized major languages. Many different languages are spoken in rural areas. The languages of northern India are mainly based on the **Indo-European** language family. Languages used in the south are mainly from the **Dravidian** language family. A few regions that border Tibet in the north use languages from the **Sino-Tibetan** language family.

Urban versus Rural

Rural and urban life within the Indian Subcontinent varies according to wealth and opportunity. While concentrated in specific areas across the landscape, in general the population in rural areas is discontinuous and spread thinly. In urban areas, the populations are very concentrated with many times the population density found in rural areas. India has six world-class cities: Kolkata, Mumbai, Delhi, Chennai, Bangalore, and Hyderabad. There are many other large cities in India; in 2010, India had forty-three cities with more than a million people each.

India's interior is mainly composed of villages. In rural villages, much of the economy is based on subsistence strategies, primarily agriculture and small cottage industries.

Figure 9.27: Farmer Tilling a Field with Oxen in Rural India. Ananth BS – Annadatha – CC BY 2.0.

The lifestyle is focused on the agricultural cycles of soil preparation, sowing, and harvesting as well as tending animals, particularly water buffalo, cattle, goats, and sheep. About 65 percent of the population lives in rural areas and makes a living in agriculture. About 35 percent of the population—which is equal to the entire US population—is urbanized. India is rapidly progressing toward urbanization and industrialization. Changes in technology, however, tend to be slow in dispersing to the rural villages. More than half the villages in India do not have road access for motor vehicles. For residents of those villages, walking, animal carts, and trains are the main methods of transportation. Agricultural technology is primitive. Diffusion of new ideas, products, or methods can be slow. Modern communication technology is, however, helping connect these remote regions.

India's cities are dynamic places, with millions of people, cars, buses, and trucks all found in the streets. In many areas of urban centers, traffic may be stopped to await the movement of a sacred cow or a donkey or bullock cart loaded with merchandise. Indian cities are growing at an unsustainable rate. Overcrowded and congested, the main cities are modernizing and trying to keep up with global trends. Traditionally, family size was large. Large family size results in a swell of young people migrating to urban areas to seek greater opportunities and advantages. In modern times, family size has been reduced to about three children, an accomplishment that did not come easily because of the religious beliefs of most of India's people. If current trends continue, India will overtake China as the most populous country in the world in about fifty years.

The level of official governmental control is usually different in an urban setting from what it is in the rural areas. There may be more police or military personnel in areas of heavy traffic or in urban areas that need extra control. A central feature of many Indian cities is an older central city that represents the protected part of the city. In Delhi, for example, New Delhi represents the new construction of government buildings that was begun during the British occupation of the region as part of the British Empire. Old Delhi represents the old markets, government buildings, palaces, fortresses, and mosques that were built during the Mogul Empire, between the mid-1500s and the mid-1800s. These older parts of the cities, particularly the markets, are bustling with activities, merchants, shoppers, cab drivers, and pedal and motor rickshaws. Rickshaws are either bicycle-driven cabs or cabs based on enclosed motor scooters.

In urban areas, there is a socioeconomic hierarchy of a small group of people who are wealthy and can afford all the amenities we associate with modern life—electricity, clean water, television, computers, and the like. One of the things that characterize modern Indian cities is an expanding middle class. Many young people see the kinds of material goods that are available in the West and are creating job markets and opportunities to allow them to reach or maintain this type of lifestyle. One of the major markets to support this burgeoning middle class is the information technology field, as well as outsourcing in many of the cities of peninsular India.

India is a country with considerable contrast between the wealthy urban elites and the poor rural villagers, many of whom move to the cities and live in slums and work for little pay. Low labor costs have enabled Indian cities to industrialize in many ways similar to Western cities, complete with computers, Internet services, and other modern communications services. India's growing middle class is a product of educational opportunities and technological advancements. This available skilled labor base has allowed India's industrial and information sectors to take advantage of economic opportunities in the global marketplace to grow and expand their activities. Development within India is augmented by outsourcing activities by American and European corporations to India. Service center jobs created by business process outsourcing (BPO) are in high demand by skilled Indian workers.

India's Economic Situation

In the past decade, India has possessed the second fastest growing economy in the world; China is first. India's economy continues to rapidly expand and have a tremendous impact on the world economy. In spite of the size of the economy, India's population has a low average per capita income.

Figure 9.28 Mumbai (Bombay), the Economic Capital and Largest City in India. Ajay Goyal – city father – CC BY 2.0.

Approximately one-fourth of the people living in India live in poverty; the World Bank classifies India as a low-income economy. India has followed a central economic model for most of its development since it declared independence. The central government has exerted strict control over private sector economic development, foreign trade, and foreign investment. Through various economic reforms since the 1990s, India is beginning to open up these markets by reducing government control on foreign investment and trade. Many publicly owned businesses are being privatized. Globalization efforts have been vigorous in India. There has been substantial growth in information services, health care, and the industrial sector.

The economy is extremely diverse and has focused on agriculture, handicrafts, textiles, manufacturing, some industry, and a vast number of services. A 60 percent majority of the population earns its income directly from agriculture and agriculture-related services. Land holdings by individual farmers are small, often less than five acres. When combined with the inadequate use of modern farming technologies, small land holdings become inadequately productive and impractical. Monsoons are critical for the success of India's agricultural crops during any given season. Because the rainfall of many agricultural areas is tied to the monsoon rains of only a few months, a weak or delayed rainfall can have disastrous effects on the agricultural economy. Agricultural products include commercial crops such as coffee and spices (cardamom, pepper, chili peppers, turmeric, vanilla, cinnamon, and so on). An important product for perfume and incense is sandalwood, harvested primarily in the dense forests of the state of Karnataka, in southwestern India. Bamboo is an important part of the agricultural harvest as well. Of course, rice and lentils provide an important basis for the local economy.

Over the last two decades, information technology and related services are transforming India's economy and society. In turn, India is transforming the world's information technologies in terms of production and service as well as the export of skilled workers in financial, computer hardware, software engineering, and software services. Manufacturing and industry are becoming a more important part of India's economy as it begins to expand. Manufacturing and industry account for almost one-third of the gross domestic product (GDP) and contribute jobs to almost one-fifth of the total workforce. Major economic sectors such as manufacturing, industry, biotechnology, telecommunications, aviation, shipbuilding, and retail are exhibiting strong growth rates.

A large number of educated young people who are fluent in English are changing India into a "back office" target for global outsourcing for customer services. These customer services focus on computer-related products but also include service-related industries and online sales companies. The level of outsourcing of information activity to India has been substantial. Any work that can be conducted over the Internet or telephone can be outsourced to anywhere in the world that has high-speed communication links. Countries that are attractive to BPO are countries where the English language is prominent, where employment costs are low, and where there is an adequate labor base of skilled or educated workers that can be trained in the services required. India has been the main destination for BPO activity from the United States. Firms with service work or computer programming are drawn to India because English is a lingua franca and India has an adequate skilled labor base to draw from.

Tourism has always been an important part of India's economy and has been focused on the unique natural environments as well as historical cities, monuments, and temples found throughout the country. Of particular importance are the Mogul-period tombs, palaces, and mosques in Delhi, **Agra**, and **Jaipur**, India's "Golden Triangle" of tourism. India is a country of contrasts. Scenic beauty abounds from the Eastern and Western Ghats to the high mountains of the Himalayas. The monsoon rains provide abundant agricultural crops for densely populated regions such as the Ganges River basin. On the other hand, places such as the Thar Desert are sparsely inhabited. There is a wide gap between the wealthy elite and the massive numbers of people who live in poverty. Mumbai has some of the largest slums in Asia, yet it is the financial capital of India, teeming with economic activity.

As incomes rise for the middle class in India, the price of automobiles becomes more accessible. On the downside, an escalation in the numbers of motor vehicles in use tends to lead to an escalation in the levels of air pollution and traffic congestion. Similarly, an expansion of transportation systems increases the use of fossil fuels. India is a major competitor for fossil fuels exported from the Persian Gulf and other Organization of Petroleum Exporting Countries (OPEC) sources. The continued industrialization and urbanization in India foretells an increase in demand for energy. Rising energy costs and demand, combined with economic growth, have caused a serious problem for India. Many areas will be without power as they are shut off the power grid for hours or days, a process known as load-shedding. This allows industry and manufacturing to use the energy resources during peak times. In general, India is poor in natural gas and oil resources and is heavily dependent on coal and foreign oil imports. India is rich in alternative energy resources, such as solar, wind, and biofuels; however, alternative energy resources have not been sufficiently developed.

Vehicle Manufacturing

Two examples of India's growing economic milieu are motor vehicle manufacturing and the movie industry. India's vehicle manufacturing base is expanding rapidly. Vehicle manufacturing companies from North America, Europe, and East Asia are all active in India, and India also has its own share of vehicle manufacturing companies. For example, Mumbai-based **Tata Motors Ltd.** is the country's foremost vehicle production corporation and it claims to be the second-largest commercial vehicle manufacturer in the world. Tata Motors is India's largest designer and manufacturer of commercial buses and trucks, and it also produces the most inexpensive car in the world, the Tata Nano. Tata Motors manufactures midsized and larger automobiles, too. The company has expanded operations to Spain, Thailand, South Korea, and the United Kingdom. The company is an example of an Indian-based international corporation that is a force in the global marketplace. In 2010, India was recognized as a major competitor with Thailand, South Korea, and Japan as the fourth main exporter of autos in Asia.

Figure 9.30: The Nano, Made in India The Nano is considered the world's most inexpensive car. Wikimedia Commons – CC BY 2.0.

The Indian Cinema

Cinema makes up a large portion of the entertainment sector in India. India's cinema industry is often referred to as "Bollywood," a combination of Bombay and Hollywood. Technically, Bollywood is only the segment of the Indian cinema that is based out of Bombay (Mumbai), but the title is sometimes misleadingly used to refer to the entire movie industry in India.

Figure 9.31: Bollywood is a major film production company located in Mumbai (Bombay). The film industry in India produces almost twice as many movies as the United States. Indian production scenes can be dramatic and expressive. Feminist julie – Bollywood – CC BY-NC-ND 2.0.

Bollywood is the leading movie maker in India and has a world-class film production center. In the past few years, India has been producing as many as one thousand films annually. The highest annual output for the US film industry is only about two-thirds that of India. According to the Guinness Book of World Records, India's city of Hyderabad has the most extensive film production center in the world. The Telugu film industry operates the studio in Hyderabad.

Indian films are produced in more than a dozen languages and appeal to a wide domestic and international audience. Indian movies range from long epic productions with stories within stories to dramas, musicals, and theatrical presentations. Their popularity extends beyond South Asia. Indian movies with modest dress, lack of explicit sexual scenes, and a focus on drama are popular in places such as Egypt, the Middle East, and other African countries. Movie stars are energetically promoted and enjoy celebrity in India, as is the case with the entertainment industry in the United States and Europe. The cinema is part of the cultural experience in Indian society. Urban life in India reserves a large presence for the entertainment industry, particularly the Indian film industry. One of the prime artistic endeavors in urban India is movie posters depicting all the glory of the latest Bollywood movie. Most of these colorful posters are painted by hand and they tend to be large; some are several stories high.

India: East and West

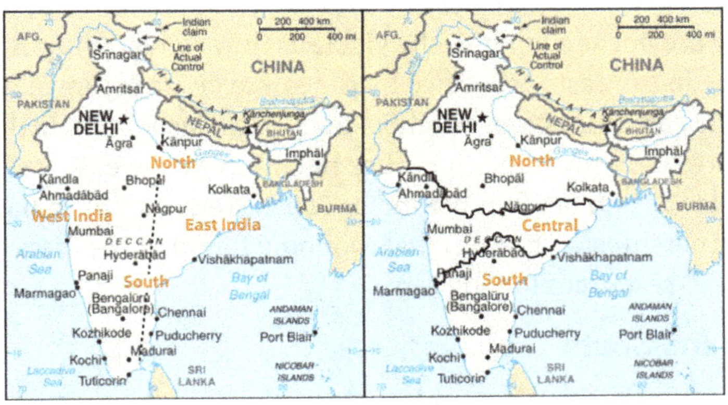

Figure 9.32: India can be divided either along north/south dimensions or along east/west dimensions. CIA World Factbook – public domain.

South Asia's physical geography—an overview of its physical features—was described at the beginning of the chapter. India makes up the largest physical area of the South Asia realm. Another way of looking at the physical and human landscapes of India is to study spatial characteristics. Additionally, the economic side of the equation can be illustrated by dividing India between east and west according to economic development patterns. To do this, on a map of India draw an imaginary line from the border with Nepal in the north, near Kanpur, to the

Polk Strait border with Sri Lanka in the south. This division of India illustrates two sides of India's economic pattern: an economically progressive West India and an economically stagnant East India.

The progressive western side of India is anchored by Mumbai and its surrounding industrial community. Mumbai is the economic giant of India with the country's main financial markets, and has been a magnet for high-tech firms and manufacturing. Mumbai's port provides access to global markets and is solidly connected to international trade networks. Auto manufacturing, the film industry, and computer firms all have major centers in the large urban metropolitan areas of the west. Large industrial cities such as Bangalore and Hyderabad have established themselves as high-tech production centers, attracting international business in the computer industry and the information sector. Chemical processing has been ongoing in **Bhopal**, which is noted for an environmental disaster, a gas leak in 1984 that resulted in the deaths of as many as ten thousand people. The nation's capital is located in New Delhi, which borders the massive city of (Old) Delhi. The western half of India has been progressing along a pattern with a positive economic outlook that views the global community outside of India as a partner in its success.

The eastern half of India has not been as prosperous as the west in its economic growth. The renowned city of Kolkata has traditionally anchored the eastern sector, but its factories have deteriorated into rustbelt status with aging and outdated heavy industries. The high-intensity labor activities of textile and domestic goods manufacturing are not as economically viable as they were in the past. The stagnant economic scene in the east is signified by the low average income levels of many of the states in the eastern region. Neighboring Bangladesh offers little in support of economic growth, and Myanmar, another neighbor to the east, has its own set of problems and lacks support for East India. The eastern half of India does not have the strong partnerships with the global economy found in the west and thus relies more on internal resources for survival.

India: North and South

There are differences in the geographic patterns between the northern and southern halves of India as well as between the eastern and western halves—depending on the criteria used to compare them. Climate patterns, for example, are more diverse in the north, with a wide range of temperatures throughout the seasons. Winter temperatures in the mountainous north are cold and summer temperatures in the Thar Desert can be extremely high. Southern India has a more moderate range of temperatures throughout the year. The far north has high mountains. The south has only the low-lying Eastern and Western Ghats. The north has the extensive Ganges River basin. The south has different drainage networks based on the plateaus of the region.

Besides physical aspects, there are cultural differences between the north and south as well. India is a complex societal mix of many ethnic groups, languages, and traditions. Spatially separating the country into vernacular regions is not conducive to agreeable results. Still, there are some recognizable trends that have been stereotyped or commonly stated between the northern and southern parts of India. The north is portrayed as a faster-paced society, with more edge and competitiveness. The south has been portrayed as more relaxed and less insistent.

As the section on languages illustrated, Indo-European languages are mainly spoken in the north and Dravidian languages are predominantly spoken in the south. Hindi is more commonly the lingua franca of the north, while English is more frequently the lingua franca of the south. People in the north are of Indo-Aryan descent, while the people in the south have a Dravidian heritage. Hinduism dominates all of India, but the north has a wider diversity of religions, such as Sikhism, Buddhism, and Islam, practiced by a large number of people. The south has a substantial Christian population along its west coast.

Food is an important aspect of the culture of societies, and there are clearly distinctions between the cuisine of the north and of the south in India. Indian cooking is primarily vegetarian, emphasizing aspects of Hinduism. However, many dishes, particularly in North India, contain goat, chicken, lamb, fish, and other meats. Beef is not eaten by Hindus, while pork and some species of fish are not traditionally eaten by Muslims. North India has more wheat-based products and less rice. Their dishes are prepared with spices and herbs, including black and chili peppers. Northern Indian food is characterized by its use of dairy products (yogurt; milk; paneer, or homemade cheeses; and ghee, or clarified butter). Onions, ghee, and spices are the common base for different types of salans or curries (gravies). Griddles are used for preparing different types of flat breads—chapattis, naan, and kulcha. Rice, lentils, and chickpeas are a staple part of the diet in North India.

Figure 9.33: A meal of fish curry with rice and peppers can be found in South India. Rovingl - Fish Curry - CC BY 2.0.

Food in the southern parts of India includes more rice as a staple, and seafood (fish and prawns) is common along the coastal areas. Coconut oil is used as a basis for cooking. Sambar, a stew made of peas and vegetables, is an important staple of the region as are rice and idlis, which are a type of cake or bread made from steaming fermented black lentils. Chili peppers are also common in South Indian cooking.

Biodiversity and the Environment

Earlier sections have introduced the issues of population growth and resource depletion in South Asia. India has its share of the same environmental problems. Water pollution along the Ganges is severe and affects the largest concentration of people in India. India is the second-largest consumer of coal in the world, coal that is mainly burned to produce electricity. Burning coal adds significantly to air pollution. A rise in the number of vehicles in use, combined with few emission controls, also adds to the air pollution in urban areas. Deforestation continues in many rural areas, as was noted in earlier sections about Pakistan and Bangladesh.

India has a number of rare animal species that need habitat if they are going to survive. A few of the larger animals include the Indian Rhinoceros, Clouded Leopard, Indian Leopard, Snow Leopard, Asiatic Lion, Bengal Tiger, Asian Water Buffalo, Asian Elephant, Stripped Hyena, and the Red Panda. Many species are endangered or threatened along with many other lesser-known organisms. The high human population growth throughout South Asia places a strain on the natural habitat of wild animals. Habitat loss caused by human development makes holding on to the wide array of biodiversity difficult.

Figure 9.34: The Indian Leopard is a near-threatened species that once lived throughout South Asia. Sid Lammata – Indian Leopard – CC BY-NC 2.0.

India has instituted measures designed to preserve its biodiversity. The Indian government has created sanctuaries for threatened or endangered species. National parks were established before India declared independence and were substantially expanded in recent decades. In 1972, The Wildlife Protection Act was instituted to create critical habitat for tigers and other rare species. There are hundreds of protected wildlife areas and fifteen biosphere reserves in India. Four of the biospheres were created in conjunction with the World Network of Biosphere Reserves.

The Indian government has established protected areas throughout the country, many of which are in the highland regions and the northern mountains. For example, the Gir Wildlife Sanctuary, including an area preserved for Asian Lions, is located on the Kathiawar Peninsula north of Mumbai, which juts out into the Arabian Sea. India is the only place left with Asian Lions in the wild. Tigers, elephants, rhinos, and leopards can be found in the sanctuaries. The country has about ninety-two national parks, which are also home to rare wildlife species, and more than three hundred fifty wildlife sanctuaries of all sizes. There are about twenty-eight tiger reserves in India. The country also has a number of marine reserves and protected areas along its coastlines.

The efforts of the Indian government to protect the country's biodiversity constitute an admirable environmental undertaking. The government has stepped up law enforcement efforts to combat poaching, which is a major cause of the decrease in numbers of rare species. Poachers kill animals such as tigers, leopards, elephants, and rhinos for their hides, horns, or body parts, which are sold on the black market in Asia for large sums of money. Many of the rare, threatened, or endangered species of India would not have a chance of survival without the government efforts to protect and provide for them. Balancing finding resources for rapid human population growth with wildlife management will continue to be a challenge in the years ahead for India and all countries of the planet.

Key Takeaways

- Colonialism had a tremendous impact on South Asia and its people. Colonial development patterns were implemented to control the people and to extract resources, not necessarily to benefit the realm.
- India has a wide disparity between its poor rural areas with agricultural economies and its wealthier bustling cities with expanding business sectors.
- Various urban centers of India have positioned themselves well to take advantage of the global economy and expand their manufacturing and industrial base. India is becoming a major manufacturing country for vehicles and high-tech industries.

- There are noticeable economic differences between the more progressive Western India and the stagnant economic conditions of Eastern India. There is also a noticeable cultural difference between the North and the South in India in the categories of language, ethnicity, food, and society.
- The Indian government has created national parks, wildlife sanctuaries, and game reserves to help protect rare, threatened, or endangered species.

Discussion and Study Questions

1. Outline the main ways in which British colonialism impacted South Asia.
2. What are the three main language families in India? What is the lingua franca?
3. List the main qualities that are different between the rural and urban areas of India.
4. How did British colonizers transport resources from the hinterland to the port cities for export back to Great Britain? How has this system changed since 1947?
5. Explain the various ways in which the rapid population growth is impacting India.
6. Why is India a major target for BPO?
7. List various ways the Indian film industry impacts India and the world.
8. How is economic development different between Western India and Eastern India?
9. Outline some cultural differences between the North and the South in India.
10. How has the government of India worked to protect the biodiversity of the natural environment? What are some of the animals that are being protected?

Geography Exercise
Identify the following key places on a map:

- Agra
- Bangalore
- Bombay (Mumbai)
- Bhopal
- Calcutta (Kolkata)
- Delhi
- Goa
- Hyderabad
- Jaipur
- Kathiawar Peninsula
- Madras (Chennai)
- New Delhi

9.5. RELIGIONS OF INDIA AND SOUTH ASIA

The region of India and its neighbouring countries is the birthplace of several ancient religions. Some individuals in the world persist in their adherence to animist beliefs, independent of the major global faiths. The most ancient religions of India are Hinduism and Buddhism. Significant religions in the region encompass Islam, Christianity, Sikhism, Zoroastrianism, Jainism, and the Baha'i faith. India is occasionally designated as a Hindu state, however the validity of this designation is questionable. India is best characterized as a secular nation with around 80 percent of its inhabitants adhering to Hindu customs. Islam is the second most prevalent religion, adhered to by approximately 13 percent of the global population. Christianity is the third-largest religion in India, adhered to by approximately 3 percent of the populace. Sikhism constitutes approximately 2 percent of India's population. Buddhism and Jainism are two minority faiths originating from South Asia. Additionally, there still Indians who adhere to animist religions that predate all the aforementioned religions, particularly in isolated regions.

Figure 9.35: Islamic Architecture in Hindu India The Taj Mahal was constructed as a mausoleum for the wife of the ruler Shah Jahan in 1653 when the Muslim Mogul Empire controlled northern India. The Taj Mahal is located at Agra, India, and is a UNESCO World Heritage Site. Particlem – Taj Mahal – CC BY 2.0.

Hinduism

Hinduism is one of the world's oldest major religions still practiced. Its origins can be traced to ancient Vedic civilizations in India approximately three thousand years ago. The religion is found mainly in India, and it has the third-highest number of believers of religions in the world. Hinduism does not originate from a single teacher but from many traditions. The Hindu belief system consists of a number of schools of thought, with a wide variety of rituals and practices.

Hinduism has a vast body of written scripture that discusses theology, mythology, and philosophy as well as providing important guidance on the practice of dharma, religious or right living. These texts include the *Vedas* and the *Upanishads.*

Other important scripts include the *Tantras*, the *Agamas*, the *Purānas*, and the epics of the *Mahābhārata* and the *Rāmāyana*. The *Bhagavad Gitā* is a small part of the *Mahābhārata* that is a conversation between one of the Pandava brothers and the god Krishna, concerning the meaning of life and worthiness. This is often thought to be a summary of the spiritual teachings of the Vedas.

Figure 9.36 Shiva Statue in Bangalore, India Andrea – Lord Shiva – CC BY 2.0.

Predominantly, Hinduism follows the teachings of many gods or goddesses, frequently including a Supreme Being. While there are hundreds, if not thousands, of gods and goddesses, many are thought to represent different aspects of the same individual or Supreme Being. These individuals can be recognized by items that they are holding as well as by the vehicle or avatar that carries them. The three main deities and most widely venerated of the Hindu faith are **Shiva** the Destroyer, **Vishnu** the Preserver, and **Brahma** the Creator. There is a continuous cycle in which the original creation was accomplished by Brahma, Shiva destroys the universe, and Vishnu will recreate or preserve that universe from destruction. Different Hindu traditions have venerated each of the three main deities as the all-encompassing Supreme Being.

The polytheistic traditions of Hinduism consider a large number of deities or spiritual entities. Since there is no one creed or unified systems of beliefs, Hinduism has been referred to as more of a religious tradition than a religion. It has been said that Hinduism cannot be defined, but is instead experienced. This understanding allows a variety of beliefs to be included in the vast array of Hindu religious practices. There is actually freedom for the individual as to the form of worship or individual beliefs. The religion in general is more of a tradition and lifestyle with different avenues of practice. This allows for the diversity of spiritual deities or their manifestations within one Hindu faith.

Hinduism is an extremely diverse religion, making it extremely difficult to define set doctrines that are accepted by all denominations. Within the wide spectrum of religious traditions are general concepts that are common to Hindu beliefs. Hindus believe in Dharma (code of conduct or duty), Samsara (reincarnation/rebirth), Karma (personal actions and choices), and Moksha (salvation) by belief in God and through an individual path of faith. Reincarnation is a cycle of death and rebirth for a soul to transmigrate through until it reaches Moksha. Karma governs how the soul is reincarnated. Actions in this life determine the soul's life cycle for the next life. Positive and upright works will draw one closer to God and a rebirth through reincarnation into a life with a wider consciousness or higher caste level. Evil or bad actions take the soul farther from God and into a lower form of worldly life or caste level. In the Hindu faith the eternal natural law applies to all life forms. The cycle of death and rebirth for the soul is necessary to reach the ultimate goal of reaching the universal divine spirit. Yoga is often used as a practice or path for following the traditions.

Pilgrimages are common in Hindu practice. Holy sites or temples are located throughout India and are regular destinations for the Hindu faithful. Pilgrimage is not required but is routinely conducted by a good number of Hindus. Besides many holy temples, a variety of cities and other holy places are pilgrimage destinations for Hindus. Varanasi, one such city, is considered by many as the holiest city of Hinduism, although other cities also hold this distinction. Located on the Ganges River, Varanasi is home to a large number of temples and shrines. The most visited shrine in Varanasi is one in honor of a manifestation of Shiva. Hindu festivals are held in Varanasi throughout the year, many along the banks of the Ganges. Varanasi is also one of the holiest places in Buddhism; it is said to have been designated by Gautama Buddha as one of four prime pilgrimage sites.

Varanasi boasts more than one hundred ghats that provide access to the Ganges River. These ghats are not to be confused with the Eastern and Western Ghats that are highland ranges located along the coasts of India. For this application, a ghat is a term for a set of steps leading to the water. Some ghats are used for bathing, some for religious rituals, and others for the cremation of the dead. More than one million Hindu pilgrims visit Varanasi annually. Mother Ganga, as the river is referred to in Hinduism, is considered holy by many Hindu followers. Devotees ritually bathe in the river or take "holy" water from it home to ill family members. Some Hindus believe that the water can cure illnesses. Others believe that bathing in the Ganges will wash away your sins. The nonspiritual truth is that the Ganges is a highly polluted waterway. The water is not considered safe for human consumption by most universal health standards.

Figure 9.5.379.5.37: Ghat in Varanasi, India, where Hindu Faithful Access the Ganges River. Rupert Taylor-Price – People washing at India ghats – CC BY 2.0.

Hindu Marriage Act

India's 1955 Hindu Marriage Act addresses the issues of marrying outside of one's religion or caste. The act proclaims that all Jains, Buddhists, Sikhs, and Hindus of any sect, creed, or caste level are considered as Hindus and can intermarry. A non-Hindu is allowed to marry a Hindu, with the provision that specific legal stipulations must be followed, regardless of the ceremony. Every Hindu marriage has to be witnessed by the Sacred Fire, in deference to Agni, the fire deity. To complete the marriage, the bride and groom must together encircle the Sacred Fire seven times.

Figure 9.38 Hindu Wedding Couple in India. .. – Hindu Weddings – North Indian – CC BY-NC 2.0.

The Hindu Caste System

The exact origins of Hinduism and the caste system are unknown, but powerful nomadic Aryan warriors appeared in northern India about 1500 BCE. The Aryans conquered most of India at the time, including the Dravidian groups of central and southern India. They organized society into separate groups or castes. Every person was born into an unchanging group or caste that remained his or her status for the rest of his or her life. All lifetime activities were conducted within one's own caste. The caste a person was born into was considered to be based on what they had done in a past life. The caste system has evolved differently in different parts of Asia. Each Hindu branch has its own levels of castes, and thousands of sublevels have been established over time. In Hinduism, the basic system originated around five main caste levels:

1. **Brahmin:** priests, teachers, and judges
2. **Kshatriya:** warrior, ruler, or landowner
3. **Vaishya:** merchants, artisans, and farmers
4. **Shudra:** workers and laborers
5. **Dalits (Untouchables or Harijan):** outcasts or tribal groups

The Dalits (Untouchables or Harijan) traditionally worked in jobs relating to "polluting activities," including anything unclean or dead. Dalits have been restricted from entering Hindu places of worship or drinking water from the same sources as members of higher castes. They often had to work at night and sleep during the day. In many areas, Dalits needed to take their shoes off while passing by upper-caste neighbors. Dalits could leave their Hindu caste by converting to Christianity, Buddhism, or Islam. The Indian government has implemented a positive affirmative action plan and provided the Dalits with representation in public offices and certain employment privileges. This policy has received harsh opposition by upper-caste groups. Technically, the caste system is illegal under current Indian law. Nevertheless, the opportunities that are available to the upper castes remain out of reach to many of the lowest caste. In some areas, education and industrialization have diminished the caste system's influence. In other areas, Hindu fundamentalists have pushed for a stronger Hindu-based social structure and opposed any reforms.

Traditional socioeconomic status tends to be more important in rural areas, where the caste system is more formally adhered to. If you live in a community of millions of people, caste affiliations tend not to be important, but in a smaller, more rural community, these relationships and the status they hold can be very important, especially as many of the castes are associated with traditional village tasks, such as religious leaders, politicians, farmers, leather workers, or other activities.

Buddhism

Around 535 BCE in northern India, a prince by the name of Siddhartha Gautama broke from the local traditions that shaped Hinduism and taught religious salvation through meditation, the rejection of earthly desires, and reverence for all life forms. Siddhartha is recognized as the first Buddha. He taught that through many cycles of rebirth a person can attain enlightenment and no longer have a need for desire or selfish interests. Enlightenment is being free from suffering and is reaching a state of liberation often referred to as Nirvana. Buddhism is considered a "dharmic" faith that concerns following a path of duty for a proper life. According to Buddhism, life is dictated by karma, which connects our actions with future experiences. Buddhism spread across the Indian Subcontinent after the sixth century BCE and became the region's dominant religion within 1,500 years. However, since that time, the religion has diminished in the Indian Subcontinent, although it has seen some revival under the influence of Buddhist scholars. Buddhism predominates in the northernmost areas of India.

Figure 9.40: Statue of Buddha at Bodh Gaya, India. Man Barlett - Untitled - CC BY 2.0.

Buddhism is the majority religion in Bhutan, Sri Lanka, Tibet, and most of Southeast Asia. It was prominent in China, Mongolia, and North Korea before their governments adopted Communist ideology. Communist governments officially announced that their countries were nonreligious, although many people still followed religious systems. Various branches of Buddhism have developed, with many schools established within each branch. Buddhism can be divided geographically into southern, northern, and eastern Buddhism. Scholars and Buddhists practitioners may arrive at various methods of classification of the various schools of thought within the Buddhist faith; the geographic basis of recognition provides only one way to understand the variations within the religion. One feature common throughout all branches of the faith is that Buddhism does not have caste levels.

The southern branch of Buddhism is known as Theravada Buddhism (*the Teaching of the Elders*), most prominent in Southeast Asia and Sri Lanka. This branch, referred to as the oldest branch of Buddhism, attempts to follow the original Buddha's teachings. Meditation and concentration are seen as keys to enlightenment. Spiritual forces do exist but it is up to the individual to attain his or her own path toward awakening.

The northern variety of Buddhism is associated with the Vajrayana Buddhist tradition and is often called Tibetan Buddhism. It has its strongest allegiance in Tibet, Western China, Bhutan, Nepal, and parts of Mongolia. Rooted in the Buddha's original teaching, Northern Buddhism seeks to break out of the cycle of birth, death, and rebirth. Spiritual techniques are often used, along with the main principle of meditation. Vajrayana Buddhism is considered by some to be a branch of Mahayana Buddhism.

In the east, the main form of Buddhism is the Mahayana tradition, which is most common in Japan, Taiwan, Korea, parts of Vietnam, and eastern China. There are various forms of the eastern traditions, including Zen Buddhism. Meditative in nature, there is a strong emphasis on universal compassion, altruism, and selflessness. Considered by many to be compatible with other religions, it is often touted as more a lifestyle than a religion. The meditative activities are often said to focus on calming the body and mind, which can provide a positive outcome for anyone seeking inner direction, even those following other religions.

All branches of Buddhism teach nonviolence, honesty, selflessness, tolerance, and moral living. Buddhism holds to the Four Noble Truths and the Eightfold Path (The Middle Way) to enlightenment. Suffering is a standard component of humanity. Only through the Eightfold Path to enlightenment is freedom from suffering possible. Enlightenment comes through wisdom, ethical conduct, and meditation. Buddhism has become the world's fourth main religion, with most of its followers in Asia.

The Four Noble Truths and the Eightfold Path

The Four Noble Truths

1. Suffering exists.
2. Suffering arises from attachment to desires.
3. Suffering ceases when desire ceases.
4. Freedom from suffering is possible by practicing the Eightfold Path.
5. The Middle Way or Eightfold Path

Attainable through wisdom

1. Right view
2. Right intention

Attainable through ethical conduct

3. Right speech
4. Right action
5. Right livelihood

Attainable through meditation

6. Right effort
7. Right mindfulness
8. Right concentration

Other Religions of South Asia

Islam

Islam diffused to South Asia from the Middle East through traders and merchants shortly after its creation. The Mogul Empire dominated northern India for centuries and is an example of the Islamic presence in the region. India has the largest Muslim minority in the world. In 2010, India had an estimated population of 140 million Muslims, which made up about 14 percent of the population. India has the third-largest Muslim population in the world, after Indonesia and Pakistan. Most Muslims in India live in the north along the Ganges River Basin and in Nepal. There is also a large Muslim population in eastern India near the border with Bangladesh. Kashmir, in northern India, also continues to hold a significant Muslim population.

Christianity

Christianity is India's third-largest religion, practiced by about 3 percent of the population. Christianity is a monotheistic religion following the teachings of Jesus Christ that originated in the Middle East. Tradition has it that Christianity first came to India through first-century CE missionary activity linked to the Apostle Thomas, and later arrived to other parts of India through the activities of western missionaries from 1500 CE onward. A major Christian stronghold is the state and city of Goa on the southwest coast, a colony of Portugal from the 1500s to the mid-1900s. There is also a strong Christian presence in eastern India, in the region bordering Myanmar.

Jainism

Jainism or Jain Dharma is a spiritual, religious, and philosophical tradition in India that dates back to about the ninth century BCE. Jains (followers of Jainism) believe that their religion's origins extend back to the distant past. A Jain is a follower of Jinas (the saints), who are humans who have rediscovered the dharma (or the way) and have become fully liberated. These Jinas can then teach this spiritual path to other people. A major characteristic of Jainism is the emphasis on the consequences of physical and mental behavior. There are about five million Jains in India, and others around the world.

Great care must be taken while going about one's daily life, as Jains believe that everything is alive and that many beings (including pests such as insects) possess a soul. All life is considered worthy of respect and all life is equal and deserves protection, especially the life of the world's smallest creatures. While in India, you can recognize Jains, because many of the strictest adherents will wear masks to prevent themselves from inhaling insects and thus destroying the insects' souls. Jains are a religious minority, with around five million followers in modern-day India.

All followers of Jainism are vegetarians. Their diet is part of the practice of nonviolence at the heart of their religion. They will eat only food items such as fresh fruit, vegetables, cereals, and legumes. Most root vegetables such as potatoes and onions are avoided by the more devout because of the harm that would be done to the plants themselves or other organisms in the soil when the roots are pulled. Additionally, Jainists will not eat honey, consume any food that may have fermented overnight, or drink water that has not been filtered.

Sikhism

Sikhism was previously discussed in the section on the Punjab, the region at the center of the Sikh community. Sikhism is a monotheistic religion centered on justice and faith. In Sikhism, salvation can be obtained through devotion to God and through disciplined meditation. There is a high importance placed on the principle of equality between all people in the Sikh religion. There should be no discrimination on the basis of gender, creed, caste, or ethnicity; every person is equal. The writings of former Sikh gurus are the basis for the religion. The center of the Sikh religion is found in the Golden Temple, in the city of Amritsar, in the Punjab. This is where Sikhs gather to unite in the faith and associate with each other. There are about twenty-six million Sikhs in the world, and about three-fourths of them live in the Punjab state of India.

Figure 9.41: Baha'i Faith's Lotus Temple in New Delhi India has the largest population of Baha'i in the world. Wikimedia Commons – CC BY 2.0.

Baha'i Faith

The Baha'i Faith is found in many large urban areas of the Indian Subcontinent, particularly New Delhi, where a large temple complex is found. This temple is commonly known as the "Lotus Temple" based on its shape, which looks like a large lotus flower. Two million Baha'is live within India, which has the largest population of Baha'is in the world. Iran has the second-highest Baha'i population in the world. There is also a major Baha'i temple and center in Haifa, Israel. The Baha'i Faith was founded by Baha'u'llah in Persia (Iran) during the nineteenth century CE. This religion focuses on the spiritual oneness of humanity and the unity of the other major world religions.

Zoroastrianism

Zoroastrianism originated in Iran (or Persia); in fact, it was once the dominant religion of Greater Iran, but it has now dwindled to about two hundred thousand Zoroastrians around the world, with concentrations in Iran, India (primarily Mumbai), and Pakistan (primarily Karachi and Lahore). Zoroastrianism follows the teachings of the prophet Zoroaster (also known as Zarathustra or Zarthosht). This philosophy acknowledges the divine authority of Ahura Mazda (Mazdaism) as proclaimed by Zoroaster. One of the tangible forms of this religion is the use of fire as a purifying agent in ceremonies. The temples are commonly known as "fire temples." Additionally, because of a prohibition of burials of bodies in the ground, Zoroastrians allow natural exposure of bodies to the elements in structures known as "Towers of Silence." The greatest numbers of followers of Zoroastrianism can be found in India, with additional numbers in southern Pakistan.

Key Takeaways

- Hinduism is one of the world's oldest still-practiced religions. There is no one specific path in the religion. Hinduism is more of a religious tradition based on core concepts than it is a formal religion.
- The caste system is a Hindu practice of placing people in social layers with similar occupations, privileges, and status. The untouchables are the lowest caste.
- Buddhism was created around 535 BCE from the traditions that shaped Hinduism by Siddhartha Gautama, who taught religious salvation through meditation, the rejection of earthly desires, and reverence for all life forms. There is no caste system in Buddhism, which has a number of branches that vary throughout Asia.
- Islam and Christianity are the second- and third-largest religions in India but did not originate in South Asia. Jainism and Sikhism got their start in South Asia and are still practiced by millions of people. The Baha'i Faith and Zoroastrianism are also active in India.

Discussion and Study Questions

1. What are the major religions in India in terms of population percentage?
2. What are four of the main concepts of Hinduism? How is yoga related to Hinduism?
3. Who are the three main deities in the Hindu tradition? What are each of these three associated with?
4. Is the caste system a centripetal force or a centrifugal force? In other words, does it divide or unite India?
5. What is the relationship between Hinduism and the Ganges River? What are ghats used for?
6. What is the caste system based on? How does one become a member of a caste level?
7. Does the United States have a caste system? What other countries or groups have caste-type levels?
8. What is the difference between Hinduism and Buddhism? Who started Buddhism, and when?
9. What are the three main branches of Buddhism, based on geographic location?
10. What are the general aspects of Jainism? How is this religion different from Islam?

9.6. CONCISE SUMMARY

The Himalayan mountain ranges delineate the northern boundary of South Asia. Nepal is situated along this border and serves as a buffer state between India and China. Nepal exhibits a significant population growth rate. The majority of its population is employed in agriculture. Deforestation is a significant environmental issue that leads to landscape erosion. Nepal, being landlocked and impoverished, faces challenges in sustaining a stable government and providing sufficient public services. South Asia was under British colonization for ninety years. Colonialism introduced an organized administration, a railway transportation infrastructure, and significant port cities utilized for the exportation of goods from the interior. The political boundaries of South Asia were delineated by British conquerors, influenced by religious affiliations and economic benefits. The British empowered Sikhs from Punjab to govern the Hindu and Muslim communities. English is extensively utilized as a lingua franca.

Conflicts persist in the mountainous region of Kashmir and the tropical island of Sri Lanka. The isolated region of Kashmir in the northern section of the territory is partitioned among Pakistan, China, and India. All three nations possess nuclear armaments. The predominant Buddhist demographic in Sri Lanka is the Sinhalese, primarily located in the southwest, where they govern the majority of the country. The Hindu minority Tamil people of Sri Lanka is located in the Jaffna Peninsula in the northeast. The Tamils seek an independent nation and have territorial claims over the island.

Port cities in South Asia serve as hubs for international trade and growth. A significant discrepancy exists between the rural impoverished and the rich elite. India has been cultivating a robust economy founded on an expanding information sector, healthcare, and manufacturing. Motor vehicles and computing technology are rising in India and competing on a global scale. Pakistan's economy is hindered by rapid population expansion and Islamic militancy.

Pakistan and Bangladesh were previously governed as a single entity. Bangladesh was once known as East Pakistan. These Muslim nations exhibit exceedingly high population densities and possess agrarian economies. The Indus River traverses Pakistan, whereas the Brahmaputra and Ganges rivers flow through Bangladesh. Monsoon flooding poses a significant threat to Bangladesh, while earthquakes have inflicted considerable destruction in Pakistan.

Hindu and Buddhist faiths initially emerged in South Asia. India possesses the largest population of Hindu adherents globally. The caste system has established socioeconomic strata within the society, which are being influenced by significant urbanization rates. Buddhism comprises several branches that can be regionally categorized as eastern, northern, and southern. Bhutan and Sri Lanka possess Buddhist majorities. South Asia is the origin of Sikhism and Jainism. Islam is prominent in South Asia: Pakistan is the second-largest Muslim nation globally, India possesses the third-largest Muslim population, and Bangladesh is also a Muslim-majority country. South Asia hosts a Christian minority among several other minority religious groups.

10 Chapter

EAST ASIA

10.1. INTRODUCTION: AN OVERVIEW

East Asia is a vibrant and diverse region that encompasses countries such as China, Japan, South Korea, North Korea, Mongolia, and Taiwan. Known for its rich cultural heritage, East Asia has a profound historical legacy that influences its modern identity. The region is characterized by a blend of ancient traditions and cutting-edge technology, with bustling cities like Tokyo and Shanghai juxtaposed against serene landscapes like Japan's cherry blossoms and China's Great Wall. Economically, East Asia is a powerhouse, driving global trade and innovation, while socially, it grapples with issues of modernization, environmental challenges, and demographic shifts. The intricate tapestry of languages, philosophies, and customs makes East Asia a fascinating area for exploration and study.

Figure 10.1 East Asia and Its Neighbors

Physical Geography

East Asia is surrounded by a series of mountain ranges in the west, Mongolia and Russia in the north, and Southeast Asia to the south. The Himalayas border Tibet and Nepal; the **Karakoram Ranges, Pamirs**, and the **Tian Shan Mountains** shadow Central Asia; and the **Altay Mountains** are next to Russia. The **Himalayan Mountains** are among the highest mountain ranges in the world, and Mt. Everest is the planet's tallest peak. These high ranges create a rain shadow effect, generating the dry arid conditions of type B climates that dominate **western China.** The desert conditions of western China give rise to a large uninhabitable region in its center. Melting snow from the high elevations feeds many of the streams that transition into the major rivers that flow toward the east.

Figure 10.2 Flags of the East Asian Nations and Territories

Created by tectonic plate action, the many mountain ranges are also home to earthquakes and tremors that are devastating to human livelihood. The Indian tectonic plate is still pushing northward into the Eurasian plate, forcing the Himalayan ranges upward. With an average elevation of fifteen thousand feet, the Tibetan Plateau is the largest plateau region of the world. It has high elevations and type H climates. The plateau is sparsely populated and the only places with human habitation are the river valleys. Lhasa is the largest city of the sparsely populated region. Sometimes called "the Roof of the World," the Tibetan Plateau is a land of superlatives. The small amount of precipitation that occurs often comes in the form of hailstorms mixed with wind. Its landscape is generally rocky and barren.

The vast arid regions of western China extend into the **Gobi Desert** between Mongolia and China. Colder type D climates dominate the Mongolian steppe and northern China. The eastern coast of the Asian continent is home to islands and peninsulas, which include Taiwan and the countries of Japan and North and South Korea. North Korea's type D climates are similar to the northern tier of the United States, comparable to North Dakota. Taiwan is farther south, producing a warmer tropical type A climate. The mountainous islands of Japan have been formed as a result of tectonic plates and are prone to earthquakes. Since water moderates temperature, the coastal areas of East Asia have more moderate temperatures than the interior areas do. A type C climate is dominant in Japan, but the north has a colder type D climate. The densely populated fertile river valleys of central and southeastern China are matched by contrasting economic conditions. Rich alluvial soils and moderate temperatures create excellent farmland that provides enormous food production to fuel an ever-growing population.

Most of China's population lives in its eastern region, called **China Proper**, with type C climates, fresh water, and good soils. China Proper has dense population clusters that correspond to the areas of type C climate that extend south from Shanghai to Hong Kong. Around the world, most humans have gravitated toward type C climates. These climates have produced fertile agricultural lands that provide an abundance of food for the enormous Chinese population. To the south the temperatures are warmer, with hot and humid summers and dry, warm winters. The climates of China Proper are conducive for human habitation, which has transformed the region into a highly populated human community. The **North China Plain** at the mouth of the **Yellow River** (*Huang He River*) has rich farmland and is the most densely populated region in China.

Northwest of Beijing is **Inner Mongolia** and the Gobi Desert, a desert that extends into the independent country of Mongolia. Arid type B climates dominate the region all the way to the southern half of Mongolia. The northern half of Mongolia is colder with continental type D climates. In the higher elevations of the highlands in western Mongolia, there is a section of type H highland climates. Its climate and location identify Mongolia as a landlocked country in the northern latitudes with a low level of precipitation. The areas of type D climate that extend north from Beijing through Northeast China at times receive more precipitation than northern Mongolia. Northeast China features China's great forests and excellent agricultural land. Many of China's abundant natural mineral resources are found in this area. Balancing mineral extraction with the preservation of agricultural land and timber resources is a perennial issue.

Lying north of the Great Wall and encompassing the autonomous region of Inner Mongolia is the vast Mongolian steppe, which includes broad flat grasslands that extend north into the highlands. North China includes the Yellow River basin as well as the municipalities of Beijing and Tianjin. Areas around parts of the Yellow River are superb agricultural lands, including vast areas of loess that have been terraced for cultivation. Loess is an extremely fine silt or windblown soil that is yellow in color in this region. Deciduous forests continue to exist in this region, despite aggressive clearcutting for agricultural purposes. The Great Wall of China rests atop hills in this region.

Most of western China is arid, with a type B climate. Western China has large regions like the **Takla Makan Desert** that are uninhabited and inhospitable because of hot summers and long cold winters exacerbated by the cold winds sweeping down from the north. In a local Uyghur language, the name *Takla Makan* means "*You will go in but you will not go out.*" To the far west are the high mountains bordering Central Asia that restrict travel and trade with the rest of the continent. Northwestern China is a mountainous region featuring glaciers, deserts, and basins.

Figure 10.3: China and Its Main Climate Regions. Updated from map courtesy of University of Texas Libraries.

The central portion of China Proper is subtropical. This large region includes the southern portion of the **Yangtze River** (Chang Jiang River) and the cities of **Shanghai** and **Chongqing**. Alluvial processes give this area excellent agricultural land. Its climate is warm and humid in the summers with mild winters; monsoons create well-defined summer rainy seasons. Tropical China lies in the extreme south and includes **Hainan Island** and the small islands that neighbor it. Annual temperatures are higher here than in the subtropical region and rainfall amounts brought by the summer monsoons are at times very substantial. This area is characterized by low mountains and hills.

River Basins of China

There are two major river systems that provide fresh water to the vast agricultural regions of the central part of China Proper. The Yellow River (Huang He River) is named after the light-colored silt that washes into the river. It flows from the Tibetan highlands through the North China Plain into the Yellow Sea. Dams, canals, and irrigation projects along the river provide water for extensive agricultural operations. Crops of wheat, sorghum, corn, and soybeans are common with vegetables, fruit, and tobacco grown in smaller plots. The North China Plain has to grow enough food to feed its one thousand people per square mile average density. This plain does not usually produce a food surplus because of the high demand from the large population of the region. Beijing borders the North China Plain. Its nearest port, **Tianjin**, continues to expand and grow, creating an economic center of industrial activity that relies on the peripheral regions for food and raw materials. Cotton is an example of a key industrial crop grown here.

The Yangtze River (Chang Jiang River) flows out of the Tibetan Plateau through the Sichuan Province, through the Three Gorges region and its lower basin into the East China Sea. Agricultural production along the river includes extensive rice and wheat farming. Large cities are located on this river, including Wuhan and **Chongqing. Nanjing** and Shanghai are situated near the delta on the coast. Shanghai is the largest city in China and is a growing metropolis. The **Three Gorges Dam** of the Yangtze River is the world's largest dam. It produces a large percentage of the electricity for central China. Oceangoing ships can travel up the Yangtze to **Wuhan** and, utilizing locks in the Three Gorges Dam, these cargo vessels can travel all the way upriver to Chongqing. The Yangtze River is a valuable and vital transportation corridor for the transport of goods between periphery and core and between the different urban centers of activity. Sichuan is among the top five provinces in China in terms of population and is dependent on the Yangtze River system to provide for its needs and connect it with the rest of China.

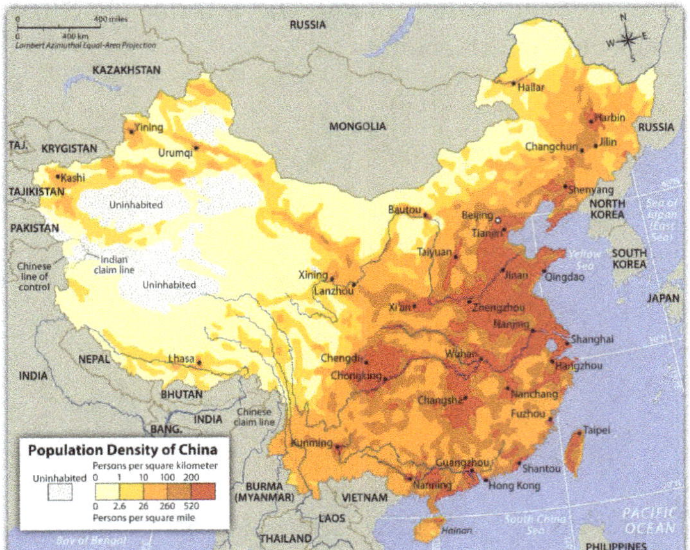

Figure 10.4: China and Its Main Population Regions

The region of eastern China that is favorable to large populations is called China Proper. River basins historically produce abundant food, which in turn leads to concentrated populations.

Northeast China was formerly known as Manchuria, named after the Manchu ethnic group that had dominated the region in Chinese history. Two river basins create a favorable industrial climate for economic activity. The lower **Liao River Basin** and the **Songhua River Basin** cut through Northeast China. The cities of **Harbin** and **Shenyang** are industrial centers located on these rivers. This region is known as the **Northeast China Plain.** It has extensive farming activities located next to an industrial landscape of smokestacks, factories, and warehouses. Considerable mineral wealth and iron ore deposits in the region have augmented the industrial activities and have created serious environmental concerns because of excessive air and water pollution. In its zenith in the 1970s, this was China's main steel production area, but the region is being reduced to a rustbelt since many of China's manufacturing centers are now being developed in the southern regions of China Proper.

The southernmost region of China Proper is home to the **Pearl River Basin**, an important agricultural and commercial district. Though smaller in size than the Yangtze River Basin, major global urban centers are located on its estuary, where the mouth of the river flows into the South China Sea. The system includes the Xi River, Pearl River, and their tributaries. As the third-longest river system in China, these rivers process an enormous amount of water, and have the second-highest volume of water flow after the Yangtze. **Guangzhou**, Macau, and Hong Kong are the largest cities located here, alongside the rapidly expanding industrial center of **Shenzhen.** As mentioned earlier, Macau was a former Portuguese colony and Hong Kong was a former British colony.

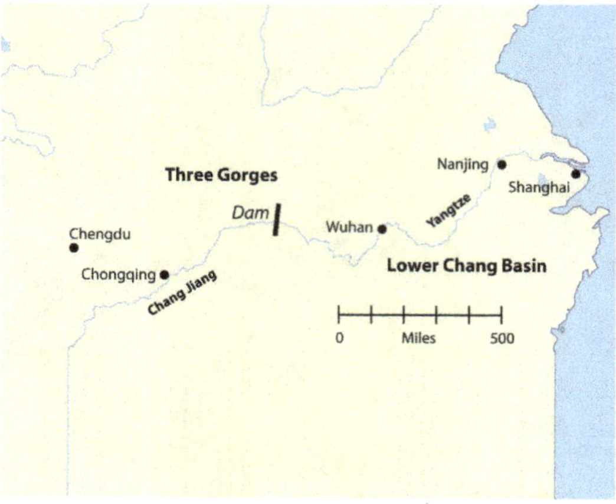

Figure 10.5: Xi-Pearl River System An estuary is a wide area at the mouth of a river where it meets the sea. Hong Kong is located on the eastern side of the Pearl River Estuary, and the former Portuguese colony of Macau is located on the western side of the waterway. Wikimedia Commons – CC BY 2.0.

These urban areas are now hubs for international trade and global commerce. Guangzhou is one of the largest cities in China, along with Shanghai, Beijing,

Wuhan, and Tianjin. Cantonese heritage and traditions form a foundation for the cultural background of the people who live here.

Three Gorges Dam (The New China Dam)

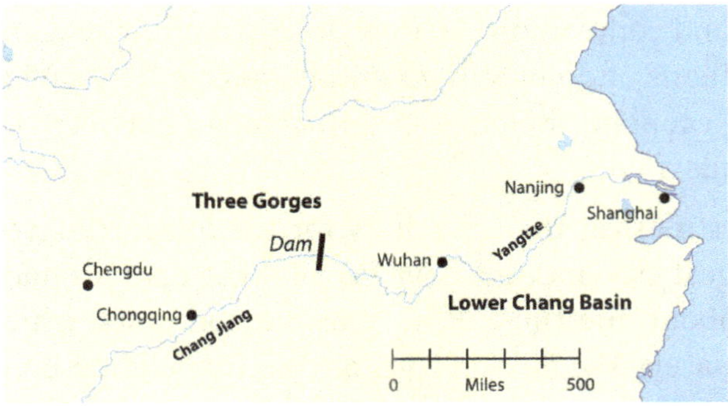

Figure 10.6 Three Gorges Dam Region

The Three Gorges Dam on the Yangtze River is known in China as the New China Dam. Its hydroelectric production system is the largest on Earth. The river system is the world's third longest, after the Nile and the Amazon. Ideas for this project go back to the days just after the last dynasty fell. Plans and development began in the decades before 1994, when the construction of the dam began. The main purposes of the dam are to control the massive flooding along the Yangtze, produce hydroelectric power, and increase shipping capacity along the river.

The Yangtze River flows through three deep gorges where a dam has been constructed to stabilize flooding, produce electricity, and support river transportation.

- Dam length: 7,661 feet
- Dam height: 610 feet
- Dam width (at base): 377 feet
- Physical construction began: December 14, 1994
- Construction cost: estimated thirty-nine billion US dollars
- Estimated surface area of reservoir: 403 square miles
- Estimated length of reservoir: 375 miles
- Capacity of thirty-two generators totaling 22,500 MW (equivalent to about twenty nuclear power plants the size of the Watts Bar 1, the newest US nuclear reactor)

Before construction of the dam, flooding along the Yangtze cost thousands of lives and billions of dollars in damage. In 1954, the river flooded, causing the deaths of more than thirty-three thousand people and displacing an additional eighteen million people. The giant city of Wuhan was flooded for three months. In 1998, a similar flood caused billions of dollars in damage, flooded thousands of acres of farmland, resulted in more than 1,526 deaths, and displaced more than 2.3 million people. The dam was rigorously tested in 2009, when a massive flood worked its way through the waterway. The dam was able to withstand the pressure by containing the excess water and controlling the flow downstream. The dam saved many lives and prevented billions of dollars in potential damage. The savings in human lives and in preventing economic damage are projected to outweigh the cost of the dam in only a few decades.

The dam produces most of the electricity for the lower Yangtze Basin, including Shanghai, the largest city in China. Five years of the dam producing electricity has already paid for about one-third of its construction costs, which is equivalent to burning approximately 150 million tons of coal (depending on coal quality). This reduces the emission of millions of tons of carbon dioxide, sulfur, and nitrogen oxides into the atmosphere, which reduces air pollution and does not contribute to climate change. Heavy freight traffic on the Yangtze was the norm even before the dam was built; in fact, it has the highest rates of transport of any river. The building of the dam has augmented the amount of freight traffic.

Figure 10.7: China's Three Gorges Dam on the Yangtze River is the largest dam in the world. Wikimedia Commons – CC BY 2.0.

All the positive attributes of the Three Gorges Dam have contributed to the economic development of China. This is a testimony to the engineering and technological capacity of the nation. However, this project has also created its own problems and negative impacts on culture and the environment. By 2008, the number of people forced to relocate from the flooding of the reservoir had reached 1.24 million. Historic villages and hundreds of archaeological sites were flooded. Thousands of farmers had to be relocated to places with less productive soils. Compensation to the farmers for relocation was forfeited because of corruption and fraud. Sadly, much of the scenic beauty of the river basin is now under water.

Animal species like the critically endangered Siberian Cranes, who had wintered in the former wetlands of the river, had to find habitat elsewhere. The endangered Yangtze River Dolphin has been doomed to extinction because of the dam and the amplified river activity. The dam restricts the flushing of water pollution and creates a massive potential for landslides along its banks, exacerbating the potential for the silting in of the reservoir and the clogging of the dam's turbines. The dam also sits on a fault zone and there is concern that the massive weight of the water in the reservoir could trigger earthquakes that may destroy the dam, with catastrophic consequences. Large development projects tend to have an enormous impact on the people and the environment that inhabit their shadow. The building of the Three Gorges Dam has created controversy, with strong arguments on both sides of the issues. To further complicate the situation, other large dams are being proposed or are under construction along the same river.

Chinese Dynasties and Colonialism

The earliest Chinese dynasty dates to around 2200 BCE. It was located in the rich North China Plain. Organized as a political system, Chinese dynasties created the Chinese state, which provided for a continuous transfer of power, ideas, and culture from one generation to another. From 206 to 220 CE, the **Han Dynasty** established the Chinese identity; Chinese people became known as People of Han or Han Chinese. The last dynasty, the **Qing (Manchu) Dynasty**, which ruled between 1644 and 1911, claimed control of a region including all of China, Mongolia, Southeast Asia, and Korea. Dynastic rule ended in China in 1911.

Figure 10.8: The Great Wall of China Construction started in the seventh century BCE and, including all its branches, the wall is about 5,500 miles long. Matt Barber – Great Wall of China – CC BY 2.0.

Europeans colonized the Americas, Africa, and South Asia, and it was only a matter of time before technology, larger ships, and the European invasion reached East Asia. European colonialism arrived in China during the Qing Dynasty. China had been an industrialized state for centuries; long before the empires of Rome and Greece were at their peak, China's industrial cities flourished with the concepts of clean drinking water, transportation, and technology. Paper, gunpowder, and printing were used in China centuries before they arrived in Europe. The Silk Road, which crossed the often dangerous elevations of the high mountain passes, was the main link between China and Europe.

European colonial powers met tough resistance in China. They were kept at bay for years. Meanwhile, the Industrial Revolution in Europe, which cranked out mass-produced products at a cheap price, provided an advantage over Chinese production. British colonizers also exported opium, an addictive narcotic, from their colonies in South Asia to China to help break down Chinese culture. By importing tons of opium into China, the British were able to instigate social problems. The first **Opium Wars** of 1839–1842 ended with Britain gaining an upper hand and laying claim to most of central China. Other European powers also sought to gain a foothold in China. Portugal gained the port of Macau. Germany took control of the coastal region of the rich North China Plain. France carved off part of southern China and Southeast Asia. Russia came from the north to lay claim to the northern sections of China. Japan, which was just across the waterfront from China, took control of Korea and the island of **Formosa** (now called Taiwan). Claims on China increased as colonialism moved in to take control of the Chinese mainland.

Though European powers laid claim to parts of China, they often fought among themselves. China did not produce heavy military weapons as early as the Europeans did and therefore could not fend them off upon their invasion. Chinese culture, which had flourished for four thousand years, quickly eroded through outside intrusion. It was not until about 1900, when a rebellion against foreigners (known as the **Boxer Rebellion**) was organized by the Chinese people, that the conflict reached recognizable dimensions. The Qing Dynasty dissolved in 1911, which also signified an end to the advancements of European colonialism, even though European colonies remained in China.

Three-Way Split in China

European colonialism in China slowed after 1911, and World War I severely weakened European powers. The Japanese colonizers, on the other hand, continued to make advancements. Japan did not have far to travel to resupply troops and support its military. In China, a doctor by the name of **Dr. Sun Yat-sen** promoted an independent Chinese Republic, free from dynastic rule, Japan, or European colonial influence. Political parties of Nationalists and Communists also worked to establish the republic. Dr. Sun Yat-sen died in 1925. The Nationalists, under the leadership of **Chiang Kai-shek**, defeated the Communists and established a national government. Foreigners were evicted. The Communists were driven out of politics.

Figure 10.9: Taiwan Currency Both the Republic of China (ROC) and the People's Republic of China (PRC) consider Dr. Sun Yat-sen to be a famous Chinese historical figure. Richard winchell – Museum of Australian Currency Notes – CC BY-ND 2.0.

Nationalists, Communists, and Japan conducted a three-way war over the control of China. Japan's military took control of parts of Northeast China, known as Manchuria, and were making advancements on the eastern coast. Nationalists defeated the Communists for power and were pushing them into the mountains. The Chinese people were in support of the two parties working together to defeat the Japanese. **The Long March** of 1934 was a six-thousand-mile retreat by the Communists through rural China, pursued by Nationalist forces. The people of the countryside gave aid to the efforts of the Communists. The Chinese were primarily interested in the defeat of Japan, a country that was brutally killing massive numbers of China's people in their aggressive war.

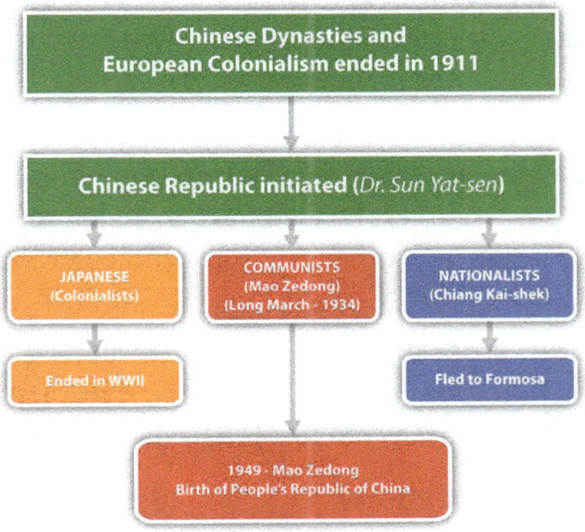

Figure 10.10: The Three-Way Split of China and the Emergence of Communist China in 1949

In 1945, the defeat of Japan in World War II by the United States changed many things. Japan's admission of defeat prompted the end of Japanese control of territory in China, Taiwan, Korea, Southeast Asia, and the Pacific. By 1948, the Communists, who were becoming well organized, were defeating the Nationalists. Chiang Kai-shek gathered his people and what Chinese treasures he could and fled by boat to the island of Formosa (Taiwan), which in 1945 had just been freed from Japan. Taiwan was declared the official Republic of China (ROC). The Communists took over the mainland government. In 1949, Communist leader **Mao Zedong** declared the establishment of the People's Republic of China (PRC) with its capital in Peking (Beijing). Japan was devastated by US bombing and defeated in World War II; its infrastructure destroyed and its colonies lost, Japan had to begin the long process of rebuilding its country. Korea was finally liberated from the Chinese dynasties and Japanese colonialism but began to experience an internal political division. Political structures in the second half of the twentieth century in East Asia were vastly different from the political structures that had been in place when the century began.

Key Takeaways

- China is the largest country in physical area and in population in East Asia. The realm is isolated by the high mountains in the west, which cause a major rain shadow and desert conditions in the western regions. Mongolia is the only landlocked country. The other countries and territories are located along the Pacific Rim.
- Robust population growth has been supported by adequate food production in the major river valleys and coastal regions of East Asia. Coastal areas receive adequate precipitation and allow access to fishing for human activities.
- The Three Gorges Dam was constructed on the Yangtze River to control flooding, generate electricity, and support shipping. The dam is the largest in the world. Downsides of the dam have included the relocation of human settlements, erosion, and other environmental concerns.
- Colonialism infiltrated China and challenged its last dynasty. By 1911, both the dynasty and European colonialism were in demise. In a three-way battle for power in China, the Communists emerged in 1949 to take control. The Nationalists fled to Formosa to form their own government.

Discussion and Study Questions

1. Outline the countries or territories that make up the region known as East Asia.
2. How is East Asia separated from the rest of Asia? How did this keep the realm isolated for many centuries?
3. What is the only landlocked country in East Asia? Describe its physical features.

4. What are the main climate types in each of the countries in East Asia?
5. Where are the four main river basins in China Proper? How do they contribute to China's development?
6. What are the three main benefits of the Three Gorges Dam? What are three of the negative impacts?
7. What was significant about the Han and Qing Dynasties? When did dynastic rule end in China?
8. How did the British attempt to break down Chinese culture? What was China's response?
9. What was the relationship between Japan, Korea, and China before World War II?
10. What was the three-way split in China about? What happened to each of the three groups?

Geography Exercise

Identify the following key places on a map:

- Altay Mountains
- Chang Jiang River
- China Proper
- Chongqing
- Formosa
- Gobi Desert
- Guangzhou
- Hainan Island
- Harbin
- Hong Kong
- Huang He River
- Inner Mongolia
- Karakoram Ranges
- Liao River
- Macau
- Nanjing
- North China Plain
- Northeast China Plain
- Pamirs
- Pearl River
- Shanghai
- Shenyang
- Shenzhen
- Songhua River
- Taiwan
- Takla Makan Desert
- Three Gorges Dam
- Tianjin
- Tian Shan
- Tibet
- Western China
- Wuhan
- Xizang
- Yangtze River
- Yellow River

10.2. EMERGING CHINA

China is the world's largest Communist country. Isolated from Europeans and Central Asians by the Himalayas and other high mountain ranges, Chinese culture has endured for thousands of years. Rich in history, adventure, and intrigue, a tour through China would reveal a people with a deep, longstanding love of the land, traditions as old as recorded history, and a spirit of commerce and hard work that sustains them to this day. China is about the same land size as the United States, although technically it is slightly smaller than the United States in total area, depending on how land and water areas are calculated. China only has an eastern coast, whereas the United States has both an eastern and a western coast. If you recall the climate types and the relationship between climate and population, you can deduce the location of the heavily populated regions of China.

The form of Communism promoted by Mao Zedong was not the same as the type of Communism practiced in the Soviet Union. Various Communist experiments were forced upon the Chinese people, with disastrous results. For example, in 1958, the **Great Leap Forward** was announced. In this program, people were divided into communes, and peasant armies were to work the land while citizens were asked to donate their pots and pans to produce scrap metal and increase the country's industrial output. The goal was to improve production and increase efficiency. The opposite occurred, and millions of Chinese died of starvation during this era.

Another disaster began in 1966 and continued until Mao died in 1976. The **Great Proletarian Cultural Revolution** wreaked havoc on four thousand years of Chinese traditional culture in a purge of elitism and a drive toward total loyalty to the Communist Party. Armies of indoctrinated students were released into the countryside and the cities to report anyone opposing the party line. Schools were closed, universities were attacked, and intellectuals were killed. Anyone suspected of subversion might be tortured into signing a confession. Violence, anarchy, and economic disaster followed this onslaught of anti-Democratic terror. Estimates vary, but most sources indicate that about thirty million people lost their lives during the Mao Zedong era through purges, starvation, and conflict.

During the early decades of Communism in China, all travel into or out of China was severely restricted by the so-called Bamboo Curtain. The United States had backed the nationalist movement during the Chinese Civil War and continued to support Chiang Kai-shek in Formosa (Taiwan). The United States did not recognize Communist China; the US embassy was in Taiwan, not in Beijing. As China was experiencing its disastrous experiments with Communism, the core economic areas of the world were advancing with commercial technology and high-tech electronics and sending rockets to the moon. China lagged behind in its industrial activities and became a country based on agriculture. A visit to China by US president Richard Nixon in 1972 signaled the opening of diplomatic relations with the United States and was also viewed as a Cold War move against the Soviet Union.

- **1949 – Communist China**
- 1958 – Great Leap Forward (*Backward*)
- 1966 – Great Proletarian Cultural Revolution
- 1972 – Richard Nixon visits China (*open door*)
- 1980s – Four Modernizations (*Deng Xiaoping*)
- 1990s – Growth of Enterprise and Industry
- 2000+ – A New Century (*Trade with the West*)

The Chinese Communist Party's approach when it took power was to institute a "planned economy." A planned economy, sometimes called a command economy, stands in marked contrast to a market economy. In a planned economy, the government controls all aspects of the economy, including what goods and products should be produced, how much of each should be produced, how products will be sold or distributed and for what (if any) price, who should have jobs and what jobs they should have, how much people will be paid, and all other decisions related to the economy. In a planned economy, businesses are nationalized; that is, businesses are owned by the government rather than by any private entity. By contrast, in a market economy, businesses are privately owned, and most decisions are driven by consumer and investor behavior.

The decade of the 1980s was a transition for China in that there was a shift of focus from China's Communist economy to a more market-oriented economy. The economic collapse of the Soviet Union in the early 1990s coincided with the opening of China to trade with the West. In 1992, China announced that it would transition to a socialist market economy, a hybrid of a Communist-planned economy and a market economy. A series of statements by China's political leaders suggested that in order for China to enjoy a more mature form of socialism, greater national wealth was needed. They further indicated that socialism and poverty should not be considered synonymous and that the country was ready to turn its attention to increasing the wealth and quality of life of its citizens. During the next decade, China experienced an enormous growth in its economy. At the beginning of this century, China was ranked in the top five of the world's largest trading nations, joining ranks with the United States, Germany, Japan, and France. The sheer size of China's population contributes to the magnitude of its economy. This does not, however, mean that most of China's population has a high standard of living.

Political Units

China can be divided into regions utilizing various criteria: political regions, economic regions, natural regions, and climatic regions to name a few. The

Communist government of the People's Republic of China (PRC) is made up of a number of types of political units. The main region of China Proper includes twenty-two provinces, including the island province of **Hainan** in the south. The island formerly known as Formosa, now called Taiwan, is considered by China to be its twenty-third province but in actuality remains under its own government—the Republic of China (ROC). Mainland China includes five autonomous regions, each with a designated minority group; four municipalities; and two special administrative regions (SARs) that hold considerable autonomy. All but Taiwan are included in the region called mainland China, except the SARs **Hong Kong** and **Macau.**

The autonomous regions exist as a kind of a compromise between China and the regions that would prefer to be totally independent and that have large non-Han or ethnic minority populations. These five autonomous regions are **Tibet**, **Guangxi**, **Xinjiang**, **Inner Mongolia**, and **Ningxia.** It should be noted that "autonomous" is how China would describe these regions, but that description is arguable. In reality, the autonomous regions have very little legal ability to govern themselves and, in fact, have in some ways less autonomy than the provinces to pass their own legislation.

Four of China's cities have governance structures that are roughly on par with that of the provinces. The directly controlled municipalities are **Beijing**, **Tianjin**, **Shanghai**, and **Chongqing** and encompass large geographic areas well beyond the city limits, including towns, villages, and rural areas.

The two SARs of China are Hong Kong and Macau. Hong Kong was part of the British Empire and later a British protectorate until its governance reverted to China in 1997. Known informally under Chinese law as "one country, two systems," Hong Kong and Macau are guaranteed a great degree of autonomy for at least the first fifty years after the transfer from Britain to China. China is responsible for defense and some portions of foreign affairs, while Hong Kong and Macau are responsible for most other matters of state, including their legal systems, law enforcement, immigration laws, monetary systems, customs policies, and others.

China's Communist government is controlled by the Chinese Communist Party, which is protected and guaranteed power by the country's constitution. Compared to the more open democratic political systems of Western Europe and North America, China is considered an authoritarian state, complete with heavy restrictions on the Internet, the press, freedom of assembly, reproductive rights, and freedom of religion. Economic trade with the West has created more openness with travel and business, which has begun to challenge or erode the hard-line politics of decades past.

Figure 10.12 China and Its Administrative Divisions, Including Autonomous Regions, Municipalities, and SARs

The People Population

With four thousand years of culture to build on, China continues to press forward into the twenty-first century. In 2010, China had more people than any other country in the world, about 1.33 billion. Most of its people live in China Proper, in the eastern regions of the country. China Proper has the best agricultural lands in the country, the most fertile river basins, and the most moderate climates. For perspective, China has over one billion more people than the United States, with most of those people living in the southeastern portion of China. During Mao's time, there was little concern for population growth, but in recent decades China has implemented measures to deal with its teeming population.

The government of China has taken the responsibility to address its growing population and the country's ability to provide for the needs of the additional populace. In 1978, China implemented the one-child-only statute, limiting family size to one child. The policy allows for exemptions under certain conditions. Couples living in rural areas and people of minority status are two examples of exempted conditions and may be permitted more than one child, especially if the first child is a girl. Peripheral administrative regions like Macau and Hong Kong are exempt. The one-child-only policy was implemented in an attempt to address environmental, economic, and social issues related to population growth. This policy has helped reduce China's potential population by hundreds of millions of people, but the controversial policy has not been easy to implement or make effective. There are growing concerns regarding the policy's negative impact on society. In response to the one-child-only policy, there have been reports of female infanticide and a higher number of abortions. However, surveys taken by independent agencies indicate that about three-fourths of China's population supports it.

Economic incentives pressure families to abide by the one-child-only policy. Government benefits and social programs offer incentives that may be lost if couples have more than one child. Enforcement of the policy has been left to the provincial authorities with varying levels of success. Parents eager to have a boy to carry on the family name might abort a female baby. Though now illegal, the use of ultrasound equipment to determine the gender of a fetus is widespread. Some provinces in China have a severe shortage of women because of the policy; men in provinces where women are scarce may have to migrate to find a wife. In China there are more boys born than girls; the ratio averages more than 10 percent more boys than girls, with some provinces reaching more than 25 percent. This imbalance creates cultural issues that may have a negative impact on traditional society. Conventional society is challenged when there are a large number of unmarried young men in an area that may not have adequate employment opportunities for them or available women for potential marriage.

Figure 10.13: Young Girl and Her Grandfather in Southern China The one-child-only policy eliminates brothers, sisters, cousins, aunts, and uncles in an extended family. Susan Sermoneta – proud grandpa – CC BY-NC-ND 2.0.

More than 50 percent of China's population lives in rural areas, meaning there is potential for high levels of rural-to-urban migration. The core industrial cities located in China Proper attract migrants in a periphery-to-core migration pattern. China hopes that this urbanization and industrial activity will also support their population control methods and fuel the industrial labor base. China's urbanization increased at an unprecedented rate from 17 percent in 1978 to 47 percent in 2010. This rural-to-urban shift has been one of the largest in human history. It still continues; many workers shift from holding temporary employment in the cities to returning to their families in the countryside between jobs.

Ethnic Groups

Han Chinese is the largest ethnic group in China, with about 90 percent of the country's population. Some of the largest minority groups include Zhuang (Tai) in the south, Manchu in the northeast, Mongolian in the north central region, Uyghur in Western China, and Tibetan in Tibet, although many other minority groups exist. The main language in China, Mandarin, is spoken in different ways in different parts of the country, particularly in the north and the south. The number of languages in China (over 290) roughly corresponds to the number of ethnic groups in the country. In Western China, where the percentage of Han Chinese is quite low, most of the population is Uyghur, a group that tends to be Muslim. There are also Kazahks, Kyrgyz, and Tajiks from Central Asia, who are also predominantly Muslim. In 2010, the precise number of Muslims in China was not known, but government reports indicated there were about twenty-one million. Other reports indicate there may be three times that many. Minority groups like the Uyghurs have often experienced discrimination by the Chinese government, which has taken measures to marginalize minority groups to keep them in check.

Figure 10.14: China and Its Main Ethnolinguistic Regions. Updated may from University of Texas Libraries.

As a Communist country, there is no official religion in China, nor is any supported by the Chinese government. Before Communism took control, most of the people followed a type of Buddhism. Other belief systems included Taoism and the teachings of Confucius. Christianity does exist in China, but it is illegal to proselytize or recruit converts. Despite that, the Christian population in China has grown rapidly in recent decades. The exact number is unknown, but the *CIA World Factbook* estimates 3 to 4 percent of the population to be Christian, or somewhere between thirty-nine and fifty-two million adherents. There are many other sects and religious groups in China that have gained attention in recent years. Among these are the followers of Falun Gong, which emerged in the 1990s and has elements of Buddhism and Taoism in its beliefs. The group claims to be more of a science than a religion. The Chinese government believes otherwise and banned the Falun Gong in 1999 after the group staged a large protest against the Communist Party. The crackdown by the government included many arrests and stiff prison sentences. The group is currently banned by the government.

Economic Summary

Over the past twenty years, China has transitioned its economy from a closed system characterized by a centrally planned, government-controlled market to one that embraces open commerce and a flexible production framework. These economic reforms have facilitated capitalistic tendencies in production, encouraged greater participation in private enterprise, and augmented international investment in the Chinese economy. China has abolished collective farms, enhanced agricultural output, adopted a more open stance towards free enterprise and international trade and investment, and experienced rapid economic growth. The facilitation of commerce and engagement with the global community has enabled China to converge with global standards in information and communication technology, while the proliferation of computer and Internet usage in China has created numerous openings and employment prospects across several sectors. China's establishment of special economic zones (SEZs) has accelerated urban and economic development in coastal cities, driving a significant rural-to-urban migration.

Figure 10.15: Beijing, a metropolitan center of Chinese culture, is the capital of China. Freeway development has expanded in recent years as higher incomes have allowed for automobile purchases. As a result, traffic congestion and air pollution are part of China's current environment. Jon Phillips – Beijing – CC BY 2.0.

The fast-paced growth of the Chinese economy in the past decade has brought with it some negative consequences. The exploitation of resources and the heavy utilization of the environment have resulted in serious soil erosion and air pollution. The water table in many parts of China has decreased because of heavy demands on the nation's water supply. Arable land is being lost to erosion and inadequate land-use practices. Rural areas have not received consideration or resources equal to the coastal cities, so conditions remain poor for most rural people. Half of China's population earns the equivalent of a few dollars per day, while a fortunate few earn high salaries. Unemployment is at an elevated level for tens of millions of migrants who shift from location to location, looking for work. There is also an unfortunate degree of corruption within the government and state-run offices.

Compared to Western countries, China is an authoritarian state that does not allow labor unions, free speech, freedom of religion, or freedom of the press. There has been more openness in China's economic reforms and in travel, but other strict rules of the state remain. There is no minimum wage law for factory workers, who work long hours and do not receive benefits or sick leave. There are fewer safety requirements or government regulations for security. China is trying to have the best of both worlds: the efficiency of an authoritarian government and an efficient market-driven capitalist economy. Sustaining the largest standing army in the world, China is geared to become a global superpower. The next great world conflict could be a cultural war between the United States and China that would involve economic, political, and human issues. To work together, the United States and China would be well served to increase their understanding of each other.

Growth of Enterprise and Industry

During the 1980s, following the death of Mao Zedong, China went through a transition period. The new leader, Deng Xiaoping, realized that in order for China to compete in the world market, its economy would have to be modernized. The challenge was to open up to the outside without the outside placing pressure on the Communist system inside of China. The so-called Bamboo Curtain, which referred to the restriction of movement of goods and people across Chinese borders, diminished. To attract business and tap into the global market economy, China established SEZs along the coast at strategic port cities. SEZs attracted international corporations who wanted to manufacture goods cheaply, while China's population of 1.3 billion people provided an enormous labor pool and consumer market. China's modernization efforts paid off in the 1990s, when world trade increased and US trade with China exploded.

By the year 2000, China had profited greatly because of its expanded manufacturing capacity. The coastal cities and the SEZs had become core industrial centers, attracting enormous numbers of migrants—most of them poor agricultural workers—looking for work in the factories. Compared to other jobs, factory jobs are prized employment opportunities. Rural-to-urban shift has kicked in and China's urban growth is occurring at unprecedented rates. Chinese people are moving to the cities looking for work, just like people in many other areas of the world. The SEZs encourage multinational corporations to move their overseas operations to China and take advantage of the lower labor and production costs. China benefits from the new business opportunities and by the creation of jobs for its citizens. SEZs operate under the objectives of providing tax incentives for foreign companies, exporting market-driven manufactured products, and creating joint partnerships so that everyone benefits. In the past decade, four main cities have been designated as SEZs, along with the entire province of Hainan Island to the south. Many coastal cities were also designated as development areas for industrial expansion.

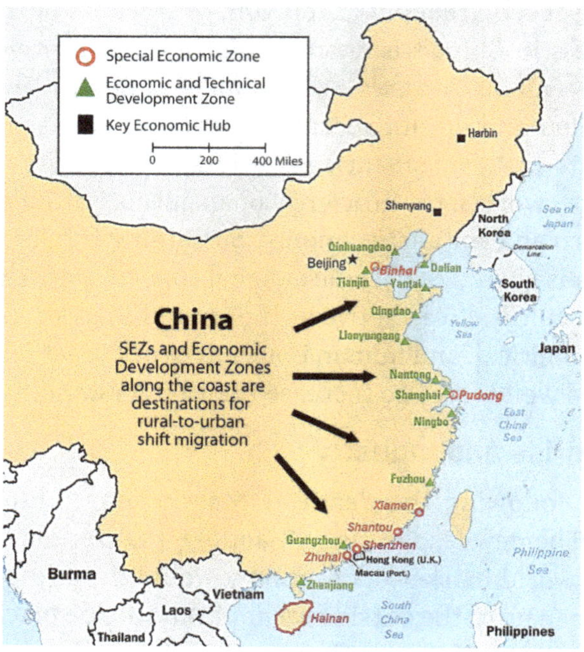

Figure 10.16: Coastal development and SEZs along the coast of China are prompting growth of the rural-to-urban shift. Updated from map courtesy of University of Texas Libraries.

In 2006, the president of China called for an expansion of the economy to focus on more innovative activities. The government allowed additional funding for more research in the areas of genetics and gene therapy, computer software development, semiconductors, and alternative energy technologies. China has a high-level energy demand to fuel its expanding economy and provide for a large population, so alternative energy development has been expanding into wind and solar power for both domestic and commercial purposes. China's industrial sector is advancing to compete in the global marketplace and muscle into the high-tech economy once dominated by core areas of Japan, Western Europe, and the United States.

Geopolitics

The political environment of China cannot be separated from its geography. To understand the larger political picture of China's position in the world, one has to understand China's geography. In terms of its land area, China is a large country. It includes an immense peripheral region surrounding China Proper consisting of minority populations and buffer states. The core of China is the Han Chinese heartland of China Proper. This is where most of the population lives and where the best agricultural land for food production is located. The surrounding periphery consists of autonomous regions and minority populations, including Tibet, Western China, Inner Mongolia, and Northwestern China. Tibet is a buffer state between China and India; high mountains provide a buffer between India and Tibet. Nepal, Bhutan, and Kashmir are also buffer states between China and South Asia. High mountain ranges extend all along China's western boundary.

Western China has a majority Muslim population closely aligned with ethnic groups in Central Asia. Inner Mongolia and the independent country of Mongolia create a buffer between China and Russia. China's actual border with Russia is remote and distant. Northeast China was historically called Manchuria and is home to the Manchus, who are not Han Chinese. Even the isolated sovereign country of North Korea acts as a buffer between China and the world superpower of the United States through its surrogate South Korea. The border with Southeast Asia is mountainous with difficult terrain and very little access. All these buffer regions insulate the Han heartland from the outside. China's powerbase is actually a territorial island insulated and isolated from the rest of the world. Technology and trade has opened links between the Han Chinese heartland and the global marketplace.

In the scope of geopolitics (the study of the geography of international relations), the core of China is protected on all fronts except one, the eastern coast. Coastal China is accessible to anyone with a ship. Coastal China is where Japan and the European colonialists such as France, Great Britain, Portugal, and Germany all came to carve up China for their imperial advancements. Also located along the coast is the island of Taiwan with the independent government of the ROC. Taiwan has been on friendly terms with the United States, which has posed a threat to the Chinese mainland. The coast makes China vulnerable. The challenge for China has been to neutralize the coastal threat and reduce its vulnerability. This is where the SEZs make a difference: the Chinese economy has relied on the SEZs to attract foreign corporations to conduct their business with the Chinese in these coastal cities. The economic world has forged ties with the Chinese along China's coastal waters. The extensive coastal development creates a situation where it is now in everyone's best interest to make the coastal region of China safe and prosperous.

In the geopolitical sense, China has historically operated under three main premises. The first premise is that China must secure a buffer zone in the periphery, which includes the regions along China's borders. Second, the country needs to continue to maintain unity within the Han Chinese majority that anchors the core region of China Proper. The third component of China's geopolitical strategy is to protect and secure its vulnerable coastal region. The country needs to find a way to marginalize outside influence and keep the heartland protected from invaders. These three geopolitical directives were designed before the postindustrial economy emerged with the information age, but the main themes of securing and protecting China remain.

China has additional political issues within the core region of China Proper. To maintain a unified Han Chinese powerbase, there should not be uneven levels of equity in the standards of living within the core region. Manufacturing has rapidly increased in the coastal cities of the SEZs and other development areas. The standard of living in the coastal cities has improved, which has created a disparity between the coastal cities and the rural regions of the interior. Factory workers can earn the equivalent of fifty cents an hour, while their counterparts working in the fields in agriculture are only making the equivalent of a few dollars per day. As a Communist country, this was one of the basic foundational principles within a socialist society—the even distribution of wealth. It kept everyone on the same playing field, at least in theory. The more open capitalist economy is changing this and challenging the equality issue.

China is also economically disproportionately reliant on its ability to maintain a high level of export manufacturing as its method of gaining wealth. The country does not have a type of geography that holds adequate agricultural resources to provide for its vast population. Manufacturing requires good political and economic relationships with the international community. In summary, China's eastern coast development is a risk that China is forced to take. The risk is a trade-off. On one hand, the eastern coast is the most geopolitically vulnerable side of China, but on the other hand, it is where the most extensive economic gains have been made in recent years. China's Bamboo Curtain has disappeared; the country is no longer isolated from the rest of the world. Economic development has created a strong degree of dependency on other countries through the doorways of China's coastal cities. China is becoming dependent on the outside world for oil, raw materials, and technology. Multinational corporations are dependent on China for low-cost manufactured goods. China is becoming more integrated within the network of the global economy.

Key Takeaways

- China has embarked on a number of serious transitions since the Communist republic was declared in 1949. Since that time, the country has shifted from a command economy to a market economy and has opened up trade to the West. This has opened up more opportunities for the Chinese people as well.
- China has a number of autonomous regions and municipalities. Many of these political units are also home to minority ethnic groups. Han Chinese make up about 90 percent of the people of China.
- China's large population has prompted the government to implement a one-child-only policy, which has resulted in a decrease in population growth but has also caused an imbalance in the percentages of boys and girls being born in many of the provinces.
- The creation of SEZs along the coast, and an open-trade policy, have rapidly developed China's manufacturing sector. This trend has attracted multinational corporations to move their operations to China to tap into the low labor and overhead costs and receive tax incentives from the Chinese government. The result has been an increase in rural-to-urban shift in China's population.
- China's historical geopolitical situation was to secure a buffer around its heartland, unify the Han majority, and protect its venerable coastal region. Economic trade and interaction with the global economy has made China dependent on other countries for its continual economic success.

Discussion and Study Questions

1. What did the Bamboo Curtain refer to? How has China changed with regard to its Bamboo Curtain?
2. What are SEZs? Where would we find them in China? Why are they located where they are located?
3. Which region of China has the highest population density? Why do they live here rather than in other areas of China?
4. What policy has China implemented to help curb population growth? What are some of the results of these policies?
5. What are the five autonomous regions based on in China? Where are they located?

6. Why did China shift from a command economy to a market economy?
7. What are the main ethnic groups, languages, and religions in China?
8. Considering current conditions, what are the major development patterns in China?
9. Under what four main principles do the SEZs operate?
10. What where the three main historical objectives in China's geopolitical situation?

Geography Exercise

Identify the following key places on a map:

- Beijing
- Chongqing
- Guangxi
- Hainan Island
- Hong Kong
- Inner Mongolia
- Macau
- Ningxia
- Tianjin
- Tibet
- Shanghai
- Western China
- Xinjiang

10.3. CHINA'S PERIPHERY

The heartland of China has consistently been included within China Proper. Adjacent to China Proper are several autonomous regions, affiliated territories, and sovereign nations. These outer regions insulate China Proper from the external world. They also furnish China with either access to commercial partnerships or access to essential raw materials for development. Every outlying governmental unit possesses distinct geographic and cultural attributes, sometimes accompanied by a minority population that is not Han Chinese.

Hong Kong and Shenzhen

Hong Kong is a former British colony and includes a number of islands on the southern coast of China. Hong Kong includes an excellent port, **Victoria**—one of the best in Asia. Victoria's deepwater port is on the interior side of Hong Kong Island and is protected from the sea, allowing ship access to an extensive sheltered harbor area. To the north of Hong Kong Island is a peninsula called **New Territories**, which the British agreed to lease from China in 1897 for a period of ninety-nine years. Bordering the New Territories on the mainland is the rapidly growing industrial city of Shenzhen.

Hong Kong was a major entry point into China for Britain during the colonial era. The colony was a British outpost that created a doorway for British expansion into

China. The port allowed for an early trade relationship to become established. The trade relationship was broken during the Korean War in the 1950s. Britain and China found themselves on opposite sides of the war. China was a newly created Communist country and sided with North Korea; Great Britain was a capitalist country and sided with South Korea. The boundary between the busy port of Hong Kong and China was closed. This left Hong Kong with a world-class port but no access to its main markets in China. Instead of closing the port and shutting down shipping, the business community of Hong Kong took advantage of its unique situation and first-rate port to create a world trade center. Ships would import raw materials, which would be processed or manufactured into finished products. The products would then be shipped back out at a profit.

Figure 10.17: The Special Administrative Region (SAR) of Hong Kong, with Shenzhen across the Border from the New Territories. Map courtesy of University of Texas Libraries .

From the theory of how countries gain wealth, recall that manufacturing was traditionally the method with the highest value-added profits. Hong Kong is a good example of this in action; that is, manufacturing was a successful method of gaining wealth for Hong Kong. First, low-level goods such as clothing and textiles were produced with Hong Kong's cheap labor and low costs. As business increased, higher-level technical goods were produced, such as radios and other electronic products. Incomes increased. Soon, Hong Kong was a business and banking center trading with all global markets. International trade attracted a strong financial network that enhanced the free market system in Hong Kong. Business in Hong Kong attracted attention in Beijing. Chinese leader Deng Xiaoping, who had begun China's open door policy in the 1980s, viewed Hong Kong as a major access point to establish trade and commerce with. In the 1990s, the special economic zone of **Shenzhen** became an important development area for Chinese industries and multinational corporations. China once again used Hong Kong as a major trade corridor so that about one-fourth of all the country's imports and exports were being shipped through the port of Hong Kong.

During the 1990s, as China actively invested in its industrial sector, it sought ways to attract the business generated by Taiwan and Hong Kong, two economic tigers. Opportunity came knocking in 1997 when the ninety-nine-year lease of the New Territories to Britain expired. China did not want to renew the lease. A deal covering all of Hong Kong was struck between China and Great Britain. Britain would relinquish all its claims to Victoria, the port, and all of Hong Kong if China would allow the area to remain non-Communist and under its own autonomy for fifty years. China agreed, and Britain left Hong Kong in 1997. Hong Kong became associated with mainland China as a special autonomous region, but remained capitalist and democratic in its operations. This opened the door for Taiwan and other trading partners to increase trade with China through Hong Kong. Shenzhen, a special economic zone (SEZ) across the border from Hong Kong in China, was ready to capitalize on its accessibility to the port and the enormous trade that had been established by Hong Kong. Shenzhen became one of the fastest growing cities in the world and has become a center of manufacturing and trade for the global economy. It grew from a moderately sized city of about three hundred fifty thousand in the early 1980s to a city of ten and half million by 2010, and is still growing. One of Shenzhen's largest trading partners is the Walmart Corporation, as factories in the city produce goods for export to the United States and sale in Walmarts around the country. Shenzhen has established a port of its own and is a magnet for international trade.

More than 95 percent of the seven million people living in Hong Kong are ethnically Chinese. The people have strong ties to mainland China but highly value their separate and independent economic and political status. Hong Kong is a major financial and banking center for Asia and has been working with the Chinese government to provide private banking services for Chinese citizens. The small land size of Hong Kong makes it a high-priced real estate destination. The cost of living is high, and space is at a premium and expensive. Nevertheless, Hong Kong attracts millions of visitors per year and has established itself as a tourism hub for people desiring to visit southern China. Tens of millions of tourists each year use Hong Kong as a base or stopover point to enter China's southern provinces. Hong Kong offers visitors immense shopping possibilities in a safe and modern environment that is attractive to people from all over the world. Cantonese is the official language, but English is widely spoken in Hong Kong because of Britain's influence and because of world trade relationships.

Figure 10.18: Victoria Harbor in Hong Kong, with its deepwater port and solid bedrock island, allows oceangoing vessels to dock right next to high-rise business and office complexes. Shashank Khanna – Victoria Harbor, Hong Kong – CC BY-NC 2.0.

Across the Pearl River Estuary to the west is the former Portuguese colony of Macau. The arrangement between China and the British government over Hong Kong provided a pattern that was also applied to Macau. At the end of 1999, Portugal relinquished its claim to Macau and the colony was turned over to the Chinese government under an agreement similar to the agreement between Britain and China regarding Hong Kong. Macau was enabled to retain its autonomy and free-market economy for fifty years as it becomes a special administrative region (SAR) of China. Macau is a much smaller territory than Hong Kong, and home to only about half a million residents.

Taiwan (ROC)

Taiwan is separated from the Chinese mainland by the Taiwan Strait. In 2010, the population of the island, which is about the size of the US state of Maryland, was twenty-two million. The island is mountainous and has rugged national parks in its interior, while most of the people in Taiwan live in bustling cities along the coast. Taiwan's push for good education and technical skills for its citizens has paid off in the form of high incomes and an industrialized economy. After World War II, when China raised its Bamboo Curtain, the United States cut its trade relations with the Communist mainland. However, the United States did trade a great deal with the Nationalists in Taiwan. The deep tensions that exist between China and Taiwan are easy to understand given the history that led to Taiwan's current existence. One of the flash points in China-Taiwan relations has been the "One China" policy. Originally asserted by Beijing, the principle contends that there is only one indivisible China, which encompasses all of China, including Taiwan, Hong Kong, Macau, and Tibet.

During the Cold War, the sizeable population of mainland China was attractive to US corporations that sought trade with China. In 1972, for the benefit of the United States and to counter the influence of the Soviet Union, US president Richard Nixon organized a visit to Beijing that opened the door to US-China diplomatic relations. There were conditions placed on the open door agreement that followed. One condition that mainland China imposed was for the US government to recognize the Communist government in Beijing as the official China. To do this, the United States had to move their US embassy from **Taipei**, the capital city of Taiwan, to Beijing and recognize only the Beijing government as the official China. The United States agreed to this policy. As a result, no government official in the US State Department, cabinet level or higher, can officially visit Taiwan; at the same time, no equal-level diplomat from Taiwan can officially visit the United States. This policy placed Taiwan in an ambiguous position: Is Taiwan an independent country or is it a part of mainland China?

Mainland China has always demanded that in order for countries to have good relations with the People's Republic of China (PRC), they must not recognize Taiwan (Republic of China, or ROC). This has caused intermittent problems for

Figure 10.19: Taiwan (ROC) is located just across the Taiwan Strait from Mainland China (PRC). CIA World Factbook – public domain.

the United States and other countries. The termination of diplomatic relations between the United States and the ROC was a huge blow for Taiwan, even though the United States continued to conduct trade with the island. The United States has always supported Taiwan's economic independence. In 2007, the United States sold military arms to Taiwan. As a result, China began to deny routine access to its ports for US ships. Taiwan's status on the world stage has fluctuated. For about twenty years after the Nationalists set up a government in Taiwan, the ROC held a seat in the United Nations (UN), and the UN recognized Taiwan's claim as the legitimate government of China. In 1971, the UN changed its position and gave the seat to Beijing. The UN indicated it could not give a seat to Taiwan because Taiwan is not recognized as a legal nation.

In 2000, Taiwan renewed its efforts to be internationally recognized as a sovereign nation and announced tentative plans to formally secede from China. China responded with many threats and conducted military exercises across the strait from Taiwan. The military exercises were seen as another threat, essentially a show of military might, meant to intimidate Taiwan into retreating from its secessionist position. The United States supported Taiwan by parking a naval fleet nearby, which incensed the mainland government. In 2005, China passed a law authorizing the use of military force to stop any secessionist actions by Taiwan.

Taiwan's economy continues to use its skilled labor base. In years past, Taiwan's factories produced textiles and light clothing. As incomes and skills advanced, production shifted to electronics and high-end goods, which it has since been exporting around the world. Taiwan has achieved a high standard of living for its people. When trade opened up through Hong Kong with mainland China, many Taiwanese industries shifted production to the SEZs along China's coast. Labor in SEZs is less expensive, leading to a corresponding rise in profits. Unfortunately, this has resulted in higher unemployment levels in Taiwan.

Figure 10.20: Street in Taiwan, ROC, Illustrating Modern Businesses and Use of Scooters for Transportation. J Aaron Farr – Taiwan Streets – CC BY 2.0.

In the past few years, Taiwan's focus has turned from secession to trade. In 2008, Taiwan elected a government that, though nationalist in its ideology, is committed to strengthening trade and investment with China. Taiwan entered the World Trade Organization (WTO) in 2001 using the name Chinese Taipei. That move started a cascade of actions that improved "cross-strait" trading conditions, and within just a few years, direct trade between China and Taiwan has become extensive. The question of Taiwan's political status is as yet unresolved and is clearly still volatile, but trade and investment between China and Taiwan are robust and growing. Taiwan has been able to independently compete in world-class sporting events, including the Olympics, under the name Chinese Taipei.

Autonomous Region of Tibet

Located in mountainous southwestern China, Tibet is classified as one of China's autonomous regions, a disputed political arrangement. It is debatable whether Tibet or any of the other Chinese regions are actually autonomous. The legal structures actually allow for very little self-governance, and most new initiatives require approval from the Chinese central government. Tibet had been an independent entity through much of its history and governed itself as a Buddhist theocracy. Its theocratic political system established the Dalai Lama as both the head of state and the religious leader of the Tibetan people.

Tibet has had a complicated history with China. This is as true of its early history as it is today. In its early history, Tibet was an independent kingdom with its unique type of Buddhism as its state religion. It was during this era that a system of Lamaism, a hierarchy of monks or other religious leaders, took hold. China's influence in Tibet waxed and waned during the later years of the kingdom, in the tenth century. The Mongols subsumed Tibet into their empire during the first part of their conquests in the thirteenth century. However, in the fifteenth century, the Mongols gave considerable local authority to the Dalai Lama, making him the spiritual leader as well a powerful political figure.

Tibet came under the control of China during the Qing Dynasty. When imperialist rule ended in China in 1911, Tibet began to once again assert its independence. Chinese, British, and Tibetan officials met and came up with an agreement that partitioned Tibet into Inner Tibet, which would be controlled by China, and Outer Tibet, which would be independent. China later indicated its intention to control all of Tibet, a move greatly resented by the Tibetan people.

When Communists took control of China in 1949, the Dalai Lama was originally allowed control over domestic affairs while China would control all other governmental functions. The Chinese government then took steps to greatly reduce the powers of the Dalai Lama and the **Panchen Lama** (spiritual leader). With the intention of spreading Communism to Tibet, China destroyed monasteries, religious symbols and icons, and other long-practiced ways of life were threatened. The Tibetans considered the Chinese political dominance a cruel and invading force. Monasteries were burned, monks were beaten or imprisoned, and Buddhism was brutally suppressed.

Figure 10.21: Potala Palace in Lhasa, Tibet, rests at 12,100 feet above sea level. This is the former seat of government of the Dalai Lama in Tibet and is a UNESCO World Heritage Site. Wikimedia Commons – CC BY-SA 3.0.

Tibetans revolted in 1956. Backed by the United States, the revolt continued even as China indicated it would suspend the transformation of Tibet. China brutally crushed the revolt in 1959, leaving tens of thousands of Tibetans dead or imprisoned. The Dalai Lama and thousands of Tibetans fled to Dharamsala, India, where they established a government in exile. At this point, China had a firm grip on Tibet, and in 1965 the reorganization of Tibet into a Chinese socialist region began in full force.

During the 1980s and early 1990s, China expressed some willingness to relent in its authoritarian control of Tibet. However, they demanded that the Dalai Lama renounce claims of Tibetan independence, which he refused. Demonstrations and continued violence occurred throughout this period. Chinese-Tibetan relations were further damaged when the Panchen Lama died mysteriously shortly after criticizing socialist reform in Tibet. The Panchen Lama is the next-highest Buddhist official under the Dalai Lama. There is now a controversy with the Chinese government over the selection of the eleventh Panchen Lama. China and Tibet both claimed the authority over choosing the Panchen Lama and each named a different person to take on the role. The situation illustrates the conflict over the minds and the people of Tibet.

Dalai Lama

The Dalai Lama serves as the spiritual and political leader of all Tibetans. The Dalai Lama is regarded as the present incarnation of former Buddhist Masters, known as Tulkus. The Masters have evolved through the cycle of death and rebirth, selecting their own moments of resurgence to assist in educating humanity on the art of living. The Dalai Lama has consistently been regarded as the head of state for the Tibetan government, with its capital situated in Lhasa. Tenzin Gyatso is the reincarnation of the thirteenth Dalai Lama. On November 17, 1950, at the age of fifteen, he ascended as the fourteenth Dalai Lama of Tibet. This transpired merely one month subsequent to China's incursion into Tibet. The Dalai Lama initially endorsed the 1951 Seventeen Point Agreement under military duress from China. In 1959, he departed Tibet and entered exile in India. China has suppressed all efforts of Tibetan resistance. During his time in India, the Dalai Lama has founded the Tibetan government in exile and is striving to safeguard Tibetan culture among the multitude of exiles who have accompanied him.

Figure 10.22 The Fourteenth Dalai Lama and Tibetan Leader in Exile Wonderlane – His Holiness the Great 14th Dalai Lama – CC BY 2.0.

In 2008, a peaceful protest led by Buddhist monks turned violent. Chinese authorities responded with a harsh crackdown. The world's attention was focused on China at the time because the Olympic Games were under way in Beijing. China has often been a target for protests because of its human rights violations against Tibetans and other minority groups. Little attention was drawn to Tibet during the 2008 Olympic Games. China wanted to showcase its advancements in industrial development and culture and did not want the world to focus on the issues with Tibet or any other part of China.

The population of Tibet is only about three million. They mainly live in the mountain valleys, below seven thousand feet. Lhasa is Tibet's main urban center as well as its capital. In the past couple of decades, China has softened its treatment of Tibetans and restored some of the monasteries. China has also moved thousands of ethnically Chinese people into Tibet to shift the ethnic balance of the population. A major rail line now directly connects Beijing with Lhasa, Tibet. Lhasa will soon have a majority Chinese population with cultural similarities that reflect more the attributes of China Proper than Tibet. The Tibetans who live there are being compromised simply by population dilution.

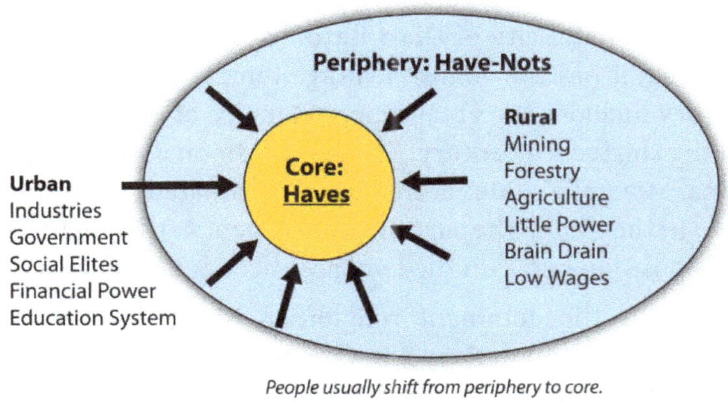

Figure 10.23: The core-peripheral spatial relationship can be appropriately applied to Tibet and its relationship with China. Tibet is a peripheral region with raw materials that provide benefits to the core areas of China, which require the raw materials for industrial development.

As indicated previously, China wants to maintain the mountainous, rural region—with little inhabitable land area and very few people—as a buffer state between India and the Han Chinese heartland of China Proper. Tibet is also becoming more important to the government of Beijing and the business people in Shenzhen and Shanghai. To understand why this is, consider the core-peripheral spatial relationship. What do the core and the periphery usually have to offer each other? Consider also how countries gain wealth. Tibet can play an important role in the larger picture of the Chinese economy. The mountains of Tibet potentially hold vast amounts of natural resources that could support the growing industrial economy in China, which is based on manufacturing goods for the global export markets. China will continue to see to it that Tibet becomes economically integrated with the rest of China in spite of the cultural differences that may exist.

Mongolia

Mongolia ranks as the world's nineteenth-largest country in terms of square miles. Mongolia shares a similar geography with much of Kazakhstan, which is the world's largest landlocked nation; Mongolia is second-largest. Despite Mongolia's large land area (slightly smaller than the US state of Alaska or the country of Iran), its population is only about three million, and the country is the least densely populated country in the world. Mountains, high plains, and grass-covered steppe cover much of Mongolia, a country that receives only between four and ten inches of rain per year, precipitation that usually comes in the form of snow. The **Gobi Desert** to the south, extending from southern Mongolia into northern China, receives even less precipitation. **Inner Mongolia** is a sparsely inhabited autonomous region south of Mongolia that is governed by the Chinese government.

Mongolia's modern capital city of **Ulan Bator** is home to about one-third of the people of Mongolia; it has the coldest average temperature of any world capital. Mongolia's history includes the great Mongol Empire of Genghis Khan, which was established in the thirteenth century. The Soviet Union used Mongolia as a buffer state with China. Mongolia's Communist party dominated politics until the Soviet Union collapsed in the 1990s. The current government in Ulan Bator has to contend with Mongolia's position between the two large economies of Russia and China.

Tibetan Buddhism is the dominant religion in Mongolia and is practiced by about 50 percent of the population. The Communist influence is evident in that approximately 40 percent of the population considers itself nonreligious. Most of the people are of Mongol ethnicity. Today, about 30 percent of the people are still seminomadic and migrate seasonally to accommodate good grazing for their livestock. A round yurt, which can be constructed at whatever location is selected for the season and disassembled for mobility, is the traditional dwelling. Mongolian culture and heritage revolve around a rural agrarian culture with an extensive reliance on the horse. Archery, wrestling, and equestrian events are some of the most popular sporting activities.

Figure 10.24: Men Wrestling on the Rugged Grasslands of Mongolia Wrestling is a major sport of the country. Wikimedia Commons - public domain.

Mongolia's economy has traditionally been centered on agriculture, but mining has grown in recent years to be a major economic sector. Mongolia has rich mineral resources of coal, molybdenum, copper, gold, tin, and tungsten. Being landlocked cuts Mongolia off from the global economy, but the large amounts of mineral reserves are in demand by core industrial areas for manufacturing and should boost the poor economic conditions that dominate Mongolia's economy. China has increased its business presence in Mongolia and has been drawing Mongolia's attention away from its former Soviet ally to become a major trading partner.

Key Takeaways

- Hong Kong was a British colony with an excellent port that provided access to China during the colonial era. Hong Kong became an economic tiger and regained its importance for trade with the mainland during the open door policy of China in the 1980s. The colony was returned to China in 1997 under an autonomous arrangement.
- Taiwan created an independent government after the three-way split in China. The separation from the mainland government has created strained relations between Beijing and Taipei. In recent years, trade and economic exchanges have brought the two Chinas together and they have become more dependent on each other.
- Tibet established itself as a Buddhist theocracy before the 1949 declaration of the Communist PRC. The Dalai Lama ruled as both political and religious leader of Tibet until China took control. The Dalai Lama fled to India, where he set up a government in exile to continue to promote Tibetan culture and support Tibet.
- Mongolia is an independent landlocked country with a small population and a large land area. It receives only modest amounts of rainfall. The country was a historic buffer state between the Communist countries of the Soviet Union and China. Mongolia is working to capitalize on its natural resources to engage the global economic situation.

Discussion and Study Questions

1. What makes the port of Hong Kong significant in both physical characteristics and relative location?
2. Why was Hong Kong important to China? How did Hong Kong become an economic tiger?
3. Where is Shenzhen? Why did this city emerge as one of the fastest growing cities in the world?
4. Why does Taiwan have a different government from mainland China? What type of government does it have?

5. Why is it that a US president or any cabinet-level US official cannot visit Taiwan but can visit Communist Beijing?
6. How did Taiwan develop a high standard of living for its people whereas mainland China lagged behind?
7. Who is the Dalai Lama? Where did the Dalai Lama rule? Where is the Dalai Lama today?
8. What is Tibet's relationship with China? Why does China want or need Tibet?
9. How was Mongolia viewed by its neighbors during the Cold War? How does China view Mongolia today?
10. What is Mongolia's physical geography like? What has been the traditional lifestyle in Mongolia?

Geography Exercise

Identify the following key places on a map:

- Gobi Desert
- Hong Kong
- Inner Mongolia
- Lhasa
- Macau
- Mongolia
- New Territories
- Shenzhen
- Taipei
- Taiwan
- Tibet
- Ulan Bator
- Victoria

10.4. JAPAN AND KOREA (NORTH AND SOUTH)

Japan

Japan comprises islands situated in the Pacific Rim, east of China and across the Sea of Japan from the Korean Peninsula. The majority of the archipelago, comprising over three thousand islands, is situated approximately north of 30° latitude. The country comprises four primary islands: Shikoku, Kyushu, Hokkaido, and Honshu. The islands include elevated mountains formed by volcanic activity. Numerous volcanoes are active, including the renowned Mount Fuji. The total land area of all Japan's islands is approximately comparable to that of the US state of Montana. The mountainous islands of Japan span two climatic zones. Except for the northern region, all of Japan experiences a type C climate. Hokkaido, an island in northern Japan, possesses a type D climate and receives sufficient snowfall for downhill skiing. Japan was the host of the Winter Olympics in 1998. Mount Fuji, situated to the west of Tokyo, is extensively photographed because to its symmetrical volcanic cone.

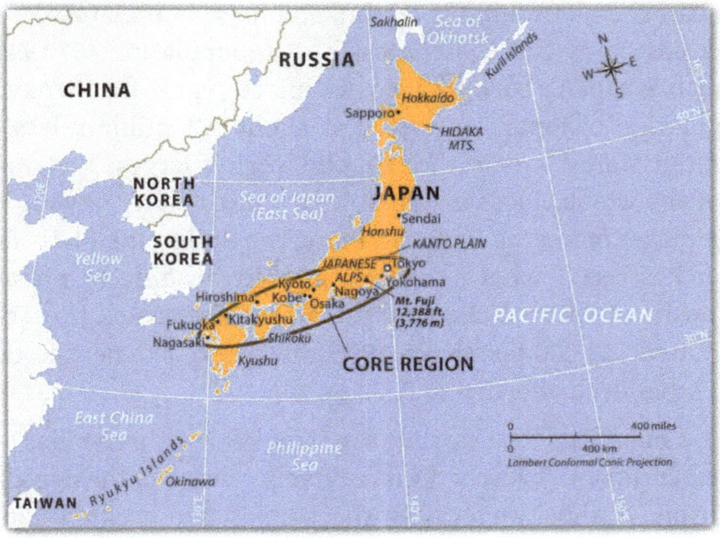

Figure 10.25: Japan's Four Main Islands and Its Core Urban and Industrial Areas

Cities like Mexico City, São Paulo, and Bombay, which have vast slum communities, each claims to be the world's largest city but lacks firm census data to verify its population. The largest metropolitan urban area in the world that can be verified is Tokyo, with a population of 26.7 million. The Tokyo metropolitan area is located in an extensive agricultural region called the **Kanto Plain** and includes the

Figure 10.26: Expanse of the Metropolitan Area of Tokyo, with Mt. Fuji in the Background. Michelle – Tokyo Tower view & Mt. Fuji – CC BY 2.0.

conurbation of Tokyo, **Yokohama,** and **Kawasaki**. The second-largest urban area in Japan is located in the **Kansai District**, and includes the cities of **Kobe, Osaka,** and **Kyoto**. Japan is a mountainous region, and most of its large cities are located in low-lying areas along the coast. Most of the population (67 percent) lives in urban areas such as Japan's core area, an urbanized region from Tokyo to Nagasaki. The 2010 population of Japan was listed at about 126 million, less than half the size of the United States. It is ironic that the world's largest metropolitan area is built on one of the worst places imaginable to build a city. Tokyo is located where three tectonic plates meet: the Eurasian Plate, the Philippine Plate, and the Pacific Plate. Earthquakes result when these plates shift, leading to possibly extensive damage and destruction. In 1923, a large earthquake struck the Tokyo area and killed an estimated 143,000 people. In 1995, an earthquake near Kobe killed about 5,500 people and injured another 26,000.

In March 2011, a magnitude 9.0 earthquake struck forty-three miles off the eastern coast of northern Japan. The earthquake itself caused extensive damage to the island of Honshu. A shockwave after the earthquake created a tsunami more than 130 feet high that crashed into the eastern coast of Japan causing enormous damage to infrastructure and loss of life. Hundreds of aftershocks were recorded; at least three registered over 7.0 in magnitude. This is the strongest earthquake to ever hit Japan in recorded history. It resulted in more than 15,500 deaths and wreaked serious damage across Japan in the value of billions of dollars. Thousands of additional people remain missing. Nuclear power plants along the coast were hit hard by the tsunami, which knocked out their cooling systems and resulted in the meltdown of at least three reactors. The nuclear meltdowns created explosions that released a sizeable quantity of nuclear material into the atmosphere. This is considered the worst nuclear accident since the 1986 reactor meltdown at the Soviet plant at Chernobyl, located north of Kiev in present-day Ukraine. The next big earthquake could happen at any time, since Japan is located in an active tectonic plate zone.

Development

The country of Japan is an interesting study in isolation geography and economic development. For centuries, shogun lords and samurai warriors ruled Japan, and Japan's society was highly organized and structured. Urban centers were well planned, and skilled artisans developed methods of creating high-grade metal products. While agriculture was always important, Japan always existed as a semiurban community because the mountainous terrain forced most of the population to live along the country's coasts. Without a large rural population to begin with, Japan never really experienced the rural-to-urban shift common in the rest of the world. Coastal fishing, always a prominent economic activity, remains so today. The capital city of Tokyo was formerly called **Edo** and was a major city even before the colonial era. Japan developed its own unique type of urbanized cultural heritage.

Encounters with European colonial ships prompted Japan to industrialize. For the most part, the Japanese kept the Europeans out and only traded with select ships that were allowed to approach the shores. The fact that European ships were there at all was enough to convince the Japanese to evaluate their position in the world. During the colonial era, Britain was the most avid colonizer with the largest and best-equipped navy on the high seas. Britain colonized parts of the Americas, the Middle East, Africa, South Asia, and Australia and was advancing in East Asia when they encountered the Japanese. Both Japan and Great Britain are island countries. Japan reasoned that if Great Britain could become so powerful, they should also have the potential to become powerful. Japan encompassed a number of islands and had more land area than Britain, but Japan did not have the coal and iron ore reserves that Britain had.

Around 1868, a group of reformers worked to bring about a change in direction for Japan. Named after the emperor, the movement was called the Meiji Restoration (the return of enlightened rule). Japanese modernizers studied the British pattern of development. The Japanese reformers were advised by the British about how to organize their industries and how to lay out transportation and delivery systems. Today, the British influence in Japan is easily seen in that both Japanese and British drivers drive on the left side of the road. The modernizers realized that to compete on the high seas in the world community they had to move beyond samurai swords and wooden ships. Iron ore and coal would have to become the goods to fuel their own industrial revolution. Labor and resources were valuable elements of early industrialization in all areas of the world, including Asia.

Figure 10.27: Japan was a major colonial power in the Asian Pacific Rim. Japan colonized parts of Russia, China, Taiwan, Philippines, Southeast Asia, and Korea.

Japan began to industrialize and build its economic and military power by first utilizing the few resources found in Japan. Since it was already semiurban and had an organized social order with skilled artisan traditions, the road to industrialization moved quickly. Japan needed raw materials and expanded to take over the island of Formosa (Taiwan) and the Korean Peninsula in the late nineteenth and early twentieth centuries. Japan was on its way to becoming a colonial power in its own right. They expanded to take control of the southern part of **Sakhalin Island** from Russia and part of Manchuria (Northeast China) from the Chinese. Japanese industries grew quickly as they put the local people they subjugated to work and extracted the raw materials they needed from their newly taken colonies.

Figure 10.28: This photo, taken after the atomic bomb blast of World War II in 1945, illustrates all that was left standing of a Shinto shrine in Nagasaki. Pingnews.com – Public Domain: Shinto Shrine in Atomic Ruins, Nagasaki – CC BY-SA 2.0.

The three-way split in China revealed that the Japanese colonizers were a major force in China even after the other European powers had halted their colonial activities. Japan's relative location as an independent island country provided both quick access to their neighbors and also protection from them. By World War II, Japan's economic and military power had expanded until they were dominating Asia's Pacific Rim community. The Japanese military believed that they could invade the western coast of North America and eventually take control of the entire United States. Their attack on Pearl Harbor, Hawaii, in 1941, was meant to be only a beginning. History has recorded the outcome. The United States rallied its people and resources to fight against the Japanese in World War II. The Soviet Union also turned to fight the Japanese empire. The end came after atomic bombs, one each, were dropped by the United States on the Japanese cities of **Hiroshima** and **Nagasaki**. The terms of Japan's surrender in 1945 stipulated that Japan had to give up claims on the Russian islands, Korea, Taiwan, China, and all the other places that they had previously controlled. Japan also lost the **Kurile Islands**—off its northern shores—to the Soviet Union. The islands have never been returned. Japan offered Russia an enormous amount of cash for them but the matter remains unresolved. The Kurile Island chain continues to be controlled by Russia. Japan also agreed not to have a military for offensive purposes. Japan was decimated during World War II, its infrastructure and economy destroyed.

Economic Growth

Since 1945, Japan has risen to become Asia's economic superpower and the economic center of one of the three core areas of the world. Japanese manufacturing has set a standard for global production. For example, think of all the automobiles that are Japanese products: Toyota, Honda, Mitsubishi, Subaru, Nissan, Isuzu, Mazda, and Suzuki. How many Russian, Brazilian, Chinese, or Indonesian autos are sold in the United States? The term "economic tiger" is used in Asia to indicate an entity with an aggressive and fast-growing economy. Japan has surpassed this benchmark and is called the economic dragon of Asia. The four economic tigers competing with Japan are Singapore, Hong Kong, Taiwan, and South Korea. Japan came back from total devastation in 1945 to become a world economic superpower.

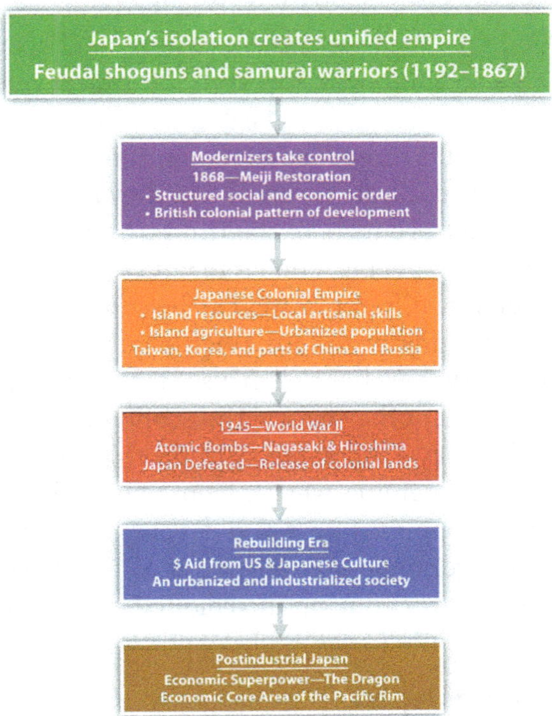

Figure 10.29: Timeline of Japanese Development

Japan's recovery in the last half of the century was remarkable. If an island nation like Japan could accomplish this rapid growth in its economy, why couldn't other similar-sized countries accomplish the same level of growth? What was it that caused the Japanese people to not only recover so quickly but rise above the world's standard to excel in their economic endeavors? The answer could be related to the fact that Japan already had an industrialized and urbanized society before the war. The United States did help rebuild some of Japan's infrastructure that had been destroyed during World War II—things like ports and transportation systems to help bring aid and provide for humanitarian support. However, the Japanese people were able to not only recover from the devastation of World War II but rise to the level of an economic superpower to compete with the United States. Japan used internal organization and strong centripetal dynamics to create a highly functional and cohesive society that focused its drive and fortitude on creating a manufacturing sector that catapulted the country's economy from devastation to financial success.

The same dynamics can be applied to Germany after World War II. Germany was completely destroyed by the war. The Allied powers decimated Germany's infrastructure and resources. Today, Germany is the strongest economy in all of Europe. Part of the reason Germany came back to become so economically successful was its industrialized and urbanized society. Germany also was able to access resources. Again, think of all the automobiles that are German products: Mercedes, Volkswagen, Porsche, BMW, and Audi. The pattern of economic growth was similar for Germany and Japan. The loss of colonies after World War II prompted Japan to look elsewhere for its raw materials. Extensive networks of trade were developed to provide the necessary materials for the rapidly growing manufacturing sector. The urbanized region around Tokyo, including the cities of Yokohama and Kawasaki, swelled to accommodate the growth of industry. Japan's coastal interior has a number of large cities and urban areas that rapidly increased with the growing manufacturing sector.

The basis for Japan's technological and economic development may be related to its geographic location and cultural development. Japan's pattern of economic development has similarities to the patterns of other Asian entities that have created thriving economic conditions. Hong Kong and Taiwan are both small islands with few natural resources, yet they have become economic tigers. Taiwan exports computer products to the United States, but Russia, with all its natural resources, millions of people, and large land area, does not. Explaining how or why countries develop at different rates can be complex, because there may be many interrelated explanations or reasons. One element might be centripetal cultural forces that hold a country together and drive it forward. The ability to manage labor and resources would be another part of the economic situation. The theory of how countries gain wealth may shed some light on this issue. Japan used manufacturing as a major means of gaining wealth from value-added profits. This is the same method the Asian economic tigers also developed their economies. Japan developed itself into a core economic country that took advantage of the peripheral countries for labor and resources during the colonial era. Japan took advantage of every opportunity that presented itself to become a world manufacturing center.

Modern Japan

Japan is a homogeneous society. In 2010, all but 1 percent of the population was ethnically Japanese. Japan resembles a nation-state, where people of a common heritage and aspirations hold to a unified government. This provides a strong centripetal force that unites the people under one culture and one language. However, religious allegiances in Japan vary, and there are a good number of people who indicate nonreligious ideologies. Shintoism and Buddhism are the main religious traditions. Shintoism includes a veneration of ancestors and the divine forces of nature. There is no one single written text for Shintoism; the religion is a loosely knit set of concepts based on morality, attitude, sensibility, and right practice. It is possible for a Shinto priest to conduct a wedding for a couple, whereas the funeral (hopefully taking place much later) for either of the spouses might be conducted by a Buddhist priest. The Buddhism that is practiced in Japan is more meditative in nature than mystical.

Modernization has brought new opportunities and problems for the Japanese people. With a highly urban and industrial society, personal incomes in Japan are high and family size is quite low, at about 1.2 children. The replacement level for a given population is about 2.1 children, which means Japan has a declining population. The population is aging and there are not enough young people to take entry-level jobs. The family has been at the center of Japanese tradition, and the elderly have been venerated and honored. As the country has become more economically developed, higher incomes for young people have prompted a shift toward convenience and consumer amenities. Consumer goods are available, but all the oil and about 60 percent of the food products have to be imported. During the off season, fruit, for example, is expensive. Beef is at a premium price, so fish is a staple source of protein. Income levels are high, but the cost of living in places like Tokyo is also quite high.

Japan is facing a labor shortage for many low-level service jobs that had traditionally been held by young people entering the workforce. Japan is an island country that has prided itself in being 99 percent Japanese. Countries in stage 5 of the index of economic development usually enter a negative population growth pattern. According to population statistics, Japan is at the start of a serious population decline, with a negative population growth pattern. While Western Europe and the United States also have a declining percentage of young people, in those countries the declining youth cohort has been replaced by an elevated level of immigration, both legal and illegal. Culturally, this is not an attractive option for Japan, but they are forced to address the labor shortage in any way they can. Their economy will depend on the ability to acquire cheap labor and resources to compete in the global marketplace.

Centripetal Forces Hold a State Together

- Ethnic Unity and Tolerance
- Social/Economic Equity

- Just and Fair Legal System
- Charismatic Leadership
- Religious Acceptance
- Nationalism
- Common Heritage
- Common Language
- Any Other Unifying Force

As the economic dragon of Asia, Japan has had an enormous economic impact on the East Asian Community (EAC). The shortage of cheap labor has forced Japan to look elsewhere for new manufacturing ventures. China is an attractive country in this respect because of its substantial population and low standard of living for many of its citizens. Japan is looking to take advantage of its geographical location to establish a more positive trade situation with its Asian neighbors. The dynamics of its economy are uniquely similar to that of the United States and Europe. All three core economic areas are competing for inexpensive labor and resources to fuel their economies. China has been a major destination for manufacturing production because of its low labor costs and attractive financial incentives. If you recall, during World War II China and Japan were locked in a bitter war that caused harsh animosities to take root against each other. The two countries have not historically been friendly toward each other. Japan's history as a colonizer has caused serious cultural attitudes against it from its Asian neighbors. Nevertheless, economics usually drive politics, and Japan wields enormous economic power. Japan now engages in trade with China and continues to be a core economic country in the global economy.

Korea

The Korean Peninsula juts out into the Pacific Rim from northwestern Asia. The peninsula is bound by the Sea of Japan (the East Sea) and the **Yellow Sea.** North and South Korea share the peninsula. These countries have been separated by the Korean **demilitarized zone (DMZ)** since 1953. To the west and north is China and to the far north along the coast is Russia. Korea is separated from China by the Yellow Sea and the Yalu and Tumen Rivers to the north. The Yalu and Tumen rivers form the actual border between North Korea and China. Japan is located just east of the Korean Peninsula across the **Korea Strait**. The Korean Peninsula is now split between South Korea and North Korea. The capital city of North Korea is **Pyongyang**, and **Seoul** is the capital of South Korea.

The topography of the Korean Peninsula is mountainous with arable or cultivable land in high demand. Approximately 70 percent of the Korean Peninsula is mountainous. The bedrock is predominantly composed of volcanic and granitic rocks that have been severely modified by glacial processes over the past twenty-five thousand years. The highest peak in North Korea rises more than nine thousand feet. The Korean Peninsula can be thought of as four general areas:

1. Western Region with an extensive coastal plain, river basins, and small foothills
2. Eastern Region with high mountain ranges and a narrow coastal plain
3. Southeastern Basin
4. Southwestern Region of mountains and valleys

Off the southern and western coasts of the Korean Peninsula are about three thousand small and mostly uninhabited islands, all within the territory of South Korea. South Korea's largest island has an area of 712 square miles. It is the highest point of land in South Korea, at 6,398 feet above sea level.

North and South Korea have very different, yet related, environmental issues, primarily related to the degree of industrialization. With a low level of industrialism, North Korea has severe issues of water pollution as well as deforestation and related soil erosion and degradation. Other issues in North Korea include inadequate supplies of clean drinking water and many waterborne diseases. South Korea has water pollution associated with sewage discharge and industrial effluent from its many industrial centers. Air pollution is severe at times in the larger cities, which also contributes to higher levels of acid rain.

Figure 10.31: The Korean Peninsula between the North and the South The thirty-eighth parallel cuts across the peninsula in the area of the DMZ farmland. The thirty-eighth parallel on or near the DMZ is the *Cease-Fire Line* between North and South Korea (*right*). The left side of the picture is South Korea, and the right side of the picture is North Korea. Wikimedia Commons – public domain.

The physical size in square miles of North Korea or South Korea is similar to the physical area of the US state of Kentucky. North Korea is slightly larger, and South Korea is slightly smaller. For centuries, Korea was a unified kingdom that was often invaded by outsiders. After the fall of the Chinese Qing (Manchu) Dynasty, the Japanese took control of the Korean Peninsula (1910) and controlled it as a colony. Japanese colonial rule ended in 1945 when the United States defeated Japan, forcing it to give up its colonial possessions. The structure of modern Korea is a result of the ending of World War II.

The conclusion of World War II was a critical period for the Korean Peninsula. The United States and the Soviet Union both fought against the Japanese in Korea. When the war was over, the Soviet Union took administrative control of the peninsula north of the thirty-eighth parallel, and the United States established administrative control over the area south of the thirty-eighth parallel. The United States and the Soviet Union divided Korea approximately in half and eventually established governments in their respective regions that were sympathetic to each nation's own ideology. The Soviet Union administered the northern portion; the United States administered the southern region. Politics deeply affected each of the regions. Communism dominated North Korea and capitalism dominated South Korea. In 1950, with aid from China and the Soviet Union, the Communists from the north invaded the south. After bitter fighting, a peace agreement was reached in 1953 to officially divide the Korean Peninsula near the thirty-eighth parallel. Korea remains divided to this day. The United States has thousands of soldiers stationed along the Cease-Fire Line (or DMZ), which is the most heavily guarded border in the world.

North Korea

The government of North Korea (Democratic People's Republic of Korea, or DPRK) was formed with the leadership of Communist dictator **Kim Il Sung**, who shaped his country with a mix of Soviet and Chinese authoritarian rule. Having few personal freedoms, the people worked hard to rebuild a state. Using the threat of a US military invasion as a means of rallying his people, Kim Il Sung built up a military of more than one million soldiers, one of the largest in the world. People could not travel in or out of the country without strict regulation. North Korea existed without much notice until the 1990s, when things suddenly changed.

When the Soviet Union fell in 1991, North Korea lost a valued source of financial support and oil. Without fuel and funding, many of the factories closed. Unemployment rates rose significantly. Meanwhile, China adopted a more open posture and began to increase its level of trade with the West. The result was that China began to lose interest in propping up North Korea politically. North Korea was facing serious problems. Massive food shortages caused famine and starvation. Thousands of people died. Kim Il Sung died in 1994 and the country lost their "**Great Leader.**" He had ruled the country since World War II and in the end was deified as a god to be worshiped by the people.

Figure 10.32: This giant statue of North Korea's "Great Leader" Kim Il Sung stands in the capital city of Pyongyang and is called the Mansudae Grand Monument. This mammoth bronze statue was built in 1972 to celebrate the Great Leader's sixtieth birthday. John Pavelka – Manusudae Grand Monument – CC BY 2.0.

The government dictatorship continued as the **Great Leader**'s son, **Kim Jong Il**, took over the leadership role. Kim Jong Il, known as the "**Dear Leader**," took repression of his people to new levels. Televisions and radios sold in North Korea can only receive government-controlled frequencies. Cell phones and the Internet are banned. The government takes a hard line against dissenters. If you are caught speaking against the state or taking any action to support dissent, you can be arrested, fined, placed in a prison camp, or executed. People are not allowed to leave or enter the country without government approval. Only a couple hundred tourists are allowed into the country per year and are closely escorted.

Kim Jong Il has taken hard measures to stay in power and to avoid yielding to the capitalist frenzy of corporate colonialism. He has nuclear weapon capability and has used his military weapons production as a bargaining chip against the United States and other world powers. North Korea has demonized the United States as the ultimate threat and has used state-funded propaganda to indoctrinate its people. North Korea's government continually tells its people that the United States will invade at any minute and to be prepared for the worst. Propaganda has been used to create and enforce military, economic, and political policies for an ideology that supports the unification of all of Korea under Communist control.

North Korea is mainly mountainous; there is little quality farmland. While only 2 percent of the land is in permanent crops, about a third of the population works in agriculture. The best farmland in North Korea is located south of the capital city of Pyongyang. The capital is a restricted area where only the most loyal to the state can live. There is a serious shortage of goods and services. Electricity is not available on a dependable basis. Massive international food aid has sustained the people of North Korea. Outside food aid is accepted even though Kim Jong Il has continued a policy of self-reliance and self-sufficiency called Juche. *Juche* is designed to keep Korea from becoming dependent on the outside. The policy of *Juche* has been quite effective in isolating the people of North Korea and keeping dictator Kim Jong Il in power. The policy of *Juche* also holds back the wave of corporate capitalism that seeks to exploit labor and resources in global markets for economic profit.

Figure 10.33: A North Korean soldier looks across the DMZ in the Joint Security Area (*left*). The modern "63 Building" in Seoul, South Korea, stands as witness to the prosperity and advancements in the South Korean economy (*right*). Expert Infantry – Korean DMZ – CC BY 2.0; PhareannaH[berhabuk] – 63 building – CC BY 2.0.

One way that North Koreans are finding out about the outside world is through smuggled-in cell phones and VHS videotapes from South Korea. Popular VHS productions include South Korean soap operas, because of the common heritage, language, and ethnicity. The North Korean government has attempted to crack down on smuggled videotapes by going from home to home. Electricity is turned off before entering the home. Once the electricity is turned off, videotapes cannot be removed from the video player. The government agents then recover the videotape. Citizens found with smuggled videotapes are punished with steep fines or imprisonment. Desperate North Koreans have escaped across the border into northern China, where thousands of refugees have sought out better opportunities for their future. It is ironic to think that Communist China, with historically few human rights, would be a place where people would seek refuge, but China's economy is growing and North Korea's economy is stagnant, which creates strong push-pull forces on migration.

South Korea

In 1993, South Korea became a fully fledged democracy with its first democratically elected president. Seoul is the capital city and is home to almost ten million people. Seoul is located just south of the Cease-Fire Line, also known as the DMZ, which is referred to in general terms as the thirty-eighth parallel, even though it does not follow it exactly. The United States has many military installations in this area.

South Korea manufactures automobiles, electronic goods, and textiles that are sold around the world. South Korea is a democracy that has used state capitalism to develop its economy. After the World War II era, South Korea was ruled by a military government that implemented land reform and received external economic aid. Large agricultural estates were broken up and redistributed to the people. Agricultural production increased to meet the demands of the population. South Korea has much more agricultural production than North Korea and is therefore in providing food for the high population density. A once-modest manufacturing sector has grown to become a major production center for export trade.

The fifty million people who live in South Korea have a much higher standard of living than the residents of North Korea. Personal income in the north is barely equivalent to a few dollars per day, while personal income in the south is similar to that of Western countries. The economic growth of the south was a result of state-controlled capitalism, rather than free-enterprise capitalism. The state has controlled or owned most of the industrial operations and sold its products in the global marketplace. Giant corporations, which forced industrialization along the coastal region, have promoted South Korea as the world's leading shipbuilding nation. South Korean corporations include Daewoo, Samsung, Kia Motors, Hyundai, and the Orion Group. As an economic tiger, South Korea continues to reform its economic system to adapt to global economic conditions.

South Korea has announced plans to conduct a comprehensive overhaul of its energy and transportation networks. Government funding will augment efforts to create more green-based initiatives. Part of this effort will focus on lower energy dependency with environmentally friendly energy developments such as wind, solar, bike lanes, and new lighting technologies. High-speed rail service and increased capacity in electronic transmissions lines are planned as part of a next generation of energy-efficient technologies that will increase economic efficiency. These policies have been enacted to update South Korea's economy and create new products for manufacturing export.

Buddhism was introduced into the Korean Peninsula, as were many other aspects of Chinese culture that had significant effects on the Korean heritage. Buddhism has been a prominent religion in Korea for centuries. The teachings of Confucius are also widely regarded. About 30 percent of the population claims Christianity as their religious background; about 20 percent of the Christians are Protestant and 10 percent are Catholic. This is the highest Christian percentage of any Asian country. Up to half of the population makes no claims or professions of faith in any organized religion. Before 1948, Pyongyang was a notable Christian center. At that time, approximately three hundred thousand people identified as Christian. After the establishment of a Communist government in North Korea, many of the Christians fled to South Korea to avoid persecution.

Unification of North and South

What if North Korea and South Korea would become unified again? In the field of geography, there is a concept of regional complementarity, which exists when two separate regions possess qualities that would work to complement each if unified into one unit.

Figure 10.34: Reunification Buddha in South Korea, Erected to Signify the Unity of the Korean People. Paul B. – DSC_0105 – CC BY-NC-ND 2.0.

North and South Korea are the classic illustration of regional complementarity. The North is mountainous and has access to minerals, coal, iron ore, and nitrates (fertilizers) that are needed in the South for industrialization and food production. The South has the most farmland and can produce large harvests of rice and other food crops. The South has industrial technology and capital needed for development in the North. If and when these two countries are reunited, they could work well as an economic unit.

The harder question is how and under what circumstances the two Koreas could ever come to terms of unification. What about the thirty-five to forty thousand US soldiers along the DMZ? What type of government would the unified Korea have? There are many young people in South Korea that would like to see the US military leave Korea and the two sides united. The generation of soldiers that survived the Korean War in the 1950s understands the bitterness and difficulties caused by the division. This segment of the population is highly supportive of maintaining the presence of the US military on the border with North Korea. Unification is not likely to take place until this generation either passes away or comes to terms with unification. The brutal dictatorship of Kim Jong Il, with his claimed nuclear capabilities, has been a main barrier to unification. This is a political division, not technically a cultural division, even though the societies are quite different at the present time. The geography of this situation is similar to that of East and West Germany after World War II and during the Cold War. Korea may have different qualities from Germany, but unification may be possible under certain conditions, foremost being different leadership in the north.

Key Takeaways

- Japan's mountainous archipelago has provided isolation and protection for the Japanese people to maintain their unique culture and heritage. The islands were created by tectonic plate action, which continues to cause earthquakes in the region.
- Japan was a colonial power that colonized much of the Asian Pacific region. Its defeat in World War II devastated the country and liberated all its colonies. Since 1945, Japan has been developing one of the strongest economies in the world through manufacturing.
- Japan is a highly industrialized country that is in stage 5 of the index of economic development. The country's small family size and declining population are a concern for economic growth in the future.
- The Korean Peninsula is divided in two by the DMZ. North Korea is a Communist dictatorship with a low standard of living and absolutely no civil rights. South Korea is a capitalist democracy with a high standard of living and a free-market economy.

- North Korea is more mountainous and lacks arable land to grow food. South Korea has more farmland and fewer mountains. If the two countries were to be united with a common economy, they would complement each other with a balance of natural resources.

Discussion and Study Questions

1. What climate types are found in Japan? How does climate type correspond to population?
2. Why has Japan never experienced a strong rural-to-urban shift in its population?
3. What is the ethnic makeup of Japan? Why does this pattern exist in Japan?
4. Does Japan have a high or low population growth rate? What problems arise from this situation?
5. How did Japan become an economic superpower after 1945?
6. How would South Korean soap operas pose a serious threat to North Korea?
7. How did South Korea develop such a robust economy with such a small physical area?
8. Why are thousands of US soldiers positioned on the DMZ?
9. How does North Korea's policy of *juche* align with the global economy?
10. Explain how the concept of regional complementarity applies to the two Koreas.

Geography Exercise

Identify the following key places on a map:

- Edo
- Hokkaido
- Honshu
- Kansai District
- Kanto Plain
- Kawasaki
- Kobe
- Korea Strait
- Kurile Islands
- Kyoto
- Kyushu
- Mt. Fuji
- Osaka
- Pyongyang
- Sakhalin Island
- Sea of Japan
- Seoul
- Shikoku
- Yellow Sea
- Tokyo
- Yokohama

10.5. CONCISE SUMMARY

East Asia is centered around Communist China, a substantial nation that predominates the geographical area. The elevated mountains along its western boundaries have sequestered China. The periphery areas of Tibet, Western China, Mongolia, and Manchuria serve as buffer states safeguarding the central Han Chinese heartland in China Proper, which is home to the majority of the population.

To enhance foreign trade and augment its manufacturing sector, China has instituted special economic zones (SEZs) along its coastline to attract international enterprises. Shenzhen is a SEZ situated adjacent to the Hong Kong border. Urbanization and industrialization have inspired a significant migration of the people from rural to urban areas.

The world-class port and strategic geographic position of Hong Kong have significantly contributed to its emergence as an economic powerhouse. The integration of high-tech manufacturing with banking and financial services has transformed the little region into a significant commercial hub for Asia. In 1998, Hong Kong was integrated with China, facilitating extensive commerce and development with China and Shenzhen.

Taiwan is a tiny island located off the southeastern coast of China. It has evolved into an economic powerhouse characterized by advanced industry and elevated earnings. The island possesses an autonomous government distinct from mainland China and operates under a capitalist economic system. Historic tension has existed between Taiwan and Beijing since the Nationalists relocated there in 1948. The economic exchange between Taiwan and mainland China has fostered tighter relations between the two parties.

Japan comprises several big islands. Following its loss in World War II, Japan endeavored to rehabilitate and establish itself as a principal manufacturing nation and the second-largest economy globally, a position now occupied by China. Japan possesses a homogeneous society, with approximately 99 percent of its population being Japanese. The small family size and declining population are raising economic concerns regarding the shortage of entry-level personnel.

Korea is divided into an authoritarian North Korea and a capitalist South Korea. North Korea is insular and possesses a fragile economy. Their citizens are deprived of fundamental human rights. South Korea has evolved into one of Asia's economic powerhouses. South Korea possesses a robust manufacturing industry that generates advanced electronics and information technologies. Numerous Korean and American soldiers continue to be deployed along the demilitarized zone (DMZ) separating the two Koreas.

Chapter 11 SOUTHEAST ASIA

11.1. INTRODUCTION: AN OVERVIEW

Southeast Asia, a vibrant and diverse region, is known for its rich tapestry of cultures, languages, and landscapes. Stretching from Myanmar in the northwest to Indonesia in the southeast, this area encompasses a variety of nations, each with its unique traditions and histories. From the bustling markets of Bangkok to the serene beaches of Bali, Southeast Asia offers a captivating blend of urban energy and natural beauty. With its intricate history influenced by trade, colonialism, and globalization, the region continues to be a dynamic crossroads of cultures, making it a fascinating destination for travelers and scholars alike.

Physical Geography

The islands and the mainland of Southeast Asia include a wide array of physical and cultural landscapes. The entire realm is located in the tropics except the northernmost region of Burma (Myanmar), which extends north of the Tropic of Cancer. A tropical Type A climate dominates the region and rainfall is generally abundant. The tropical waters of the region help moderate the climate. Southeast Asia is located between the Indian Ocean on the west and the Pacific Ocean on the east. Bordering the many islands and peninsulas are various seas, bays, straits, and gulfs that help create the complex maritime boundaries of the realm. The South China Sea is a major body of water that acts as a separator between the mainland and the insular region. The thousands of islands that make up the various countries or lie along their coastal waters create a matrix of passageways and unique physical geography.

The three longest rivers of the realm, Mekong, Red, and Irrawaddy, are located on the mainland and have their headwaters in the high elevations of Himalayan ranges of China. The Mekong River makes its way from the high Himalayas in China and helps form the political borders of Laos and Thailand on its way through Cambodia to Vietnam where it creates a giant delta near Ho Chi Minh City (Saigon). The Red River flows out of China and through Hanoi to the Red River delta on the Gulf of Tonkin. The Irrawaddy River flows through the length of Burma providing for the core area of the country. Another major river of the mainland is the Chao Phraya of Thailand. With its many tributaries, the Chao Phraya creates a favorable core area that is home to the largest population of the country. Many other rivers can be found on both the mainland and the insular region. The rivers transport water and sediments from the interior to the coasts, often creating large deltas with rich soils that are major agricultural areas. Multiple crops of rice and food products can be grown in the fertile river valleys and deltas. The agricultural abundance is needed to support the ever-increasing populations of the realm.

Figure 11.1 Southeast Asia: The Mainland Region and the Insular Region (the Islands)

Tectonic plate activity has been responsible for the existence of the many islands and has created the mountainous terrain of the various countries. High mountain ranges can have peaks that reach elevations of over fifteen thousand feet. The high-elevation ranges of New Guinea, which are along the equator, actually have glaciers, ice, and snow that remain year-round. The island of Borneo, in the center of the insular region, is actually a segment of ancient rock that has been pushed upward by tectonic forces to form a mountainous land mass. The mountains on Borneo have been worn down over time by erosion. Mountains and highlands stretch across the northern border of the realm along the borders with India and China. The interior nature of this border makes it less accessible. Similar dynamics can be found in the interior of the islands of the insular region, where the isolation and remoteness have helped create the environmental conditions for unique flora and fauna. In the highland areas the human cultural landscape can be diverse.

Time and isolation have worked together to form the traditions and cultural ways that give local groups their identity and heritage.

Tectonic activity makes the region vulnerable to earthquakes and volcanic eruptions. The volcanic peak of Mt. Pinatubo, in the Philippines, erupted in 1991, spewing ash and smoke into the atmosphere and impacting much of the planet. An earthquake of 9.0 magnitude occurred off the coast of the Indonesian island of Sumatra in 2004 and caused widespread disaster throughout the wider region of the Indian Ocean. As many as one hundred fifty thousand deaths were reported, mainly from flooding. A thirty-five-foot-high wall of water from the tsunami devastated many coastal areas from Thailand to India.

Impact of Colonialism

Southeast Asia has not escaped the impact of globalization, both colonial and corporate. As Europeans expanded their colonial activities, they made their way into Southeast Asia. Southeast Asia was heavily influenced by European colonialism. The only area of the region that was not colonized by the Europeans was Thailand, which was called **Siam** during the colonial era. It remained an independent kingdom throughout the colonial period and was a buffer state between French and British colonizers. The Japanese colonial empire controlled much of Southeast Asia before World War II.

Some of the countries and regions of Southeast Asia became known by their colonial connection. Indonesia was once referred to as the Dutch East Indies, which was influential in the labeling of the Caribbean as the West Indies. French Indochina is a term legitimized for historical references to the former French claims in Southeast Asia. Malaya and British Borneo each had its own currency based on a dollar unit that was legal tender for the regions of the Federation of Malaya, Singapore, Sarawak, North Borneo, and Brunei. Independence from the European powers and freedom from Japanese imperialism by the end of World War II provided a new identification for the various countries of the realm. Cultural and economic ties remain between many former colonies and their European counterparts.

Figure 11.2: Southeast Asia was colonized by Europeans and later by Japan.

East Timor, a former Portuguese colony south of Indonesia, has been the most recent colony to gain independence. Timor is an island just north of Australia. The western portion is claimed by Indonesia. The whole island was annexed to Indonesia in 1975. As a result of separatist movements that entailed conflict and violence, the eastern portion was finally granted independence in 2002. Since then, East Timor has been working to establish itself as a country and is now negotiating its offshore boundary to include important oil and gas reserves.

Cultural Introduction

Southeast Asia has a population of more than six hundred million people; more than half the population lives on the many islands of Indonesia and the Philippines. The small island of Java in Indonesia is one of the most densely populated places on Earth. More than half of the two hundred forty-five million people who live in Indonesia live on the island of Java. The island of Luzon in the Philippines is also one of the more densely populated areas of the insular region. The Philippines has over one hundred million people, Vietnam has more than ninety million, and Thailand has about sixty-seven million. Local areas with high food-producing capacity are also high population centers, which would include deltas, river valleys, and fertile plains.

The ethnic mosaic of Southeast Asia is a result of the emergence of local differences between people that have evolved into identifiable cultural or ethnic groups. Though there are a multitude of specific ethnic groups, a number of the larger ones stand out with recognizable populations. On the mainland the Burmese, Thai, Khmer, and Vietnamese are the largest groups, coinciding with the physical countries from Burma to Vietnam. A similar situation can be found in the insular region. Many distinct groups can exist on the many islands of the region. The island of New Guinea, for example, has hundreds of local groups with their own languages and traditions. The large number of ethnic groups is dominated by Indonesians, Malays, and Filipinos, coinciding with the countries of Indonesia, Malaysia, and the Philippines. Each of these main groups has many subgroups that hold to their own cultural heritage in the areas where they exist. The many islands of Indonesia and the Philippines create the opportunity for diversity to continue to thrive, in spite of the globalization process that increased the interaction and communication opportunities between groups.

Indonesia is also home to the largest Muslim population in the world. All major religions can be found here. The Philippine population is predominantly Christian, but there is a minority Muslim community, including rebel insurgents. Most of people in Malaysia follow Islam. About 95 percent of the people in Thailand and more than 60 percent of the people in Laos are Buddhist. Hinduism is present in the Indonesian island of Bali and in various other locations in the region. Animism and local religions can be found in rural and remote areas. Clearly, Southeast Asia is a mix of many ethnic groups, each with its own history, culture, and religious preference.

Overseas Chinese

Southeast Asia is also home to over thirty million overseas Chinese—ethnic Chinese who live outside of China. The Chinese exodus to the realm was the greatest during the last Chinese dynasties and during the colonial era. European colonial powers enhanced this migration pattern by leveraging the use of people with Chinese heritage in their governing over the local populations in the realm. Life has often been difficult for overseas Chinese. The Japanese occupation of the realm during World War II was a time of harsh discrimination against Chinese. Japanese occupation and colonialism diminished with the end of World War II. The overseas Chinese minority retained an economic advantage because of their former colonial status and their economic connections. Chinatowns emerged in many of the major cities of Southeast Asia. The discrimination against the Chinese, fueled by religious or socioeconomic differences, often continued after World War II by the local ethnic majorities. Nevertheless, overseas Chinese in Southeast Asia have been instrumental in promoting the global business arrangements that have established the Pacific Rim as a major player in the international economy.

Key Takeaways

- Southeast Asia can be studied by dividing up the realm into two geographic regions: the mainland and the insular region. The mainland borders China and India and has extensive river systems. The insular region is made up of islands and peninsulas between Asia and Australia, often with mountainous interiors.
- France and Britain colonized the mainland region of Southeast Asia. Burma was a British colony and the rest was under French colonial rule. The Japanese took control of the region briefly before World War II ended in 1945. Siam was the only area not colonized. Siam became the country of Thailand.
- The physical geography of the mainland and the insular region is dominated by a tropical type A climate. Cooler temperatures may be found in the mountainous regions and more even temperatures ranges can be found along the coasts. Tectonic plate activity is responsible for the many earthquakes and volcanic eruptions that occur in the realm.
- Southeast Asia is ethnically, religiously, and linguistically diverse. A number of major ethnic groups dominate in the mainland and insular region but are only examples of the multitude of smaller groups that exist

in the realm. One minority group is the overseas Chinese, who immigrated to the realm during the colonial era.

Discussion and Study Questions

1. Which countries of Southeast Asia are in the mainland region? Which countries are in the insular region?
2. What are the four main rivers of the mainland region and how do they contribute to each region's economic activities?
3. How has tectonic plate activity been evident in Southeast Asia? How has the rest of the world been influenced by tectonic activity in the realm?
4. Which European countries have been the main colonizers and which countries did each colonize? How has the colonial experience influenced the realm?
5. Where are the main population centers? Why are these locations favorable to such large populations? Which countries are the most populous?
6. What are the main ethnic groups on the mainland and in the insular region? Why are there so many ethnic groups in this realm? How has physical geography contributed to the diversity?
7. What are the main religious affiliations of the realm? Which countries have the largest Buddhist, Muslim, and Christian populations? How do you think this diversity of religious beliefs could create difficulties in the workplace?
8. What is the newest country to declare independence in the realm? Which European country colonized the area? Why didn't this country gain its independence many decades ago?
9. How have the overseas Chinese influenced the realm of Southeast Asia? Why have the overseas Chinese been so influential in the economic situation of the Pacific Rim?
10. What other region of the world has similar dynamics in physical geography and colonial activities to the insular region of Southeast Asia?

Geography Exercise

Identify the following key places on a map:

- Chao Phraya River
- Gulf of Tonkin
- Ho Chi Minh City
- Indochina
- Insular Region
- Irrawaddy River
- Mainland Region
- Mekong River
- Red River
- Saigon
- Siam

11.2. THE MAINLAND COUNTRIES

Learning Objectives

1. Summarize the main economic activities of each country.
2. Understand how Vietnam was divided by civil war and the impact the war had on the country.
3. Realize how the country of Laos is addressing its rural landlocked economic situation.
4. Describe the radical conditions that led to the creation of Democratic Kampuchea.
5. Outline the physical geography of Thailand and how this country has developed its economy.
6. Comprehend the conditions in Burma. Learn why the Burmese people would be opposing the government.

Vietnam

The elongated state of Vietnam is slightly larger than Italy and about three times the size of the US state of Kentucky. In 2010 it was estimated to have a population of about ninety million people. Sixty percent of the population is under age twenty-one. This indicates that the population was only about half its current size at the end of the Vietnam War. Vietnam has two main urban core areas: Ho Chi Minh City (Saigon) in the south and the capital, Hanoi, in the north. The middle region of Vietnam is narrow, with higher elevation. Each core area is located along a major river delta. The **Red River** delta is located east of Hanoi in the north, and the mighty **Mekong River** delta is located next to Saigon in the south. These river deltas deposit silt from upstream and provide excellent farmland for growing multiple crops of rice and food grains per year.

Vietnam has a tropical Type A climate with a long coastline. Fishing provides protein to balance out nutritional needs. More than 55 percent of the population works in agriculture. Family size has dropped dramatically because of population growth and a trend toward urbanization. Rural-to-urban shift has caused the two main urban core cities to grow rapidly. Saigon is the largest city in Vietnam and has a port that can accommodate oceangoing vessels. Hanoi, the capital, is not a port city and is located inland from the nearest port of Haiphong on the coast of the **Gulf of Tonkin**.

Political Geography

An understanding of Vietnam is not complete without understanding the changes in political control the country of Vietnam has experienced. Different Chinese dynasties controlled Vietnam at different times. When France colonized Vietnam, it imposed the French language as the *lingua franca* and Christianity as the main religion. Both changes met resistance, but the religious persecution of Buddhism by the French colonizers created harsh adversarial conditions within the culture. The French domination started in 1858. The Japanese replaced it in 1940; this lasted until the end of World War II. With the defeat of Japan in 1945, the French desired to regain control of Vietnam. The French aggressively pushed into the country, but met serious resistance and were finally defeated in 1954 with their loss at the battle of Dien Bien Phu.

Figure 11.3: Southeast Asia and Vietnam. The two main cities of Vietnam are both located next to large rivers. The capital, Hanoi, in the north is on the Red River. Ho Chi Minh City (Saigon) to the south is next to the delta of the Mekong River. Updated from map courtesy of Wikimedia Commons - public domain.

In the mid-1950s, the Vietnamese began asserting their request for an independent country. The dynamics were similar to that of Korea. After 1954, Vietnam needed to establish a government for their independent country. They were not unified. The northern section rallied around Hanoi and was aligned with a Communist ideology. The southern region organized around Saigon and aligned itself with capitalism and democratic reforms.

During the Cold War, the United States opposed Communism wherever it emerged. Vietnam was one such case. Supporting South Vietnam against the Communists in the north started not long after the defeat of France. By 1960, US advisors were working to bolster South Vietnam's military power. After the assassination of John F. Kennedy, President Lyndon Johnson had to make a choice to either pull out of Vietnam or push the US military to fully engage the Communists in North Vietnam.

Not wishing Vietnam and its neighbors to "go Communist" through a domino effect—where if one country fell to Communism its neighbors would follow—President Johnson decided to escalate the war in Vietnam. By 1965, more than one half million US soldiers were on the ground in Vietnam. History has recorded the result. Just as Vietnam was divided by political and economic ideology, the Vietnam War also divided the US population. Protests were common on college campuses and public support for the war was often met with public opposition.

Figure 11.4 Portrait of Ho Chi Minh, the Communist Leader of North Vietnam during the Cold War Wikimedia Commons – public domain.

The US government, under President Richard Nixon, finally decided to pull all US troops out from Vietnam after a cease-fire was agreed upon in a Paris peace conference in 1973. More than fifty-seven thousand US soldiers had died in the Vietnam War. Two years later, in 1975, the North Vietnamese Communists invaded South Vietnam and took control of the entire country. Vietnam was unified under a Communist regime. More than two million people from South Vietnam escaped as refugees and fled to Hong Kong, the United States, or wherever they could go. Thousands were accepted by the United States, which caused ethnic rifts in US communities. The United States placed an embargo on Vietnam and refused to trade with them. The United States did not open diplomatic relations with Vietnam again until 1996. The Vietnam War devastated the infrastructure and economy of the country. Roads, bridges, and valuable distribution systems were destroyed. Vietnam could only turn to what it does best: growing rice and food for its people.

Figure 11.5: Man Hauling Cut Wood on a Bicycle Cart (Pedicab) by the Perfume River Near the City of Hue, Vietnam. Heiko Carstens – Vietnam_full_04 – CC BY 2.0.

Modern Vietnam

For the past three decades, Vietnam has been recovering and slowly integrating itself with the outside world. Its population has doubled; most of the population was born after the Vietnam War. Their main goal is to seek out opportunities and advantages to provide for themselves and their families. Vietnam has been a rural agrarian society. The two main core cities, however, are now waking up to the outside world, and the outside world is discovering them. Looking for cheap labor and economic profits, economic tigers such as Taiwan are turning to Saigon to set up light manufacturing operations. People from the rural areas are migrating to the cities looking for employment. Saigon has more than 8.5 million people and has a special economic zone (SEZ) located nearby. Rural-to-urban shift is kicking in. After 1975, the city of Saigon was renamed Ho Chi Minh City after the victorious Communist leader, Ho Chi Minh. Many of the people who live there and who live in the United States still refer to it as Saigon.

Any country that experiences rapid urbanization or economic change suffers from serious growing pains. Conflicts usually erupt over control of resources and land ownership, ethnic groups usually vie for power, and environmental damage is usually extensive. All these issues are evident in Vietnam. The dogmatic Communist government has acted to moderate both the problems and the economic growth. The future of Vietnam may be similar to most of Southeast Asia as it balances out the strong adhesive forces of local culture and the demands of a competitive global economy. The growing population will add to the demand for resources and employment opportunities. Vietnam has been a relatively poor country but it still has been able to export rice and other agricultural products. In recent years, the Communist government has implemented a series of reforms moving toward a market economy, which has encouraged economic development and international trade.

Globalization has prompted a strong rural-to-urban shift within Vietnam. The rural countryside is still steeped in its agrarian heritage based on growing rice and food crops, but the urban centers have been energized by modern technology and outside economic interest. Vietnam has enormous growth potential. The country's urban centers are shifting from stage 2 of the index of economic development into stage 3, where the urbanization rates are the strongest. The rapid rise of the global economy that is connected to Vietnam's major cities has provided jobs and opportunities that are highly sought after by the growing population. The city streets are filled with a sea of motorbikes and bicycle traffic. Cars are becoming more plentiful. Saigon has been a major destination for the export textile industry and other industries seeking a cheap labor base. Cell phones and Internet services have connected a once-isolated country with the rest of the world.

Laos

The geography of Laos centers on the Mekong River basin and rugged mountain terrain. Laos is landlocked. Vietnam shields Laos from the South China Sea to the east and Cambodia to the south. It doesn't have a port city to the outside world.

Figure 11.6: Woman in Attapeu Province, Laos, Cooking Midmorning Snacks for Schoolchildren These snacks are a corn-soy blend shipped in from the United States and sponsored by the US McGovern-Dole Act. Food is provided in schools in rural Laos to encourage attendance and enrollment in school. Thousands of children benefit from this program. Prince Roy – Attapeu – CC BY 2.0.

The mountains reach up to 9,242 feet. The Type A climate provides a rainy season and a dry season. The rains usually fall between May and November, followed by a dry season for the remainder of the year. The Mekong River flows through the land and provides fresh water, irrigation, and transportation. The country's capital and largest city, Vientiane, is located on the Mekong River. Laos is about the same size in area as the US state of Utah.

The Lao Kingdom coalesced in the 1500s and was eventually absorbed by the Kingdom of Siam, which thrived during the eighteenth and nineteenth centuries. France muscled in during the colonial era and created a French Indochina. Laos received independence from France in 1949. Laos is a rural country with about 80 percent of the population working in agriculture. Globalization has not yet been established in this country and infrastructure is less developed. Electricity is not available on a consistent basis and transportation systems are quite basic. There aren't any railroads and there are few paved roads. Clean water for human consumption is not always available. The economy is based on agriculture, with some outside investments in mining and natural resources.

Two-thirds of the people in Laos are Buddhists. Animist traditions and spirit worship have the next highest percentage of followers. Muslims and Christians make up a small percentage of the population. Lao make up the largest ethnic group and 70 percent of the population. Other ethnicities include the Hmong and mountain tribal groups, which can be found in various remote regions of the country. The remoteness and rural heritage of the many tribal people have started to attract tourism. Tourism has increased in recent years, partially due to the Chinese government allowing its citizens to travel outside their borders from China into Laos. Laos has two UNESCO World Heritage Sites: the historic town of Luang Prabang, and the southern site of Wat Phou (Vat Phu), which is an ancient Hindu temple complex.

Laos is a poor country. It has fewer employment opportunities for its citizens than other developing countries have. The one-party Communist political system of the Cold War has been decentralizing control and working to encourage entrepreneurial activities. Foreign investments are increasing in the areas of mining, hydroelectric production, and major construction projects. The World Bank and other agencies have supported efforts to improve infrastructure and provide opportunities for the people of Laos. China has been partnering with the Laotian government to help build rail transport in the country. These efforts have assisted in reducing poverty and increasing the economic and physical health of the country.

Cambodia

A Notorious History

Cambodia is about the same size in area as the US state of Missouri. The population in 2008 was estimated at 14.5 million. The Khmers created the Angkor Empire, which reached its peak between the tenth and thirteenth centuries. Preceding the colonial period, the Angkor Empire entered into a long era of decline. France took control of the region in the latter part of the seventeenth century. Japan took control of the region before World War II and then relinquished it when they surrendered to end World War II. France regained control of Cambodia after the Japanese army was defeated. Cambodia finally received independence from France in 1953.

To understand Cambodia, one has to understand its recent history. This country has undergone some of the most extreme social transitions in modern times. The **Khmer Rouge**, under the leadership of **Pol Pot**, turned society upside down, giving the country a legacy that it will carry forward as integration continues into the world community.

Between 1969 and 1973, while the United States was fighting the Vietnam War, US forces bombed and briefly invaded Cambodia in an effort to disrupt the North Vietnamese military operations and oppose the Khmer Rouge. Millions of Cambodians were made refugees by the war, and many ended up in Phnom Penh. The number of casualties from the US bombing missions in Cambodia is unknown. The US war in Vietnam thus had spilled over into Laos and Cambodia and advanced the opportunities for the Khmer Rouge regime to gain power. Pol Pot's Communist forces of the Khmer Rouge finally captured Cambodia's capital of Phnom Penh in 1975. The Khmer Rouge evacuated all cities and towns and forced the people to move to the rural areas. The country's name was changed to Democratic Kampuchea. China's Great Cultural Revolution and the Great Leap Forward disaster were influential for Pol Pot's radical experiment. Since Vietnam was supported by the Union of Soviet Socialist Republics (USSR), the Khmer Rouge looked to China for arms and support.

Pol Pot was creating an agricultural model for a new country based on eleventh-century ideals. People in urban areas were forcibly marched off into the countryside for labor in agriculture. Anyone who resisted or even hinted at dissent was killed. All traces of Westernized ideas, technology, medical practice, religion, or books were destroyed. Thousands of people were systematically killed in an attempt to bring into being a rural agrarian utopian society. The thousands upon thousands who were systematically eliminated gave rise to the term Killing Fields, meaning fields where massive groups of people were forced to dig their own graves and then were killed. The mass killings were reminiscent of those carried out by Hitler, Stalin, and Mao. Pol Pot's regime also targeted ethnic minority groups. Muslims and Chinese suffered serious purges. Professional, educated people, such as doctors, lawyers, and teachers, were also targeted for execution. According to some reports, the very act of wearing eyeglasses was a death sentence as it was a symbol of intellectualism.

In a country of eight million in 1970, more than two million people were executed or died as a result of Pol Pot's policies. The total number will never be known. Hundreds of thousands became refugees in neighboring countries.

Figure 11.7: Skulls of the Victims from the Killing Fields of Pol Pot and the Khmer Rouge in Cambodia from the 1970s. Wikimedia Commons – public domain.

By 1978, the Khmer Rouge was isolated in the countryside. Vietnamese forces controlled the urban areas. A decade of civil war and unrest followed. Paris peace talks, cease-fires, United Nations–sponsored elections and coalition governments have since helped provide political stability. Pol Pot died under unclear circumstances in 1998 while being held under house arrest. As of 1999, the Khmer Rouge elements that were still in existence had surrendered or were arrested. Many of the Khmer Rouge leaders were charged with crimes against humanity by United Nations–sponsored tribunals.

Modern Cambodia

Cambodia is working to become a democratic and open country with established trade relationships with global markets. The people have struggled to create a stable society that can rebound from their legacy of turmoil and conflict. The country's population is relatively young. More than half the population is under age twenty-five; one-third is under fifteen. The rural areas and the generations who remain there continue to lack the basic amenities of modern society. Education, electricity, and modern infrastructure are lacking. More than half the population works in agriculture. Since less than 25 percent of the population lives in cities, Cambodia is likely to experience a high rural-to-urban shift in its future.

People are returning to religious practices that were banned during the Pol Pot era. Buddhism is the dominant religion of about 95 percent of the population. Small percentages of the population also practice Christianity, Hinduism, Islam, or tribal beliefs. There are at least twenty distinct hill tribes that hold to their own traditions and cultural ways. The country has historically been self-sufficient with food, but the rapid population growth, political instability, and lack of infrastructure are challenging the future of the country.

Agriculture has been the main economic activity, though textiles (clothing manufacturing) have increased in recent years because of the low cost of labor combined with an abundant workforce. The international business sector has sought to exploit this opportunity, but multinational corporations are hesitant to invest in a country that suffers from political instability or a high level of corruption within the public and private sector.

Figure 11.8: A Small Section of the Angkor Wat Temple Complex in Northwest Cambodia Angkor Wat is said to be the largest religious structure in the world. Hindu in origin, it was converted to a Buddhist site during the country's early years. Paul Arps – Angkor temple (Angkor Wat, Cambodia 2011) – CC BY 2.0.

Cambodia has been attempting to build a sustainable economy. The textile industry is the number one source of national wealth. Sweatshops and low-tech manufacturing have begun to take root in the expanding capital city of **Phnom Penh.** Tourism is another sector that has experienced rapid growth. Though nonexistent in earlier decades, tourism has taken off. Cambodian tourism provides travelers with an experience that is more pristine and less commercialized. Tourism has been rated as the second-largest sector of the economy. One of the main sites that attract many visitors is the extraordinary ancient site of **Angkor Wat** (Angkor means "city" and Wat "temple"). This site is one of the best-preserved showcases of Khmer architecture from its early empire years. Angkor Wat is being developed as a major tourist attraction. The twelfth-century complex was first a Hindu site dedicated to Vishnu, and then it was converted to a Buddhist site. Angkor Wat has become an international tourist destination. It is one of the largest temple complexes in existence in the world and is a UNESCO World Heritage Site. The city of Angkor has been estimated to have been the largest city in the world at its peak. As many as a thousand other temples and ancient structures have been recovered in the same area in recent years.

Cambodia has pressing environmental problems. The country has the notorious designation by the UN as the nation with the third-highest number of land mines on Earth. Since 1970, more than sixty thousand people have been killed, and many more injured or maimed because of unexploded land mines in rural areas. The growing population, attempting to recover from decades of devastation, has cut down the rainforest at one of the highest rates in the world. In 1970, rainforests covered about 70 percent of the country. Today there is only about 3 percent of the rainforest left. A rise in the need for resources, along with illegal timber activities, has devastated the forests, resulting in a high level of soil erosion and loss of habitat for indigenous species. The loss of natural resources is likely to hinder the country's economic growth.

Thailand

Thailand is larger than Laos and Cambodia combined but smaller than Burma. The physical regions that make up Thailand include the mountainous north, where peaks reach up to 8,415 feet; the large southeastern plateau bordering the Mekong River; and the mainly flat valley that dominates the center of the country. The southern part of the country includes the narrow isthmus that broadens out to create the **Malay Peninsula.** The tropical Type A climate has dry and rainy seasons similar to Cambodia. The weather pattern in the main part of Thailand, north of the Malay Peninsula, has three seasons. The main rainy season lasts between June and October, when the southwest monsoon arrives with heavy rain clouds from over the Indian Ocean. After the rainy season, the land cools off and starts to receive the northeast monsoon, which is a cool dry wind that blows from November to February. Considered the dry season, its characteristics are lower humidity and cooler temperatures. From March to May, the temperatures rise and the land heats up. Then the cycle starts over again with the introduction of the rainy season. The weather pattern in the southern part of Thailand in the Malay Peninsula receives more rain throughout the year, with two rainy seasons that peak from April through May and then again from October through December.

Thailand was formerly known as the Kingdom of Siam. In 1932, a constitutional monarchy was established after a bloodless revolution erupted in the country. The name was officially changed to Thailand in 1939. The ruling monarch remains head of state but a prime minister is head of the government. Siam was never colonized by either the Europeans or the Japanese. The leaders of Siam played France and Britain against each other and remained independent of colonial domination. During World War II, the Japanese did extend influence in the region. Thailand briefly engaged the Japanese military in World War II but worked out an armistice that used the Japanese military to regain territories lost to Britain or France. At the same time, Thailand was working to support Allied efforts in the region.

About three-fourths of the population is ethnically Thai. There is a noticeable Chinese population and a small percentage of people who are ethnically Malay. There are various minority groups and hill tribes. The country's official language is Thai. Buddhism is adhered to by about 95 percent of the population. The ruling monarch is considered the defender of the Buddhist faith. Southern Buddhism is fervently practiced here. Thailand does not use the Western Gregorian calendar. Thailand uses an official calendar based on an Eastern translation of the Buddhist era, which essentially adds 543 years to the Gregorian calendar. For example, when it was 2010 AD in the West, it was 2553 BE in Thailand.

There have been clashes between Thailand's small Muslim minority groups in the south, which have been increasing since 9-11. Islamic influences have been increasing near the border with Malaysia, which is about 60 percent Muslim. The Buddhist government of Thailand has sought to keep extremist groups like Al-Qaeda from operating within its borders. A series of Muslim-inspired bombings in recent years have increased social tensions and brought more attention to the religious division in the south.

Figure 11.9: Modern City of Bangkok with High-Rise Office Buildings and Business District. Efendi Kwok – Bangkok Skyline – CC BY-NC 2.0.

Thailand has an excellent record of economic growth and has been one of Southeast Asia's best performers in the past couple of decades. Thailand is developing its infrastructure and has established measures to attract foreign investments and support free-enterprise economic activities. The recent slowdown in the global economy and internal political problems have caused a sharp downturn in Thailand's economic growth. Nevertheless, Thailand remains a strong economic force and one of the best economies in the region. The positive indicators include a strong focus on infrastructure, industrial exports, and tourism.

Urbanization rates are increasing; at least one-third of the population lives in cities. Family size has fallen to lower than two children per family, while education rates have increased. The country has also tapped into its natural resources for export profits as the world's third-largest exporter of tin and the second-largest exporter of tungsten. Light manufacturing has taken off and become a major component of the economy, accounting for about 45 percent of the gross domestic product (GDP). The country is a major manufacturer of textiles, footwear, jewelry, auto parts, and electrical components. Thailand has been the major exporter of rice in the world and has a strong agricultural base.

Thailand is a newly industrialized country and has all its bases covered to build national wealth: a balance of agriculture, extractive activities, manufacturing, and postindustrial activities (tourism). Thailand is considered the third-largest manufacturer of motor vehicles in Asia, after Japan and Korea. Vehicle producers from the United States and Asia are manufacturing large numbers of cars and trucks in Thailand. Toyota dominates the market in both truck and auto production. Truck production is augmented by Mitsubishi, Nissan, Chevrolet, Ford, and Mazda. Honda, from Japan, and the Tata Motor Corporation, from India, are expanding their operations in Thailand. Thailand is in a good position to advance its economy and shift the whole country into the next stage of development to become a major participant in the global economic marketplace.

The tourism industry has grown immensely in Thailand over the past few decades. Green and lush tropical mountain landscapes, the exquisite architecture of ancient Buddhist temples, and beautiful golden beaches along warm tropical coastlines make for an excellent tourism market. Some of the best world-class tropical beach resorts are located along the sandy and sunny shores of Thailand. The country is open to outsiders and has welcomed tourism as part of its economic equation. The relatively stable country provides a safe and exciting tourism agenda that has a global clientele. The downside of the thriving tourism industry is the sex trade. Relaxed laws on sexual activity have made Thailand a destination for people from around the world seeking "sex tours" and erotic experiences. Not surprisingly, a sharp increase in the number of individuals infected with sexually transmitted diseases has been documented. Approximately one million people in Thailand tested HIV positive in the mid-1990s. The sex industry has been big business for Thailand and at the same time has created an unfortunate negative stereotype for the overall tourism situation. There is much more to the tourism industry in Thailand than the sex trade.

The country of Thailand has the potential to recover from the global economic downturn and once again claim its role as an economic tiger of Southeast Asia. If political stability serves to enhance economic investments, the country will continue to experience economic growth. The low population growth is a model for other countries in the region. Thailand provides a good example of the theory that as a country urbanizes and industrializes, family size will usually go down. Thailand is also moving forward in the index of economic development. It is in stage 3, where there is a strong rural-to-urban shift in the population. The capital city of **Bangkok** has stage 4 development patterns and is an economic core area for the country and the region. As large as New York City, Bangkok has developed into the political, cultural, and economic center of Southeast Asia. Often referred to as the "Venice of the East" because of its city canals, Bangkok has become a global city with a population of more than eight million people officially and more than fifteen million unofficially.

Figure 11.10: Burma (Myanmar): The Irrawaddy River and the North/South Layout. CIA World Factbook – public domain.

Myanmar or Burma

The Union of Myanmar (Union of Burma) is the official name for Burma. Since 1989, the military authorities in Burma have promoted the name Myanmar as a conventional name for their state. The US government and many other governments have not recognized or accepted the name change. Some groups within Burma do not accept the name because the translation of Myanmar is also the name of an ethnic minority in Burma. The use of the name Burma or Myanmar is split around the world and within the country.

Burma is the largest country on the Southeast Asian mainland in terms of physical area. It is about the same size in area as Texas and had a population in 2010 of about fifty-four million. The country has a central mass with a southern protrusion that borders Thailand toward the Malay Peninsula. The northern border area between India and China has high mountains that are part of the Himalayas, with towering peaks extending to 19,295 feet. The **Irrawaddy River** cuts through the center of the country from north to south, creating a delta in the largest city, **Rangoon** (Yangon). Most of the country's population lives along this river valley.

There are differences in physical landscape between the north and south. The northernmost area is mountainous with evergreen forests. Cool temperatures are found in the north and warmer annual temperatures are found in the south. To the west of the Irrawaddy River and north of Mandalay the land cover is mainly deciduous forests. The eastern region from **Mandalay** to the Laos border is scrub forests and grasslands. This area is considered the dry zone, with an annual rainfall of about forty inches. The more tropical south and coastal areas can receive higher levels of precipitation. The area around the core city of Mandalay was a major focus of agricultural development before British colonialism. Dryland crops were most common. During the colonial era, the British looked to the rich farmlands of the southern Irrawaddy delta and emphasized Rangoon as the center of their exploitations. Wetland rice is a major crop of the southern Irrawaddy basin. The southwest and the southern protrusion are mainly tropical evergreen forests. There has been oil exploration along the coastal regions of the Bay of Bengal and along the **Andaman Sea.**

The country was colonized by the British and was once a part of Great Britain's empire in South Asia as a province of India. Burma was one of the most prosperous colonies of Britain until World War II, when the Japanese invaded and war devastated the region. Democratic rule existed from 1948 until 1962, when an authoritarian military dictatorship took over the country. A revolutionary council ruled the country between 1962 and 1974. This government nationalized most of the businesses, factories, and media outlets. The overall operating principle of the council was a concept called the **Burmese Way of Socialism.** This concept was based on central planning and Communist principles mixed with Buddhist beliefs.

Between 1974 and 1988, the sole political party of the country was the Burma Socialist Program Party, which was controlled by the same military general and his comrades who had been in control for decades. During this time, the rest of the world was advancing in technology and economic development and moving forward with advancements in health care and education. Burma remained an impoverished and isolated nation. A number of countries, including the United States, have trade restrictions with Burma. For decades, the authoritarian regime in Burma has been accused of serious human rights violations, which have largely been ignored by the outside world.

Protests against the military rule have always existed in Burma but have been suppressed by the armed forces and the authoritarian government. In 1962, the government cracked down on demonstrations at Rangoon University, resulting in fifteen students being killed and many others in need of medical attention. The military government has taken serious action against any antigovernment protest activities. By 1988, the people of Burma were taking to the streets with widespread demonstrations and protests against the government over claims of oppression, mismanagement, and lack of democratic reforms. A total crackdown on the people was implemented, with thousands of protesters killed. A new council led by a military general created the **State Law and Order Restoration Council** a year later. Martial law was imposed and even harsher policies were imposed on anyone opposing the government. This is when the name of Myanmar was first used for the country.

The name change and the military rule have not been universally accepted. The United States still refers to the capital city as **Rangoon**, not as **Yangon**. In 2006, military rulers moved the capital north to the city of **Naypyidaw**. The purpose of the move was to establish a *forward capital* and shift development and political energy more toward the center of the country, rather than along the coast. World nations are divided on the issues of how to deal with the changes and the military regime in Burma. The governments of some countries believe more sanctions should be implemented to force the leadership into compliance. Other countries believe sanctions are not effective against the government; that is, sanctions harm the people and do not affect the military leadership. Countries on this side of the equation believe that open trade is the best policy.

Figure 11.11: Demonstrators Marching to Express Discontent with the Government of Rangoon (Yangon), 2007 The banner, written in Burmese, refers to a national movement to promote nonviolence. A Buddhist monk is in the foreground, and the Shwedagon Pagoda is in the background. Wikimedia Commons – CC BY 2.0.

Antigovernment protests erupted in 2007 when the military-ruled government allowed prices on fuel and energy to double and triple in price. Protesters were quickly and violently dealt with and many were arrested and jailed. Later that year, thousands of Buddhist monks led a peaceful protest to gain the government's attention to make democratic changes. The demonstration ended in a renewed government crackdown. Another voice in the antigovernment demonstrations has been that of opposition leader **Aung San Suu Kyi**, a UN worker in the early 1960s and a Nobel Peace Prize winner in 1991. Her opposition to the military rule has led to imprisonment and house arrest for decades. She has been a symbol of the opposition and hope for democratic reforms. In 2010, Aung San Suu Kyi was at long last released from house arrest and allowed more freedom of movement under government restrictions.

Burma has been placed in the same category as North Korea and Somalia in terms of authoritarian rule, lack of human rights, and stagnant economy. Economic conditions are poor. The military rulers have gained control of the main income-generating enterprises in the country, including the lucrative drug trade from the prime opium growing region of the northern Golden Triangle, where Burma borders Laos and Thailand. All factors seem to indicate an increase in opium production in recent years. Precious gemstones such as rubies, sapphires, and jade are abundant in Burma. Rubies bring the highest incomes. Burma produces about 90 percent of the world's supply, with superb quality. The Valley of Rubies in the north is noted for quality gem production of both rubies and sapphires. Most of the gems are sold to buyers in Thailand. All the profits go to Burma's military rulers in the government, and since there is a high level of corruption and mismanagement within the government and business, the income from the gems produces limited economic development for the main population and discourages foreign investment in the country. Burma has become one of the poorest countries in Southeast Asia. China has emerged as the main trading partner with Burma and has been propping up the dictatorial military regime. China supplies the regime with arms, constructs many of the infrastructure projects, and supplies natural gas to the country.

Burma is ethnically diverse. Though it is difficult to verify, the government of Burma recognizes one hundred thirty-five distinct ethnic groups within its borders. It is estimated that there are over a hundred different ethnolinguistic groups in Burma. About 90 percent of the population is Buddhist. This high level of diversity can allow for strong centrifugal forces that are not generally conducive to unity and nationalism. The heavy emphasis on the national military is one of the only centripetal forces within the population, even though the military leadership is also looked at with distain by those desiring more openness and democratic conditions.

Key Takeaways

- France and Britain colonized the mainland region of Southeast Asia. Burma was a British colony and the rest was under French colonial rule. The Japanese took control of the region briefly before World War II ended in 1945. Siam was the only area not colonized. Siam became the country of Thailand.
- Vietnam was divided by a Communist north and a capitalist south during the Cold War. Vietnam is emerging from decades of isolation to provide the global economy with a large low-cost labor pool that has been attracting foreign investments by multinational corporations.
- The rural and landlocked nation of Laos has strong Buddhist traditions and an agrarian society.
- Cambodia was impacted by the Vietnam War and then by the devastation of Pol Pot's Khmer Rouge radical experiment in agrarian socialism, which killed as many as 2.5 million people. Recovery has been slow, but the textile industry and tourism have contributed to economic growth.
- The Buddhist country of Thailand has been experiencing major economic development in recent decades and has established itself as a major economic power in the region. The modern capital city of Bangkok is a major center of manufacturing and cultural activities.
- The people of Burma (Myanmar) continue to suffer under an authoritarian regime that offers few civil rights or democratic processes to its people. Poor, isolated, and militarily controlled, Burma has been at the center of many human rights violations in recent decades with little response from the international community.

Discussion and Study Questions

1. What are the main physical features and qualities of the mainland region of Southeast Asia?
2. What are the two main core areas of Vietnam? Which river is associated with each city?
3. What prompted the United States to enter into the conflict in Vietnam? Did Vietnam attack the United States?
4. What river flows through parts of Laos? What is the main economic activity in Laos?
5. What geographic aspect isolates the country of Laos and restricts its globalization efforts?

6. Who was Pol Pot? What type of society did he attempt to create? What were some of his methods?
7. What were the Killing Fields? What people were targeted to be eliminated? Why were these people killed?
8. What attracts tourism to Cambodia, Laos, or Thailand? Why aren't more tourists going to Burma?
9. What is the main religion in Southeast Asia? Who is considered the defender of this faith?
10. Who is Aung San Suu Kyi? How is she a reflection of conditions in her country?

Geography Exercise

Identify the following key places on a map:

- Andaman Sea
- Golden Triangle
- Malay Peninsula
- Mandalay
- Naypyidaw
- Rangoon

11.3. THE INSULAR REGION (ISLANDS OF SOUTHEAST ASIA)

The insular region of Southeast Asia comprises Malaysia, Singapore, Brunei, East Timor, Indonesia, and the Philippines. Among Southeast Asian nations, East Timor attained its independence most recently, as noted in the preceding lesson. Comparing these island nations reveals significant variety in all dimensions. Significant disparities exist in cultural, economic, and political dynamics, as well as in the ethnic groups constituting the predominant majorities in each context. A significant degree of linguistic and religious variety exists. The physical geography differs among the islands; some include elevated mountainous terrain, while others are low-lying and rather flat. The active tectonic plate movements in the region induce earthquakes and volcanic activity, leading to infrastructure devastation and loss of life, both of which severely affect human activities. Economic forces continue to prompt the countries of Southeast Asia to enter into trade relationships that integrate them with global networks based on dependency and reliance. The old colonial powers may no longer control them politically but may affect them economically. The new dynamics of corporate colonialism, with their economic power located in the core economic regions, still seek to exploit the countries of Southeast Asia for their labor and resources. These Asian nations are working to develop their own economies and use their own labor and resources to gain national wealth and increase the standard of living for their people. Each country has to contend with globalization forces within the international network of economic relationships.

Malaysia

Malaysia is a country made up of various British colonies that came together as a federation and then became an independent country. Britain started establishing colonies in the region in the late 1700s. The two main areas include the western colonies on the Malay Peninsula and the eastern colonies on the island of Borneo. The western settlements were part of the Malay Peninsula, which included the colonies of **Pinang** and **Singapore**. Eventually, the British took control of the eastern colonies of **Brunei**, **Sarawak,** and **Sabah** on the island of **Borneo.** In 1957, the western colonies on the mainland peninsula broke from their British colonizers and became an independent country called the **Federation of Malaya.** In 1963, the British Borneo colonies of Sarawak and Sabah joined the Federation of Malaya to form the current country, which is called **Malaysia.** In 1965, Singapore broke off from Malaysia and became an independent country. Brunei, which was still a British protectorate, became independent in 1984.

Malaysia has two main land areas separated by the South China Sea. The regions of Sarawak and Sabah, on the island of Borneo, are called **East Malaysia;** the mainland on the Malay Peninsula is called **West Malaysia.** These regions have similar physical landscapes, which include coastal plains with nearby densely forested foothills and mountains. The highest mountains, rising 13,436 feet, are in East Malaysia on the island of Borneo. Located near the equator, Malaysia has a tropical Type A climate with monsoons regularly occurring from October to February.

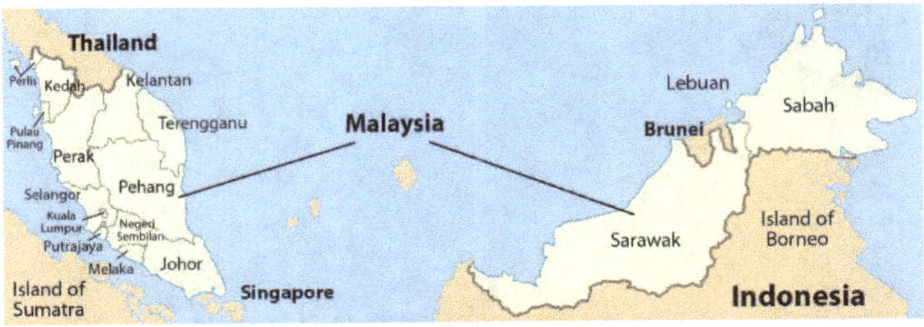

Figure 11.12: Provinces in East and West Malaysia

Diversity of Culture and Ethnicity in Malaysia

Malaysia's culture is diverse in that several major religions are practiced within its borders. Islam is considered the official religion and is supported by at least 60 percent of the population. About 20 percent of the people are Buddhists, 10 percent

Christians, and 6 percent Hindu. The remaining percentages of the population include traditional Chinese religions and local tribal beliefs. In this Islamic country, there are concerns that Muslims get preferential treatment by government programs and policies. There are even special judicial legal courts for Muslims only to work out issues regarding marriage, custody, inheritance, or other conflicting Islamic issues regarding faith and obligation. This court only hears Islamic issues and no other legal matters. There have been movements by minority extremist groups that would like to see Malaysia shift toward a true Islamic state, complete with the Sharia Criminal Code as the law of the land. The movement, however, has been cracked down on by the government. Since the 9-11 incident in the United States, there has been more concern about extremist religious views.

People of Malay ethnic background make up more than 50 percent of the population. People of Chinese descent are the second-largest group at about 24 percent. An additional 11 percent of the population is made up of indigenous groups. During British colonialism, a number of people from South Asia were brought to Malaysia. For example, Tamils were brought from India to work the plantations. Their Hindu beliefs were infused into the culture and some Tamils also converted to Christianity. Sikhs were brought from South Asia to help Britain run the country as police, soldiers, or security officers. The Sikhs who came brought their religion with them, which added to the multireligious dynamics of the country.

Figure 11.13: The Tuaran Road of Kota Kinabala City in Malaysia during a Time of Slow Traffic Notice that the cars are driving on the same side of the road as they would be in Great Britain, Malaysia's former colonizer. Jason Thien – Busy Roads of Kota Kinabalu City – CC BY 2.0.

Malaysia's diverse ethnic and cultural mix often results in strong centrifugal forces that push and pull on the societal dynamics of the country. China has been active in the business community and has established strong economic ties with regional countries that have Chinese populations. The single largest minority group in the province of Sarawak on Borneo is Chinese. As a minority group, Chinese citizens of Malaysia have felt discrimination. Since the official language is Malay and the official religion is Islam, there have been concerns about discrimination against all minority groups. Working through the cultural and ethnic diversity has been a major challenge for the country. Each minority religious or ethnic group desires to celebrate its own special holidays. For example, there is the usual New Year's celebration on January 1, and then there is the traditional fifteen-day Chinese New Year celebration celebrated at a different time of the year. Sikhs celebrate the Sikh New Year. Buddhists celebrate a holiday in honor of the life and enlightenment of Buddha. Christians celebrate Christmas and Easter. Many other holidays of significance are respected or honored by various minority groups.

Economic Development in Malaysia

Malaysia has rapidly advanced its economy in recent decades and is modernizing its infrastructure—roads, bridges, highways, and urban facilities. In the capital city, **Kuala Lumpur**, Malaysia built a modern central business district with a twin high-rise office building claimed to be the world's tallest at the time of construction. Before the global economic downturn that started in 2007, Malaysia had developed a fast-growing economy and was industrializing at a rapid rate. Malaysia has taken advantage of its location on a major shipping lane and has shifted to manufacturing as an important sector of its economy. The country has been a leader in the export of natural resources such as tin, rubber, and palm oil and has developed its agricultural and extractive sectors to gain income. The 1980s and 1990s were prosperous times for the country and it matured its manufacturing base from light textiles into electronics and heavy industries.

One aspect of the country that is looming on the horizon and may cause problems is the high population growth rate. In 2010, Malaysia's population was estimated at more than twenty-five million, with a doubling time of about forty years. Though the country is 70 percent urban, family size (fertility rate) is still at about 3.0, which indicates an increasing population growth pattern. One-third of the population is under the age of fifteen. Malaysia is one case where the general principle that if a country urbanizes and industrializes the family size will go down has not taken place fast enough. The fertility rate has dropped from 5.0 to 3.0, but it needs to get below a rate of about 2.0 if the country is going to successfully stabilize its population growth. Unless the country addresses this population growth, the demand for resources might outstrip economic progress in the future.

Singapore

Under British colonial rule, the island of Singapore was included in the Malaysian federation. It broke away and became independent in 1965. It is a small island measuring about thirty miles long at its widest point. Singapore is about two hundred forty square miles in area. Singapore's most valuable resource is its relative location. Singapore is similar to Hong Kong in its development. With a good port, Singapore is a hub for ships sailing between Europe and China. It serves Southeast Asia as an entrepôt, or break-of-bulk point, where goods are offloaded from large ships and transported to smaller vessels for distribution to the Southeast Asian community.

Figure 11.14: Singapore and Neighboring Countries

Singapore has made good strategic utilization of its geographic location by serving as a distribution center for goods and materials processed in the region. Crude oil from Indonesia is unloaded and refined here. Raw materials are shipped in, manufactured into finished products, and then shipped out to global markets. Since Singapore is small, it has had to concentrate on manufacturing goods that provide for optimal profits. As an economic tiger, Singapore has transitioned through the same stages as Taiwan, South Korea, and Hong Kong to become an economic power in Southeast Asia.

To keep labor costs low, initial manufactured products were textiles, clothing, and cheap goods. As incomes and labor skills rose, so did the complexity of the manufactured goods. The government of Singapore has targeted certain types of products to ensure a high profit margin and a global market need. This has included automation equipment, biotechnology, and high-end information technologies. Singapore doesn't manufacture automobiles but it does manufacture automation robotic components that most modern auto manufactures will purchase and use. Medical technology is expensive and is in high demand the world over. Singapore is targeting this market. The information age has spawned new technologies that are evolving rapidly and, once again, Singapore has been at the center of this industry. Singapore has been a center for the production of computer disc drives for a multitude of global corporations.

Singapore Island is a swampy place with no natural resources. All production components, food goods, construction materials, and energy must be imported. Importing everything has raised the cost of living. To compete with the other Asian economic tigers in the global marketplace, Singapore has implemented severe control measures on its operations. There are harsh penalties for criminal activities and for even misdemeanor offenses. Singapore is a safe place to live because of its strict state rules. It has an authoritarian government, which strives to create an attractive place for international corporations to operate. One of the objectives is to eliminate corruption and establish a business-friendly environment.

Figure 11.15: Singapore's Modern Development Modern high-rise office buildings fill the central business district of Singapore. The administrative district can be seen in the foreground complete with British colonial architecture, established when Singapore was a British colony. Brian Evans – Singapore – CC BY-ND 2.0.

The government of Singapore has entered into trade agreements with two of its neighbors to provide raw materials and cheap labor. A trade triangle has been established between Singapore, Malaysia, and Indonesia. Malaysia and Indonesia provide Singapore with raw materials and cheap labor; Singapore provides its neighbors with technical know-how and financial support. Everyone benefits. Singapore is an excellent example of the upper end of the economic spectrum in Southeast Asia. Countries like Laos or Vietnam would be at the opposite end, since they have a largely rural population based on agriculture that is just beginning to shift to the cities with industrialization. Singapore is already 100 percent urban with high incomes based on high-tech manufacturing and processing of raw materials. Singapore is an economic hub for Southeast Asia, complete with global airline connections and is located on a major shipping lane. Singapore's world-class port is one of the busiest in Asia. The rest of Southeast Asia is somewhere in between these two ends of the spectrum as far as economic development is concerned.

Indonesia

The country of Indonesia is the world's largest archipelago state, consisting of more than 17,500 islands, about one-third of which are inhabited. Indonesia is the sixteenth-largest country in the world by area. The combined area of all the islands and regions of Indonesia would equal about the size of the country of Mexico. The country shares land borders with the Borneo side of Malaysia, the western half of the island of Timor, and the western portion of the island of New Guinea, which is shared with the country of Papua New Guinea.

The country's location on both sides of the equator provides a tropical Type A climate, complete with a monsoon season. Average rainfall can vary from seventy to two hundred forty inches per year. The highest mountain is in **West Papua** and rises to about 16,024 feet. Indonesia is located on the Pacific Rim, where tectonic plate activity produces earthquakes and volcanic activity. The country is home to over one hundred fifty active volcanoes, including two of the most famous ones, **Krakatoa** and **Tambora**. Both had devastating eruptions in the past two centuries. One of the most violent volcanic explosions ever recorded in human history came from Krakatoa, which is located between the islands of **Java** and **Sumatra**. A series of eruptions in 1883 were heard as far away as the coast of Africa. Shockwaves reverberated around the globe seven times. Ash erupted into the atmosphere to a height of about fifty miles.

Figure 11.16: Major Islands and the Thirty-Three National Provinces of Indonesia. Wikimedia Commons – CC BY-SA 3.0.

The official death toll was 36,417, but estimates from local sources place it as high as 120,000. Global temperatures fell by about 2 °F, and weather patterns were disrupted for the next five years. Krakatoa remains active. Over the past few decades, the volcanic peak has been growing at the average rate of about five inches per week.

The tropical climate and the archipelago nature of the country provide for enormous biodiversity within the environment. Second only to Brazil in its biodiversity, Indonesia is host to an enormous number of unique plants and animals. The habitats of many of these creatures are being encroached upon by human activity. The remote islands have more of a chance of escaping habitat devastation and remaining intact, but agricultural and extractive economic activities have converted much of the natural environment into a cultural landscape that is not conducive to environmental sustainability.

Animals such as orangutans are losing their natural forests and may become extinct if current trends continue. The timber industry has brought about deforestation. Slash and burn agriculture has destroyed forest habitat, and human development patterns such as roads and urbanization have altered the ecosystems of the region. According to recent reports, Indonesia is one of the largest emitters of carbon dioxide in the world because of the high number of forest fires set each year. In 2009, the United States brokered a deal with Indonesia to forgive thirty million dollars of its debt if the country would work to protect forests on the island of Sumatra, which is home to endangered indigenous animals such as tigers, elephants, rhinos, and orangutans.

In 2010, the estimated population of Indonesia was about 245 million. Indonesia has the fourth-largest population of any country in the world, after the United States, India, and China. Indonesia also has more Muslims than any other country in the world. More than half the population of Indonesia lives on Java, the island where Jakarta, the capital city, is located. Java is the most populous island in the world, and has a population density of more than 2,400 people per square mile. Java is the size in area of the US state of Louisiana. Java has 135 million people, whereas Louisiana has 4.5 million people. Jakarta is a world-class city that is larger than New York City and encompasses a large metropolitan area, complete with many manufacturing centers, business complexes, and housing districts.

The many islands of Indonesia are home to a large number of diverse ethnic and religious groups that vary as widely as any Southeast Asian nation. There may be as many as three hundred different and distinct ethnic groups in Indonesia. Many of the ethnic groups are further divided by islands or distance. More than two hundred fifty separate languages and hundreds of additional dialects are spoken. There are an estimated seven hundred fifty languages spoken on the island of New

Guinea itself, with hundreds of them spoken on the Indonesian side of the island, in a population of less than three million. The most prevalent language group in the country as a whole is Javanese, which is spoken by about 42 percent of the population. Javanese includes the official language of Indonesian, which is taught in schools and used in business and politics as the *lingua franca* of the country. Many people speak more than one language or even a number of languages to communicate throughout the country.

Islam was diffused to Indonesia in the thirteenth century and by the sixteenth century had become the dominant religion. The Indonesian constitution allows for religious freedom, although more than 85 percent of the population follows Islam. There are at least four other religions that are officially recognized: Christianity (both Protestantism and Roman Catholicism), Hinduism, Buddhism, and Confucianism. Since Islam is followed by such a large percentage of the population, the other religions do not carry the same influence. Regional and ethnic differences play a role in the varied religious dynamics. The island of **Bali**, for example, is home to a majority Hindu population. Most of the Buddhists are ethnically Chinese, and they only make up a small percentage of the population. Christians and Muslims have had conflicts on the island of Sulawesi. It is common to find the practice of these religions less than orthodox in the more rural communities of the country.

Figure 11.17: Skyline of Jakarta, the Capital of Indonesia Jakarta has a population of about ten million and is located on the island of Java. Java has more than 135 million people and has about the same physical area as the US state of Louisiana, which has about 4.5 million people. Java has more people than any other island in the world. Global Panorama - Jakarta, Indonesia - CC BY-SA 2.0.

In spite of the diversity within the population, the country of Indonesia has established a substantial degree of nationalism as a centripetal force that holds the country together. There is a relatively high degree of stability in spite of the surface tensions or ethnic and religious conflicts that may erupt. An example of the social tensions is demonstrated in the case of Chinese citizens of Indonesia, who only make up about 1 percent of the population but impart a substantial influence over the privately owned business sector of the economy. This seemingly inequitable relationship has resulted in considerable resentment by other portions of the population, often with violent results. The many islands have become natural divisions between cultural groups.

Some of the islands—or portions of them—have attempted to break away in a devolutionary manner and become independent countries. Just as East Timor became independent, the most western province of **Aceh** on the island of Sumatra had a similar movement toward independence. West Papua on the island of New Guinea has also had an independence movement. The Aceh situation was negotiated out while the West Papua movement has been suppressed by military and political force. Many of the islands possess large amounts of natural resources, so the country of Indonesia does not want to lose these national assets that could prove valuable in gaining wealth for the future. It is not easy to create national unity with such a diverse population scattered throughout such a large archipelago.

Agriculture has been the historic base of the Indonesian economy. In 2010, it accounted for about 13 percent of the nation's gross domestic product (GDP). Agriculture is the largest employment sector—approximately 42 percent of the workforce. This equates to more than half of the population being rural. Many of the agricultural methods in rural areas are traditional; for example, farmers use water buffalo or oxen for tilling the land. The tropical climate and adequate rainfall provide for multiple crops of rice per year in many areas. Spices, coffee, tea, palm oil, and rubber are also produced in substantial quantities.

Industries are an important building block for how a country gains wealth. In the case of Indonesia, industry accounts for about 40 percent of its GDP and employs about 20 percent of its workforce. Major industries include oil, natural gas, mining, and textiles or clothing manufacturing. Indonesia's economy has been affected by global markets, but in 2005 still managed to run a trade surplus. Japan has been its main trading partner, and China has also been a major supplier of imported goods. Indonesia has been taking advantage of the trade triangle it has with its neighbors, Singapore and Malaysia, to increase its import and export trade activities.

The political background of Indonesia is similar in dynamics to many of its neighbors. Colonized by Europeans, Indonesia was previously called the **Dutch East Indies**, which explains why the islands of the Caribbean were called the **West Indies**. The Dutch colonized Indonesia in the early seventeenth century but had to relinquish possession of the archipelago to the Japanese in World War II. In 1945, after the Japanese surrendered, Indonesia declared its independence, which was finally granted in 1949 after much negotiation. The country's government quickly moved toward authoritarian rule.

During a fifty-year time period, there were only two authoritarian leaders: **Sukarno** (1949–68) and **Suharto** (1968–99). Near the end of Sukarno's rule, there were violent conflicts between Sukarno's military and the Communist Party of

Indonesia, which resulted in more than five hundred thousand deaths. Suharto's regime was credited for substantial economic growth but was also accused of serious corruption and the repression of opposition political voices. Since 1999, Indonesia has conducted free parliamentary elections and is now considered the third-largest democracy, after India and the United States.

Brunei

There are noticeable similarities between the oil-rich sheikdoms of the Persian Gulf region in the Middle East and the small sultanate of Brunei on the northern coast of Borneo. Bordered by Sarawak, the sultanate is actually two small separate regions along the coast of the South China Sea. The former British protectorate of Brunei is today a major oil and natural gas exporter. It provides a high standard of living for its small population. The compact country is about the size of the US state of Delaware. The country's population for 2009 was listed at about 388,000. Brunei is attracting immigrants seeking opportunities and advantages. It is called a sultanate because the kingdom has been ruled by sultans (rulers) from the same family for the past six centuries.

Figure 11.18: Brunei. CIA World Factbook – public domain.

The main ethnic groups in Brunei are Malay, at 66 percent, and Chinese, at 11 percent. Brunei is an Islamic State with Islam as its state religion. About two-thirds of the population is Muslim. Buddhism is the second-most popular religion. The ruling sultan is not only head of state but also prime minister of the government and leader of the Islamic faith. Similar to states in the Middle East where Islam is the official religion, alcohol is banned and the public consumption or sale of it is illegal. Prohibition against alcohol has eliminated the establishments of pubs and nightclubs. Non-Muslims and visitors to the country can legally hold small quantities of alcohol for personal consumption.

The people of Brunei have a high standard of living, with the availability of modern amenities. The government has been concerned about integrating the country into the global economy. Natural gas and crude oil bring in about 90 percent of exports and just over half of the GDP. Education and medical care is free. Food, housing, and rice farming are subsidized by the state. The state has been working to expand the economy beyond natural gas and oil. Agricultural production has been increased and unemployment has been a major focus. The wealthy emirate has also been developing its tourism sector and the financial and banking industry.

Brunei may have to take a lesson from the United Arab Emirates (UAE)—that is, to work to develop a free trade zone to attract international trade—if the country wants to continue to gain wealth once the oil and natural gas run out. It has an excellent location on the South China Sea but would have to compete with the established economic tigers of Singapore and Hong Kong as well as the other rising urban centers in the region, such as Kuala Lumpur or Bangkok.

The Philippines

Located on the eastern side of the Southeast Asian community is the archipelago state of the Philippines. With more than 7,100 islands, many volcanic peaks, and an expanse of coastal waterways, the Philippines is home to more than ninety million people in a combined land area about the size of Arizona. The Philippines were a Spanish colony. The name is taken from Spain's sixteen-century King Philip II. Spain relinquished its claim on the Philippines to the United States in 1898 after its defeat in the Spanish-American War. The people of the Philippines wanted independence at that time and fought a bitter war with the United States in which more than a million people died. The United States allowed the Philippines to become a commonwealth in 1935. The independence movement was placed on hold while the Japanese invaded and controlled the Philippines during World War II. After the war was over, the United States granted the Philippines their independence in 1946.

Environmental Forces

The islands of the Philippines are of volcanic origin. They are mainly mountainous and covered in tropical rainforest. The highest mountain, at 9,692 feet, is Mt. Apo, which is located on the southern island of **Mindanao**. The Philippines has a number of active volcanoes. The northern island of Luzon is home to the Taal Volcano, Mt. Pinatubo, and Mt. Mayon. The Pacific tectonic plate reaches the southern edge of the Philippine plate where it meets up with the Eurasian Plate. The juncture of tectonic plates creates a similar situation to that of Tokyo, which is at the opposite end of the Philippine plate. Active seismic forces result in many earthquakes. As many as twenty earthquakes a day can be registered here, though many are too weak to be noticed. In 1990, an earthquake on the island of Luzon registered at a magnitude of 7.8 and killed more than 1,621 people, causing extensive damage.

Figure 11.19: Ash Plume from Mt. Pinatubo during 1991 Eruption The island of Luzon in the Philippines has a number of active volcanoes. Ash from Mt. Pinatubo caused so much damage that it resulted in the permanent closure of major US military bases in the Philippines. UGS – public domain.

Luzon's Mt. Pinatubo volcano has been active in recent years. Before 1991, the mountain attracted little attention, was heavily forested, and was home to tribal indigenous people. The volcano had a colossal eruption in 1991 that was recorded as the second largest in a century, after Alaska's 1912 Novarupta eruption. Mt. Pinatubo began giving signs of an eruption, which were heeded by the government. Thousands of people were evacuated from the area, which saved many lives. The eruption caused billions of dollars in damage. More than eight hundred people were killed, and more than two million were directly impacted. The eruption destroyed more than eight thousand homes and the overall effects of the volcano were felt around the world.

Mt. Pinatubo's eruption forced billions of tons of magma, ash, sulfur dioxide, minerals, and particulates into the atmosphere and onto the earth's surface. The sun was blocked out, temperatures dropped, and ash piled up in nearby areas, causing extensive damage to roofs, roadways, and agricultural lands. The damage from the eruption was amplified by the fact that a full-scale typhoon hit the country on the same date, bringing torrential rainfall and wind that mixed with the ash in the air to create extremely dangerous environmental conditions. The damage had a massive impact on the entire economy of the Philippines.

The eruption severely damaged civilian infrastructure and US military bases in the region. The Subic Bay Naval Base was fifty miles to the southwest of Mt. Pinatubo's summit, while Clark Air Base was less than sixteen miles to the east. Enormous clouds of ash covered everything. As a result of the damage to the operations at the bases, the United States Air Force evacuated and moved all air base personnel and military assets to bases in Guam, Okinawa, or Hawaii. The United States ultimately abandoned Clark Air Base, while Subic Bay reverted to the Philippines. There are thirty-seven volcanoes in the Philippines, of which eighteen are still active. Mt. Mayon is the most active volcano at the present time. It has had forty-seven eruptions in recorded history. The eruption in 1993 killed sixty-eight people and caused the evacuation of sixty thousand more.

Earthquakes and volcanoes are not the only serious natural concerns of the Philippine Islands; they are also directly in the center of the Western Pacific's major typhoon belt. As many as twenty typhoons occur yearly in the area of the islands, and as many as half hit the islands directly. The 1991 typhoon Thelma/Uring killed as many as eight thousand people. The 1911 typhoon dumped over forty-six inches of rain in a twenty-four-hour period. Flooding is usually the main problem with typhoons and is the number one killer related to typhoon deaths. For more information, see Chapter 5 "Middle America", Section 5.5 "Tropical Cyclones (Hurricanes)". Typhoon activity also brings precipitation to the islands and the region. The Philippines are in the major path of typhoons in the Pacific and will continue to combat the effects of these powerful forces of nature.

Political Geography

The Philippines can be divided into three main geopolitical regions: **Luzon, Visayas,** and **Mindanao.** The northern island of Luzon is home to the nation's **national capital region** with Quezon, the largest city, and Manila, the capital. Both cities are a part of **metropolitan Manila,** which has a population of more than twenty million. The northern island of Luzon is home to half the population of the country. The central Philippines consists of the **Visayas Island group,** including the islands between the Sulu Sea and the Philippine Sea. The southern region of the country is anchored by the large island of Mindanao.

Figure 11.20: The Three Main Regions of the Philippines. CIA World Factbook – public domain.

The government of the Philippines is a constitutional republic with an elected president. With independence in 1946 came various leaders who have shaped the political landscape of the Philippines. After recovering from the devastation of World War II, the country prospered during the 1960s and showed positive economic gains. The political scene entered a difficult political era with the election of **President Ferdinand Marcos** in 1965, which turned into an authoritative dictatorship. During his time in power, the economy became sluggish and social unrest began to arise in opposition to his leadership.

Barred by law from being elected for a third time, Ferdinand Marcos declared martial law in 1972 under the premise that there was too much political conflict with Communist elements and Islamic insurgencies. Marcos ruled with his wife **Imelda Marcos** until 1986, when conditions worsened and the two were implicated in the assassination of opposition leader Benigno Aquino. Corruption, vote rigging, and the dictatorial actions of President Marcos caught up with him through mass protests, which eventually led to his removal from office. He left the Philippines for his exile in Hawaii. It was later alleged that during his twenty years in office, Ferdinand and Imelda Marcos had embezzled billions of dollars of public funds and moved them to bank accounts in Switzerland, the United States, other countries, and into fictitious money-laundering corporations. Ferdinand Marcos died of illness in 1989 in Honolulu.

Imelda Marcos returned to the Philippines to run for public office and even attempted a failed run for the presidency. Imelda was known for her thousands of shoes, which she had acquired while in power. Many of them are in a shoe museum in the Philippines. She was also known for her extravagant spending trips around the world. Several different political leaders have come to power since the Ferdinand Marcos era. Political stability has been difficult to achieve. The national leadership has faced Islamic insurgencies, attempted coups, corruption in the government, and a high national debt. These issues continue today but a modest level of stability has encouraged economic growth.

Cultural Geography

The Philippines is a diverse country with hundreds of ethnic groups. Many tribal groups as well as a large number of immigrants from Asia, Spain, and the United States have made the Philippines home. Together with Spanish influence, mixed ethnic groups have been created. They are an example of the confluence of cultures that make up the country. The Philippines is the only country in Asia where Roman Catholicism predominates, other than recently independent East Timor.

Figure 11.21: Ferdinand Magellan brought Christianity to the Philippines and converted members of the Cebuano tribe to Catholicism in 1521. Lezumbalaberenjena – Catedral de Santa Clara – CC BY-NC-ND 2.0.

Christians make up about 90 percent of the population. All but 10 percent identify themselves as Roman Catholic. A modest Muslim population is prominent in the southern island of Mindanao and neighboring islands. Islamic fundamentalism has increased the insurgency in the region, causing political and economic turmoil and conflict. People of Chinese heritage often follow Buddhism, Taoism, or Chinese folk religions. Various tribal groups still follow their cultural animist beliefs and have traditional shaman religious leaders.

The Philippines is home to more than one hundred eighty native languages and dialects. English and Filipino were declared the official languages of the Philippines in 1987. Tagalog is the main language spoken. Filipino is a version of Tagalog that is used in many of the urban areas. English and Tagalog are used in different parts of the country. The population growth rate is considerable. The Philippines will soon push past the one hundred million mark, at which point it will become a country in which 35 percent of its citizens are under the age of fifteen. Average family size is more than 3.2, which will continue to influence the economic situation of the country.

The Global Economy and Outsourcing

The modest level of political stability has caused the Philippines to become an attractive destination for global corporations who seek to **outsource** their information and technology service jobs. Any work that can be conducted over the Internet or telephone can be outsourced to anywhere in the world with high-speed communication links. Countries that are attractive to **business process outsourcing (BPO)** are countries where the English language is prominent, where employment costs are low, and where there is an adequate labor base of skilled or educated workers that can be trained in the services required. All three of these requirements are met by the labor force of the Philippines. The historical influence of the United States has provided a base of English language speakers. The country also has an adequate population base with the education or professional skills necessary to meet these demands.

Figure 11.22: Street Scene in Manila with Jeepney on the Left and a Bicycle Card Front and Center The greater metropolitan region of Manila has more than twenty million people. The city of Manila itself is one of the most densely populated large cities in the world. Stefan Munder - Manila - CC BY 2.0.

Corporate colonialism has the Phili-ppines in its business focus and is finding a good source of available labor.

In 2005, information technology and BPO amounted to about thirty-four billion dollars globally. Since 2005, that amount has increased dramatically, doubling and tripling in some countries by 2009. India has been a major destination for BPO, but the Philippines is gaining ground and increasing its infrastructure in an attempt to gain a larger share of the market. Other countries around the world are a part of this outsourcing market. This type of business activity shifts jobs from one country to another. A country might lose these types of jobs, but its corporations can remain competitive in the global marketplace if they can cut costs of operation by outsourcing their service work to a low-cost country.

Jeepneys

The term jeepney is derived from the use of early US army jeeps left over from World War II that were used as base vehicles transformed into a type of taxi. These transformed vehicles took on a cultural identity as jeepneys with their flamboyant colors and extended seating. Jeepneys are now produced for this purpose and are the most widely used public transportation mode in the Philippines. An electric version of the jeepney is being developed for a number of Asian countries.

US corporate giants like America Online, Texas Instruments, Citibank, Hewlett Packard, JPMorgan Chase & Co., and the McClatchy Company (third-largest US newspaper company) have been shifting call centers and other back-office functions to the Philippines. Other European companies like Germany's global Siemens Corporation, Royal Dutch Shell, Swedish Telecom provider Ericsson Telecommunications, and Danish shipping giant Maersk are examples of corporations that have established outsourcing centers in the Philippines. The economic savings can be considerable. BPO wages in the Philippines are one-fifth of the wages paid for the same jobs in the United States. Those same wages are double the national average wages for Philippine employees. A rise in the number of outsourced jobs is welcome news for the Philippines, whose economy is in need of a boost.

East Timor (Timor-Leste)

Timor is an island of southern Indonesia not far from Australia. The island is divided by its colonial history. The eastern half was a Portuguese colony beginning in the sixteenth century. Portuguese colonizers introduced Christianity in the form of Roman Catholicism. The western half was associated with Indonesia, which was a Dutch colony during the colonial era. The Japanese occupied the Dutch colony during World War II but had to give it up after they surrendered in 1945. Indonesia received its independence in 1949 and laid claim to the whole island of Timor. East Timor made a declaration of independence in 1975 but was occupied by Indonesia. A bitter civil war erupted. A year later, Indonesia declared it its twenty-seventh province. The civil war resulted in the deaths of as many as two hundred fifty thousand people. It wasn't until 1999 that Indonesia finally ceded its political control over East Timor. The Australian military has been instrumental in securing East Timor for independence, and has been serving as a peacekeeping force for internal security for the past decade. The United Nations (UN) recognized East Timor as a sovereign independent country in 2002. The official name of the country is listed as Timor-Leste.

Timor-Leste has a population of about 1.2 million. About 98 percent of the population is Roman Catholic. The only other predominant Catholic country in Asia is the Philippines. About 90 percent of the population still works in agriculture. The country has had a difficult time establishing a stable government and reducing conflict. Almost all its infrastructure was damaged in the civil war and rebuilding has been slow. Poor and impoverished due to the civil war over independence, the country does have some opportunity derived from the large natural gas field in the vicinity. East Timor has been working to gain control of its maritime boundaries to benefit from the offshore natural resources.

Key Takeaways

- Malaysia was a former British colony made up of various regions from both the Malay Peninsula and the island of Borneo. Modern Malaysia has diverse cultural dynamics and is modernizing its economy to compete with the core economic areas of the world.
- Singapore is an economic tiger that doesn't have natural resources but makes good use of an excellent location. High-tech manufacturing has been Singapore's main method of gaining wealth.
- Indonesia is made up of thousands of islands and hundreds of ethnic groups. Indonesia is the fourth-most populous country in the world and has the world's largest Muslim population. More than half the population lives on the island of Java.
- The Philippines has more than ninety million people on thousands of islands. The country was colonized by Spain and was then a possession of the United States before it gained independence. Roman Catholicism and the English language are common in the Philippines, both of which augment a large outsourcing industry.
- Brunei is a small Muslim emirate with high incomes because of oil revenues. East Timor is half of a small island north of Australia. It is a former Portuguese colony and just gained its independence in 2002.

Discussion and Study Questions

1. Which European country colonized each of the Southeast Asian nations discussed in this lesson?
2. What is the dominant religion in each country? Name the two Roman Catholic countries in Asia.

3. What cultural and ethnic issues does Malaysia need to address?
4. What type of economic activity has Singapore engaged in to gain wealth?
5. How does the growth triangle that Singapore is engaged in work? How does each partner benefit?
6. How does an *entrepôt* fit into the core-periphery spatial relationship in Southeast Asia?
7. What environmental problems exist in Indonesia and the Philippines? Which are natural phenomena?
8. Where are devolutionary forces active in Indonesia? How could the government address them?
9. What are the three main regions of the Philippines? Which region has the largest population?
10. What qualities or conditions are necessary for BPO?

Geography Exercise

Identify the following key places on a map:

- Aceh
- Bali
- Borneo
- East Malaysia
- Java
- Luzon
- Mindanao
- Pinang
- Quezon
- Sarawak
- Sabah
- Sumatra
- Visayas
- West Papua

11.4. CONCISE SUMMARY

Southeast Asia comprises two primary physical regions: the mainland area adjacent to China and the insular region consisting of islands or parts thereof situated between Asia and Australia. Borneo, a huge island, is divided among Brunei, Malaysia, and Indonesia.

The sole territory in Southeast Asia that evaded colonization by European powers was the Kingdom of Siam, now incorporated within modern-day Thailand. This territory, colonized by France, is frequently designated as French Indochina. Britain, Holland, Portugal, and Spain were significant colonizers in the region.

Southeast Asia exhibits diversity in both its human and physical settings. Tropical climates prevail, with mountains and coastal regions encompassing the primary land surfaces. This region has significant seismic activity, featuring numerous active volcanoes and a susceptibility to earthquakes.

All principal world religions are represented here. Indonesia is the most populated nation with a Muslim majority globally. East Timor and the Philippines are the sole largely Christian nations in Asia. Buddhism is the predominant religion in the mainland area. Malaysia and Singapore both include substantial Hindu minority populations.

Economic activity in Southeast Asia differ, with Singapore recognized as an economic powerhouse and Brunei identified as an oil-rich emirate. Thailand is emerging as a significant industrial hub, while the Philippines has established itself as a destination for outsourced information technology positions. Landlocked Laos and isolated Myanmar possess fragile economies. Vietnam and Cambodia are emerging from political isolation.

Indonesia possesses the fourth-largest population globally. Fifty percent of its population resides on the island of Java. Bali, an Indonesian island, has a Hindu majority and is a prominent tourist destination. The island of Timor is partitioned into an Indonesian western segment and the autonomous eastern segment of East Timor, a former Portuguese colony.

12 Chapter | AUSTRALIA AND NEW ZEALAND

12.1. INTRODUCTION: AN OVERVIEW

Australia and New Zealand, two remarkable nations in the Southern Hemisphere, are renowned for their stunning landscapes, rich indigenous cultures, and diverse ecosystems. Australia, the larger of the two, boasts iconic landmarks such as the Great Barrier Reef and Uluru, while New Zealand captivates with its breathtaking fjords, lush forests, and vibrant Maori heritage. Both countries offer unique experiences, from bustling urban centers like Sydney and Auckland to the serene beauty of their natural environments. Together, they embody a blend of adventure, culture, and natural wonders that continue to draw travelers from around the globe.

Figure 12.1 Australia and New Zealand

Isolation Geography

Figure 12.2: Wallace's and Weber's Lines were developed independently to account for the differences in biodiversity between the Austral realm and the Asian realm. Scientists continue to analyze the true boundary between the realms. These lines demarcate a clear environmental difference in species development between the two sides. CIA World Factbook – public domain.

The historic isolation of New Zealand and Australia from the rest of the world has caused animals and organisms that are not found anywhere else to develop in these two countries. The unique biodiversity includes **marsupials**, or animals whose young are raised in the mother's pouch, such as kangaroos, wallabies, koalas, and bandicoots. It is believed that these creatures developed separately after the continents broke away from each other more than two hundred million years ago. Many plant species are also unique to this realm. The biodiversity found here is separate from that of Asia. This has been explained by various biogeographers by drawing imaginary lines just north of Australia to indicate the line of division between the Asian realm and the Austral realm. Wallace's Line and Weber's Line are two such examples. Both examples attempt to establish the correct line of demarcation for the differences in species development between the two sides. During the ice ages, sea level was lower, and the many islands of Southeast Asia were connected by land to the mainland. Papua New Guinea was connected to Australia. Wallace and Weber believed that no land bridge connected the Asian side with the Austral side for animals to cross over. This separation caused the organisms to the south to develop independently of those in the north. For example, marsupials are not found on the Asian side of these lines but are found on the Australian side.

Colonialism

New Zealand and Australia were both inhabited before the era of European colonialism. **Aboriginal** people are said to have migrated to Australia across Southeast Asia from the mainland of Asia more than forty thousand years ago. They made Australia their home and adapted to the physical geography of the continent. For tens of thousands of years before the Europeans arrived, the

Aborigines carved out an existence in Australia and developed their cultural ways. Only about four hundred fifty thousand Aborigines remain in Australia today. New Zealand was inhabited by the Polynesian group called the Maori who established themselves on the islands in the tenth century. For hundreds of years they, too, established their culture and traditions in the region before the Europeans arrived. The Aborigines in Australia and the Maori in New Zealand were both confronted with the European invaders. From their standpoint, there was much to lose by the arrival of the Europeans. Lands were lost, new diseases killed many, and control of their methods of livelihood were taken over by Europeans. The Maori initiated a number of wars against British colonizers, but in the end the greater military power gained the advantage. At the present time, the Maori make up less than 10 percent of the population of New Zealand.

The sighting of Australia by the Dutch dates to 1606. Portuguese explorers may have discovered Australia earlier, but there are no written records. In the early 1700s, the northern and western coastlines of Australia were known as "New Holland." There were no established colonies. James Cook, a naval officer working for the British navy, commanded the good ship *Endeavor* and mapped Australia's eastern coast in 1770. He made port at Botany Bay, just south of the current city of **Sydney** and claimed the region for Britain. He named the land New South Wales. The charting of the coast resulted in continued attention being paid to the region.

Meanwhile, England had a severe problem with overcrowding of its prisons. Its problem was exacerbated by the loss of Britain's American colonies. Upon Cook's return to England, interest was generated in the concept of relieving prison overcrowding by sending prisoners to Australia. In 1787, eleven ships with seven hundred fifty convicts sailed from Great Britain to Botany Bay. Prison colonies were established in Australia. By the end of the seventeenth century, the entire Australian continent was under the British Crown. At the same time that the movement of prisoners from England to Australia was diminishing, the next wave of immigration was being fueled by the discovery of gold in the 1850s. The practice of transferring prisoners to Australia ended in 1868. The arrival of the Europeans had caused a serious demise in the Aboriginal population. Aborigines were completely decimated in Tasmania.

In 1901, the various territories and states of Australia came together under one federation called the Commonwealth of Australia. A new federal capital city of Canberra was proposed. By 1927, Canberra was ready for government activity. This commonwealth government still allowed for individual state differences. The British monarch is considered the head of state, though it is mainly a ceremonial position. There have been movements within Australia in recent years to separate from the British Crown, but they have not been approved. Australia has a democratically elected government.

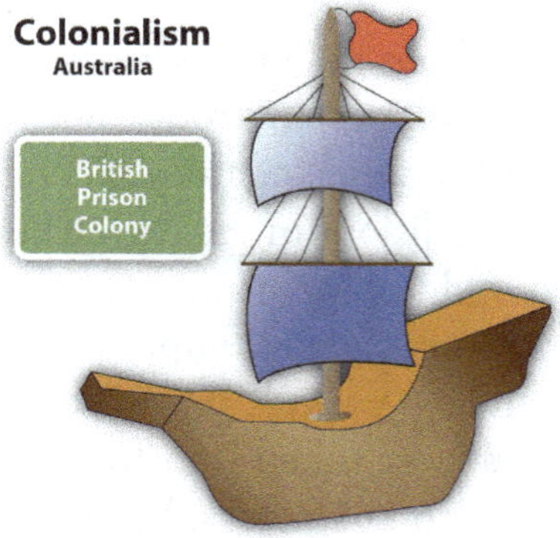

Figure 12.3: Great Britain colonized Australia by establishing prison colonies. The prison colony of Botany Bay was located near the current city of Sydney, Australia.

British naval officer James Cook mapped the coastline of New Zealand in 1769. As the colonial era emerged, Great Britain took possession of New Zealand and included it with its colony of New South Wales. In the 1840s, New Zealand became a separate crown colony. The colony developed a local parliament and a representative government. By 1893, New Zealand made headlines as the first country in the world granting all women the right to vote. As a part of the British Empire, the country was made a commonwealth nation in 1947 and has been functioning independently ever since.

Key Takeaways

- Australia is an island continent that was home to aboriginal people who have lived there for tens of thousands of years. The British colonized Australia by first creating prison colonies for convicts from Great Britain.
- New Zealand has two main islands and is home to the Maori, who were originally from Polynesia. The British colonized New Zealand and often were in conflict with the Maori.
- Australia is relatively flat with low elevation highlands and an extensive dry interior, while New Zealand has high mountains and receives adequate rainfall.
- The Austral realm was isolated by physical geography. Weber's Line and the Wallace Line were attempts to distinguish the location of the separation between biological environments.

Discussion and Study Questions

1. For what purpose was Australia first colonized? What European country colonized Australia and New Zealand?
2. How did the colonial activity impact the indigenous people?

3. How is the Austral realm isolated from the rest of the world?
4. Who are the main indigenous people of New Zealand and where did they originally come from?
5. Explain how the colonial development of Australia was similar to the colonial development of the United States.

Geography Exercise

Identify the following key places on a map:

- Arafura Sea
- Cape York
- Coral Sea
- Great Australian Bight
- Great Barrier Reef
- Gulf of Carpentaria
- New South Wales
- Sydney
- Tasman Sea
- Tasmania
- Timor Sea

12.2: AUSTRALIA

Australia, often referred to as the "Land Down Under," is a vast and diverse nation known for its stunning landscapes, unique wildlife, and vibrant culture. From the arid deserts of the Outback to the lush rainforests of Queensland, Australia boasts a wide range of natural wonders, including the iconic Great Barrier Reef and the majestic Uluru. Its cities, like Sydney and Melbourne, are vibrant hubs of art, food, and multiculturalism, reflecting the country's rich history and diverse population. With a lifestyle that embraces outdoor adventure, coastal living, and a deep connection to nature, Australia offers an unforgettable experience for both residents and visitors alike.

Physical Geography

There is an international attraction to the island continent of Australia, and the attraction has grown in intensity in the past few decades. Tourism is now the number one economic activity in Australia. Just slightly smaller in physical area than the continental United States, Australia is a large country with many resources but few people relative to its size. The Tropic of Capricorn runs right through the middle of this country. Australia hosts many unique species of plants and animals, including marsupials and a host of poisonous snakes and insects. With the advent of European colonialism, new species were introduced to the country, which regrettably caused the extinction of some of the native species but also gave Australia a wide diversity of organisms and natural conditions.

Australia is a relatively low-lying island with low relief. It is the flattest of all the continents. The various highland ranges are pronounced, but are not high in elevation. The **Great Dividing Range** is a mountain chain extending from **Melbourne** in the south to Cape York in the north. This low-lying range of highlands averages about four thousand feet and reaches an elevation of just over seven thousand feet at its highest peaks in the south. The largest river in Australia is the **Darling-Murray River** system that starts in the highland of the Great Dividing Range and flows inward through New South Wales, Queensland, Victoria, and South Australia.

The great interior of the country is home to the massive **outback.** Extending west from the Great Dividing Range, the outback encompasses most of the interior. This region receives less rainfall than along the coast and its terrain consists of deserts and semiarid plateaus with rough grasses and scrublands. The outback is sparsely populated, but is home to a number of aboriginal groups. Many of the school-age children in the outback have traditionally received their school lessons through television or radio broadcasts because of their isolation. Mining and some agricultural activities can be found in the outback. **Alice Springs** is located in the center of the continent and has been given the designation of the middle of nowhere, or the center of everything.

Figure 12.4: The Ghan (Train) Waiting at Alice Springs Station before Continuing North to Darwin The remote town of Alice Springs is located at the center of the Australian outback. Wikimedia Commons – CC BY-SA 3.0.

The deserts of Australia's interior make up a large portion of the continent. Western Australia has three large deserts: the **Gibson Desert, Great Victoria Desert**, and **Great Sandy Desert**. The **Simpson Desert** is located in the border region between the Northern Territory, Queensland, and South Australia. These deserts are not all sand; course grasses and various species of spinifex, a short plant that grows in sandy soil, also grow in the deserts. The **Great Artesian Basin** on the western edge of the Great Dividing Range receives very little rainfall. It would be classified as a desert but for its underground water resources, which support extensive farming operations. Large livestock businesses exist in Australia's interior with massive herds of cattle and sheep. The grassy plateaus and scrublands provide grazing for domesticated livestock and even wild camels.

The Great Barrier Reef, the largest barrier reef in the world, extends for 1,600 miles off the northeastern coast of Australia. It is home to a host of sea creatures and fish that draw millions of tourists each year. The reef attracts scuba divers and water enthusiasts from around the world. The reef is a main tourism attraction and brings income to the Australian economy. The Great Barrier Reef has been designated as a United Nations World Heritage Site. **Brisbane** is located on the **Gold Coast**, which gets its name from the beautiful sandy beaches. The beaches attract an important tourism market for the country.

Figure 12.5: Aerial view of Uluru (Ayers Rock), Located in the Interior of Australia near Alice Springs The rock rises 1,142 feet above the outback and is 2.2 miles long. The site is sacred to the

Aborigines and is a major tourist attraction. It is listed as a World Heritage Site. Wikimedia Commons – public domain.

A couple of large physical features of interest and significance to Australia are the two largest monoliths in the world. In western Australia, more than five hundred miles to the northeast of Perth, is Mt. Augustus National Park, which features the rock known as **Mt. Augustus.** It is considered to be the largest single rock in the world. Mt. Augustus rises 2,352 feet above the desert landscape. The single structure is about five miles long. Mt. Augustus is more than twice the size of the most famous Australian monolith of **Uluru (Ayers Rock).** Uluru is located about two hundred miles southwest of Alice Springs in the Northern Territory and is a well-known tourist attraction. Uluru rises 1,142 feet above the outback and is about 2.2 miles long. Both rocks hold significant cultural value to the aboriginal populations in Australia. They both have ancient petroglyphs, and both are considered sacred sites. Uluru has been more popularized through tourism promotions.

Climate Regions

Central and western Australia are sparsely populated. Large areas of the Northern Territory and the desert regions are uninhabited. Approximately 40 percent of

Australia's interior is desert, where Type B climates dominate. The large land mass can heat up during the summer months, triggering high temperatures. Low humidity allows heat to escape into the atmosphere after the sun goes down, so there is wide temperature variation between day and night.

Along the northern coastal region there are more tropical Type A climates. Closer to the equator and with the sea to moderate temperatures, the northern areas around **Darwin** and Cape York have little temperature variation. Temperatures in Darwin average about 90 °F in the summer and 86 °F in the winter. Spring monsoons bring additional rainfall from February to March.

Tasmania, Victoria, and the core region of the southeast have a more moderate and temperate Type C climate. The main cities, such as Sydney, Melbourne, and Adelaide, are within this area. It is not surprising that there is a direct correlation between Type C climates and the major population areas. The Tropic of Capricorn cuts across the continent, indicating that the cities are not that far south of the tropics. Average winter temperatures in June and July do not usually fall below 50 °F and average summer temperatures in January and February remain around 70 °F. Since the seasons are reversed from that of the Northern Hemisphere, many Australians go to the beach for Christmas.

Figure 12.6: Australia's Provinces and Territories and Their Respective Major Cities The two core areas, where most of Australia's population resides, are also noted. Not unexpectedly, the core areas have a dominant type C climate, following the general principle that humans gravitate toward type C climates. Updated map from Wikimedia Commons – CC BY-SA 3.0.

Population, Urbanization, and the Core-Periphery Spatial Relationship

Australia is divided politically into six states and two territories. They are the Northern Territory, Australian Capital Territory, Western Australia, Tasmania, South Australia, Queensland, and New South Wales. Australian protectorates are composed of a number of small islands around Australia. Australian core areas are conducive to large human populations. To locate the core population areas in Australia, simply find the moderate Type C climates. Australia has two core regions. There is a small core region in the west, anchored by the city of **Perth.** Most of Australia's people live in the large core region in the east along the coast. This region extends from Brisbane to **Adelaide** and holds most of the country's population.

The total population of Australia in 2010 was only about twenty-two million. There are more people living in Mexico City than in all of Australia. More than 90 percent of this population has European heritage; most of this percentage is from the British Isles. English is the dominant language. Christianity is the dominant religion of choice. The makeup of the people is a product of European colonialism and immigration.

Only about 2 percent of the current population consists of Aboriginal people, the original people of Australia. Australia's population has seen periodic growth spurts as waves of immigrants responded to national policies encouraging immigration. This was especially true after World War II. About 24 percent of the current population was born outside Australia; most come from the United Kingdom, and another large percentage comes from New Zealand. Asian countries have also contributed to the Australian population, with measurable numbers of immigrants from China, Vietnam, and the Philippines. And lastly, people from Italy and India also make up a notable proportion of Australia's immigrant population.

Australia's population is not spread evenly across the landscape, since a large portion of the country is desert. The population is concentrated mostly in the urban areas. About 90 percent of the population inhabits the cities, which are mostly in coastal areas. The largest city, Sydney, is often referred to as the New York of Australia. Sydney is positioned at the heart of the main core area, the state of New South Wales. To the south of Sydney is the Australian Capital Territory, home to the capital city of **Canberra.** Other major Australian cities include Melbourne, Perth, Adelaide, and Brisbane. **Hobart** is the

Figure 12.7: The Sydney Opera House Viewed from the Water with the City Skyline. Wikimedia Commons – CC BY 2.5.

largest city on the island of Tasmania and Darwin is the largest city in the Northern Territory.

All the large cities of Australia—with except the planned capital city of Canberra—are located on the coast. This pattern of urban distribution was a product of European colonial development. Most of Australia's population lives in the two economic core regions, so Australia exhibits a distinct core-periphery spatial pattern. The core areas hold the power, wealth, and influence while the periphery region supplies all the food, raw materials, and goods needed in the core. Australia has never had a majority rural population since its Aboriginal times. There has been little rural-to-urban shift in Australia's population. This is similar to Japan's urban development pattern.

English is the first language of the vast majority of the population. Recently enacted policies and changing attitudes toward multiculturalism have spurred growth in the number of immigrants and their descendants who speak two languages fluently—English and the language of their birthplace or national heritage. Indigenous languages have not fared so well. As many as three hundred indigenous languages were spoken by Aborigines before the Europeans arrived, and just a few hundred years later, that number now stands at about seventy. Most aboriginal languages are in danger of dying out.

Culture and Immigration

Until 1973, Australia had a collection of laws and policies known as the White Australia policy, which served to limit the immigration of nonwhite persons to Australia. While the White Australia policies limited immigration from some areas, other policies sought to expand immigration from the United Kingdom. Subsidies were offered to British citizens to relocate to Australia. Between 1830 and 1940, more than a million British citizens took advantage of the offer.

Recent census data indicate that about a quarter of the population identifies itself as Roman Catholic and another 20 percent self-identifies as Anglican (the national religion of the United Kingdom). An additional 20 percent self-identify as Protestant, other than Anglican, and about 15 percent as having no religion. Regular church attendance is claimed by at about 7.5 percent of the population. Despite modern Australia having been settled by the British, Australian law decrees that Australia will have no national religion and guarantees freedom of religion.

Sports are an important part of Australian culture, perhaps owing to a climate that allows for year-round outdoor activity. About a fourth of the population is involved in some kind of organized sports team. Football (soccer) is popular, as is true in most European countries, and rugby and cricket are popular as well. The most popular spectator sport in Australia is Australian Rules Football, also known as Aussie Rules Football, or simply "footy." This uniquely Australian game has codified rules that date back to 1858 and is a variant of football and rugby. Other forms of entertainment include television, film, and live performances of every kind. Although Australia has a number of its own television stations, there are concerns that popular culture is beginning to be dominated by American influences. Australia's large cities have extensive programs in the arts. Sydney is becoming a center for world-class performances in dance, opera, music, and theatre.

Education is well funded and internationally respected. School attendance is compulsory between the ages of six and fifteen, and the adult literacy rate has held steady at about 99 percent. Most students attend publicly funded schools, which are secular. Private schools, which charge tuition fees, do exist and are typically run by religious organizations, predominantly the Catholic Church.

Economic Geography

Most of Australia—especially the wide expanse of the arid interior known as the outback—has immense open spaces, agricultural potential or excellent resource extraction possibilities.

Figure 12.8: The agricultural region of the Barossa Valley in South Australia grows grapes and produces wine. Agricultural production is a major source of economic wealth for Australia even though only 11 percent of the population lives in rural areas. Australia's wine production is expanding to compete in the global marketplace, with France and California as major competitors. Wikimedia Commons – CC BY 2.0.

The extensive grasslands support tens of millions of domesticated animals—mainly cattle and sheep—which accounts for up to one-fifth of the world's wool production. Large agricultural businesses include thousands of acres under one operation. The western sector of the Great Dividing Range in New South Wales is an excellent region for commercial grain operations. The coastal region in

Queensland, since it is warmer and receives more rainfall, is good for sugarcane and similar crops. Sheep and cattle ranches are common in central Queensland and Western Australia. Various regions of southern Australia are excellent for grape and fruit production. Australian wine production has risen to compete with the US and European markets. Only the dry central desert regions in the center of the continent are not favorable for agriculture. In the early portion of the twentieth century, Australia gained enormous wealth by exporting food products to the rest of the world. This is still true, but the profit margin on food goods is no longer what it used to be. The country has had to look elsewhere to gain wealth.

Australia has excellent food production capabilities. It also has an excellent mineral resource base. Different types of minerals can be found in different regions throughout Australia. Western Australia has iron ore mines. The eastern region of Queensland and New South Wales has abundant coal reserves. Minerals such as zinc, copper, gold, silver, tungsten, and nickel can be found in various parts of the country, including Tasmania. Oil and gas fields can be found in the northwestern coastal waters and in the Tasman Sea east of Melbourne. The country is self-sufficient in natural gas but does have to import some petroleum products.

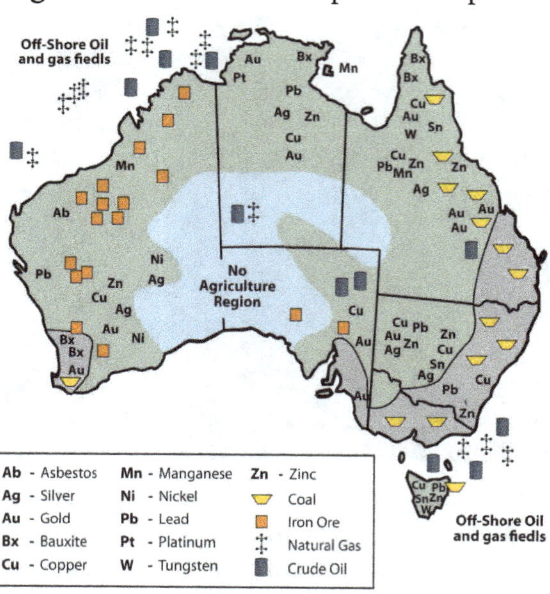

Figure 12.9: Australia's Distribution of Raw Materials. Updated from map courtesy of Wikimedia Commons – CC BY-SA 3.0.

Are any Australian-manufactured products available where you live? What products can you think of? Australia does not export many manufactured goods. Its main exports are food and raw materials. If you remember how countries gain wealth, the method with the highest valued-added profit is manufacturing. Think about Japan and the four Asian economic tigers, and how they have gained their wealth. The economic tigers have few raw materials. Where do you suppose the economic tigers and Japan get their raw materials? With Japan's enormous manufacturing capacity, it has a high demand for imported iron ore, minerals, and raw materials. Though Australia is a former British colony, Great Britain is not considered Australia's largest trading partner. Australia is closer geographically to the Asian economic community than to the European Union. Japan has become Australia's biggest trading partner. When Australia is viewed in the news, in television programs, and in Hollywood movies, it is portrayed as a country with a similar standard of living to the United States or Europe. How do Australians have such a high standard of living if they don't manufacture anything for export? To evaluate this, think about the size of the population of Australia and consider the distribution of wealth. They export an immense amount of raw materials and have a relatively small population to share the wealth.

Figure 12.10: Bondi Beach The Gold Coast of eastern Australia draws tourists from the Northern Hemisphere throughout the winter season. It is called the Gold Coast because of the long stretches of golden sand beaches, the golden tanned bodies of beach goers, and the high level of income (gold) derived from the tourism industry. Wikimedia Commons – public domain.

Australia is an attractive place to visit. The environment, the animals, and the culture make it inviting for tourism. As of the year 2002, tourism has become Australia's number one means of economic income. From the Great Barrier Reef and the Gold Coast to the vast expanse of the outback, Australia has been marketing itself as an attractive place to visit with great success. Tourism from Japan provides a large percentage of the tourist activity. Australia has moved through the initial stages of the index of economic development to become a society that is about 90 percent urban with small families and high incomes.

Mining and Aboriginal Lands

Territorial control of Australian lands has become a major issue in recent years. Large portions of western Australia and the outback have traditionally been Aboriginal lands. European colonialism on the Australian continent displaced many of the native people. Large sections of land once used by the Aborigines were taken over by the government or by private interests. Large agricultural operations and mining operations have used the lands without adequate compensation to the Aboriginal people who once controlled them. Court rulings aimed at reparation for native people have had mixed results.

There are as many as four hundred different groups of Aborigines currently in Australia that make up a total population of about four hundred fifty thousand. This is a small percentage of Australia's population but involves a large part of the physical area of the country. Their land claims include all of the Northern Territory, a large portion of western Australia, and parts of South Australia and Queensland. This is in addition to claims located within many urban areas, such as the largest city, Sydney. Mining operations on Aboriginal lands have become highly regulated. Concerns have arisen that Australia's extractive industries will diminish, causing a decline in the economy. The concern for the Aboriginal population has increased in the past few decades and the government has made attempts to mediate their political and economic issues as well as strengthen programs that address their social welfare.

Australia's Future

The economic future of Australia is complex. Though tourism has become a viable means of providing income, Australia must import manufactured products that it does not produce locally, including electronic goods, computers, and automobiles. Import dependence has increased its trade deficit. Trade agreements and protectionism have become a part of the economic puzzle of how to sustain a competitive standard of living. Australia is located next to the Asian realm. Its economy, culture, and future are becoming more Asian. Immigration has been an issue in that the government has always restricted immigration to ensure a European majority. Millions of Asian people would like to migrate to Australia to seek greater opportunities and advantages, but they are legally restricted. It is becoming more difficult for Australians to hold to their European connections with such an Asian presence. How the country will handle this situation in the future will prove interesting.

Key Takeaways

- Australia is an island continent that was home to aboriginal people before the British colonized it by first creating prison colonies for convicts from Great Britain.

- Australia is a relatively flat continent with low elevation highlands, including the Great Dividing Range along its eastern coast. The interior outback lacks precipitation and has numerous deserts.
- Two main core areas exist where Type C climates prevail and where most of the population lives: a large core area on the southeastern coast and a small core area around Perth on the western coast. The sparsely populated outback makes up the vast periphery, which has large amounts of mineral and agricultural resources.
- Aboriginal people were in Australia for forty thousand years. The British colonial activity didn't heat up until the late 1700s. Today, most of the twenty-two million people are from the British Isles and Europe. Only about four hundred fifty thousand Aborigines remain.
- Australia has few manufacturing enterprises for export profits. Tourism has become the number one method of gaining wealth, with the export of raw materials the second-largest method.

Discussion and Study Questions

1. Name the European country that colonized Australia. What was the original reason for colonizing Australia?
2. What are Australia's main political divisions? What is the vast interior called?
3. What are some of Australia's main physical features? How are they developed for tourism?
4. What are the main climate types in Australia? How does climate relate to population?
5. What type of government does Australia have? What is its capital city? Who is head of state?
6. Outline the core-periphery relationship in Australia. What distinguishes the core areas?
7. How does Australia compare with the Asian economic tigers? How do they support each other?
8. How does Australia maintain a high standard of living when it exports few manufactured products?
9. What are Australia's main exports? Who is their main trading partner?
10. What stage is the country in with regard to the index of economic development?

Geography Exercise

Identify the following key places on a map:

- Adelaide
- Arafura Sea
- Australian Capital Territory
- Brisbane
- Canberra
- Cape York
- Coral Sea
- Darling-Murray River
- Darwin
- Gibson Desert
- Gold Coast
- Great Artesian Basin
- Great Australian Bight
- Great Barrier Reef
- Great Dividing Range
- Great Sandy Desert
- Great Victoria Desert
- Gulf of Carpentaria
- Hobart
- Melbourne
- New South Wales
- Northern Territory
- Outback
- Perth
- Queensland
- Sydney
- Simpson Desert
- South Australia
- Tasman Sea
- Tasmania
- Timor Sea
- Victoria
- Western Australia

12.3. NEW ZEALAND

New Zealand, a picturesque island nation in the South Pacific, is celebrated for its breathtaking landscapes and rich Maori culture. Comprising two main islands, the North and South, New Zealand features a stunning variety of scenery, from the geothermal wonders of Rotorua to the majestic fjords of Milford Sound. Outdoor enthusiasts flock to its mountains, beaches, and forests, drawn by opportunities for hiking, skiing, and water sports. The country's unique flora and fauna, including the famous kiwi bird, further enhance its appeal. With a strong emphasis on sustainability

Figure 12.11: Southern Alps, New Zealand. Wikimedia Commons – public domain.

and a welcoming spirit, New Zealand offers a distinctive blend of adventure and tranquility, making it a captivating destination for travelers seeking both natural beauty and cultural richness.

Physical Geography

To the east of Australia across the Tasman Sea is the country of New Zealand. New Zealand is one of a number of sets of islands that make up Oceania, also referred to as the Pacific Islands, a region occupying the western and central Pacific Ocean. The Pacific Islands region is generally divided into three subregions: Micronesia, Melanesia, and Polynesia, with New Zealand being part of Polynesia. The Pacific Island region includes more than twenty-five thousand individual small islands representing twenty-five nations and territories. Most of these islands are very small. The South and North islands of New Zealand are the second- and third-largest islands, respectively. The North and South Islands of New Zealand are separated by a body of water known as the **Cook Straight,** which is only about thirteen miles wide at its narrowest point. The North and South Islands together are about the same size as the US state of Colorado. New Zealand also includes a number of smaller nearby islands.

New Zealand, like Australia, is in the Southern Hemisphere, which means that its seasons are expressed at the opposite times of the seasons in North America.

Figure 12.12: New Zealand Wellington is the capital, Auckland is the largest city in the country, and Christchurch is the largest city on the South Island. The Southern Alps extend along the western length of the south and reach elevations of twelve thousand feet. CIA World Factbook – public domain.

In other words, the warmest summer months are January and February and the cooler winter months are June and July. New Zealand lies within the Temperate Zone. There are only very moderate seasonal differences, which are slightly more pronounced in the inland areas because the inland areas lack the moderating influence of the ocean. In general, the North Island has somewhat warmer average temperatures than the South Island. In summer, average low temperatures are about 50 °F, with daytime highs around 75 °F. In the winter months, low temperatures average about 35 °F and high temperatures are about 50 °F. The occurrence of more extreme temperatures is limited to the mountainous peaks of the Southern Alps. Snow is common in these mountainous regions but rarely occurs in coastal regions.

Rainfall is heaviest on the western coasts of both islands, but especially on the South Island. The prevailing westerly winds, carrying moisture from the ocean, come in contact with the mountains of the **Southern Alps** and high precipitation results. The mountains also have the opposite effect. On the eastern side of the mountains is a rain shadow where the westerly winds blow hot, dry air and the eastern coasts are therefore substantially drier than the western coasts. Therefore, average precipitation rates vary widely across the country. The average annual rainfall in **Christchurch**, which is on the eastern coast of the South Island, is about twenty-five inches per year. **Auckland**, in the midportion of the North Island, receives twice that amount and areas on the wetter western coast receive as much as one hundred fifteen inches per year.

As an island nation, New Zealand's coastlines and oceans are some of its most important geographic features. New Zealand has one of the world's largest exclusive economic zones, an oceanic zone over which a nation has exclusive rights of exploration and exploitation of marine resources. New Zealand's exclusive economic zone covers more than one million square miles. The dramatic nature of New Zealand's landscape is well known to many moviegoers as the landscape of Middle Earth, as depicted in New Zealand film director Peter Jackson's version of J. R. R. Tolkein's *Lord of the Rings.*

Tectonic Plates and Gondwanaland

The North Island of New Zealand features a rather rugged coastline with numerous harbors, bays, and inlets. The port cities of Auckland and **Wellington** are located on two of the largest bays. The coastline of the South Island is somewhat more regular, except along the southern portion of the eastern coastline, which has deep fjords. Though the North Island has lower relief than its southern counterpart, its few mountains are volcanic in origin. The two main islands are accompanied by smaller islands around their shores. The North Island's highest peak is **Mt. Ruapehu**, which reaches almost 9,175 feet and is an active cone volcano. It is located in the south central part of the island. A range of highlands runs along its eastern side. The volcanism associated with Mt. Ruapehu results from New Zealand's location atop two tectonic plates: the Pacific Plate and the Indo-Australian Plate. The boundary of these two plates forms a subduction zone under the North Island; consequently, New Zealand experiences tens of thousands of earthquakes per year. Though most of the earthquakes do not greatly disrupt human activity, some have registered higher than 7 on the Richter scale. New Zealanders have made use of the geothermal power generated by this and the tectonic features of this area and hence, New Zealand is home to several hydrothermal power plants.

A range of mountains, the Southern Alps, divides the South Island lengthwise. Many of the peaks reach over ten thousand feet. The highest peak is **Mt. Cook**, which reaches higher than twelve thousand feet. These mountains are also formed by the area's tectonic situation. However, the two plates meet in a different way under the South Island. Rather than creating a subduction zone, the plates move laterally. The lateral movement created the South Alps. New Zealand's location places it along one of the edges of the so-called Ring of Fire, which encircles the Pacific Ocean basin. The positions and activity of the tectonic plates in this zone cause most of the world's earthquakes and have formed more than 75 percent of the world's volcanoes.

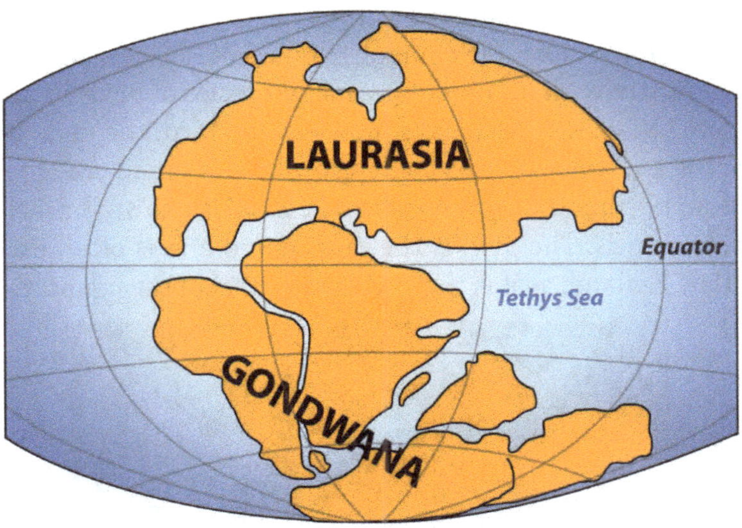

Figure 12.13: The Supercontinent of Pangaea about Two Hundred Million Years Ago The southern portion was called Gondwana (Gondwanaland). New Zealand and Australia come from Gondwana; their origin explains why their fauna and flora are so unique. Wikimedia Commons – CC BY 3.0.

New Zealand has a fascinating ecological history. New Zealand was once part of Gondwana, also known as Gondwanaland. About two hundred to five hundred million years ago, Gondwana was the southern portion of the supercontinent of Pangaea. Gondwana included most of the land in what is the Southern Hemisphere today, including Australia and New Zealand. Over time, through various forms of tectonic activity, the earth's plates shifted and the supercontinent of Gondwana broke apart, leaving New Zealand geographically isolated for millennia. The species of flora and fauna found in New Zealand either descended from the species found on Gondwana, flew there, floated there via the ocean, or were brought by people.

Biodiversity in New Zealand

New Zealand's geological history has laid the groundwork for more than 2,000 indigenous plant species, about 1,500 of which are found nowhere else in the world. The biomes of the North Island include a subtropical area, including mangrove swamps, an evergreen forest with dense undergrowth of mosses and ferns, and a small grasslands area in the central volcanic plain. The South Island biomes include extensive grasslands in the east, which are excellent for agricultural pursuits; forest areas, dominated by native beech trees in the west; and an alpine vegetation zone in the Southern Alps.

In terms of fauna, the most influential factor may be the relative absence of predatory mammals, again related to New Zealand's geological history. With few ground predators and a favorable climate, bats, small reptiles, and birds were able to thrive and flourish. Without predators, many species of birds became flightless, such as the noted Kiwi. New Zealand's most famous bird, the moa, was similar to the ostrich but is now extinct. Moa could grow to more than twelve feet high and weigh more than five hundred pounds. New Zealand is known for its large number of species of wild birds. The Kiwi is the most noted and is often used to refer to people from New Zealand, as it is the national symbol of the country.

Figure 12.14: The flightless Kiwi bird is the national symbol of New Zealand. All five species of Kiwi are endangered. Sportler – kiwi – CC BY-NC-ND 2.0.

New Zealand has a variety of landscapes that have been attractive for economic activity and tourism. The South Island is larger than the North Island and is more mountainous. The snowcapped peaks of the Southern Alps run the western length of the South Island. Large livestock-raising operations and agricultural activities can be found on the vast grasslands of the South Island. Millions of sheep and cattle are raised on the grassy highlands and the valley pasturelands. The North Island has more low-lying terrain, which is also good for agriculture and is home to a large dairy industry. The central highlands of the north offer some rugged relief and provide for a diverse physical landscape.

Cultural Dynamics and the Maori

New Zealand is home to many Polynesian groups. Its original inhabitants were the Maori, who came to the islands around the tenth century. They grew crops of gourds and sweet potatoes. Fur seals were hunted regularly, as were moa, which were hunted to extinction before the Europeans arrived. The Maori had created extensive trading networks with other island groups and developed a heritage of traditional rituals and cultural ways. The Maori culture thrived for hundreds of years and was well established in New Zealand before the arrival of the colonial ships from Europe.

Britain was the main colonizer of the islands. The British settled in to establish their presence and gain control. In 1840, the British colonizers and the Maori signed the

Figure 12.15: A Maori Man with Traditional Topknot and Tattoos Many Maori participate in performances in New Zealand both for tourism and to maintain their heritage and traditions. Steve Evans – New Zealand: Maori Culture 002 – CC BY-NC 2.0.

Treaty of Waitangi, which granted British sovereignty over the islands but allowed the Maori certain rights over tribal lands. The actual language in this treaty has been debated between the English version and the Maori version. Over time the tribal lands were codified into legal arrangements by the European colonizers. Since the Treaty of Waitangi, the situation has evolved, with subsequent land exchanges, some legal and others questionable. The Maori have complained about unfair treatment and the loss of land and rights in the process. These issues have finally reached a point of negotiation in the past couple of decades. Starting in the 1990s, treaty settlements have been made to help correct the actions of the colonial activity and compensate the Maori for the conditions they were subjugated to.

Most of the Maori have lived on the North Island. They were a real concern for the European colonizers. Claimed by Great Britain in their colonial empire, the country of New Zealand became independent of Britain in 1901. In 2010, the estimate of the population of the country was at about 4.3 million, with Europeans making up 60 percent of the population and the Maori making up about 8 percent. There are also many people who are of mixed ethnic background, including Maori and other groups. Asians, Polynesians, and other ethnic minorities make up the rest. New Zealand's main religion is Christianity and English is the official language.

The Maori have not been integrated into New Zealand society to the same extent as Europeans have. The Maori now join the ranks of other Pacific Islanders that have moved to New Zealand from Samoa, Tonga, Cook Islands, and many other places in the South Pacific. New Zealand's urban areas reflect diversity in the various cultural landscapes and ethnic communities that have established themselves in specific neighborhoods within the main cities. A common dilemma with all peoples is the draw to return to their heritage and roots, which typically results in a more traditional lifestyle with stronger cultural ways. At the same time, the modern world pulls people toward a more global and cosmopolitan culture that is steeped in modernity with changing fashions. The Maori and other ethnic groups in New Zealand find themselves facing this dichotomy of societal dynamics.

Economic Conditions

Land and climate could be said to be New Zealand's most important natural resources. Fertile soils and a mild climate, complete with thousands of hours of sunshine annually, create ideal conditions for agriculture. Grass continues to grow throughout the year, which means that sheep and other livestock can be well grazed. Wool and other agricultural products, notably meat and butter, are important exports for New Zealand's economy. Healthy forests produce timber products, which are important to the economy as well. Some of New Zealand's natural resources are found underground, including coal, natural gas, gold, and other minerals.

Wellington is the capital of the country and is located on the southern end of the North Island. Wellington is one-fourth the size of the primary city, Auckland, which has 1.2 million people and is located in the north. The major cities are located along the coastal regions and provide a connection to sea transportation. Christchurch is the largest city of the South Island and is located along the eastern seaboard on the productive **Canterbury Plain**. The soils and conditions on the Canterbury Plain are excellent for productive agriculture of all types. Coastal plains also provide access to building transportation systems of highways and railroads that are more costly to construct in the mountainous regions of the Southern Alps or the northern highlands.

Figure 12.16: Harvesting Grapes in Marlborough District, New Zealand This is at the north end of the South Island across the Cook Strait from Wellington. Agricultural products make up a large portion of the nation's exports that contribute to the country's economy. Phil Norton – Harvester – CC BY-NC-ND 2.0.

The modern cities are home to a multitude of processing centers preparing the abundant agricultural products for domestic consumption and for export products. The ever-growing populations of Asia and the rest of the world continue to place a high demand on food products and welcome New Zealand's agricultural exports. In relation to how countries gain wealth, agricultural profits are usually quite competitive and normally provide a low profit margin. New Zealand does not gain a large part of its national income from mining or manufacturing, though these industries do exist. The high standard of living that exists in New Zealand is similar to that of Australia in that the population is not very large, so that the national wealth can be distributed via the private sector economy to accommodate a relatively good lifestyle and provide for a comfortable standard of living.

New Zealand has a market economy. The mainstay of the economy is, and has been for many years, a productive agricultural sector that has been geared toward export profits. New Zealand's climate and soils help give it a place in the economy of the region through agricultural exports. A "wool boom" in the 1950s furthered the emphasis on agricultural products as tremendous profits accrued in the wool production and export industry. Today, New Zealand's economy is still heavily focused on the export of agricultural products, though the economy has diversified into other areas such as tourism and exploitation of natural resources, especially natural gas. The development of hydroelectricity generation in recent years has been important to the economy.

Key Takeaways

- New Zealand has been isolated by the separation of continents through tectonic plate action. Shifting plates continue to create earthquakes and volcanic activity in the region.
- New Zealand has high mountains, with the Southern Alps along the western coast of the South Island and highlands along the eastern side of the North Island. Adequate rainfall and good soils provide for excellent agricultural production, which has been the traditional economic activity.
- The Maori were established in New Zealand before the British colonized it. Various agreements were made to work out common arrangements, with varying degrees of success. The Maori continue to be a minority population and have not acculturated into the mainstream society of the country.
- New Zealand has historically relied on agricultural exports for national income. Shifting global markets and changes in government policies and structures have highly affected the economy of New Zealand. The country continues to work its way through the transition to a more global economy.

Discussion and Study Questions

1. What are the main physical features of the South Island and North Island of New Zealand?
2. How is the North Island different from the South Island in population and economic activities?
3. How has physical geography been helpful in the economic development of New Zealand?
4. Where are the main cities of New Zealand? What are the capital and the largest cities on each island?
5. Who were the inhabitants of New Zealand before the colonial era? Where did they come from?
6. What issues do these inhabitants of New Zealand have with the current government?
7. How are the dynamics with the Maori similar to those with the Aborigines in Australia?
8. What are the main methods that New Zealand has used to gain national wealth?

9. How has the economic situation in New Zealand changed over the past few decades?
10. How is New Zealand different in physical and human geography from Australia?

Geography Exercise

Identify the following key places on a map:

- Auckland
- Canterbury Plain
- Cook Strait
- Christchurch
- Mt. Cook
- Mt. Ruapehu
- Southern Alps
- Wellington

12.4. CONCISE SUMMARY

Australia is a sovereign nation and a somewhat low-lying island continent characterized by its extensive semiarid interior known as the outback. Two principal regions characterized by moderate Type C climates along the eastern and western coasts contain the majority of the modest population of approximately twenty-two million.

Great Britain established Australia as a penal colony. European customs and legacy dominate, with English as the primary language and Christianity as the predominant religion. The Aborigines inhabited the region for tens of thousands of years prior to the arrival of Europeans, although just approximately four hundred fifty thousand remain today.

The conventional approach to accumulating wealth in Australia has been through agricultural productivity. During the latter half of the twentieth century, the exportation of minerals and raw commodities emerged as a principal means of accumulating riches. Over the past decade, tourism has emerged as Australia's primary source of revenue. Australia's physical attributes, like the Great Barrier Reef, the outback, and its diverse fauna, render it a prime destination for global travelers.

New Zealand is approximately 1,500 miles east of Australia, separated by the Tasman Sea. The two principal islands of New Zealand exhibit significant elevation, with the Southern Alps on the South Island. The temperate climate and sufficient precipitation render New Zealand an attractive destination for agriculture and tourism. The remote settings engender a sense of isolation that impacts the expenses associated with travel and trade.

The Māori from Polynesia inhabited New Zealand prior to British colonization. The Maori experienced disputes with the British and subsequently the government around land dispossession and legal agreements. The Maori constitute fewer than 10 percent of New Zealand's 4.3 million population. The Maori predicament shares similarities with those of the Aborigines in Australia, both of whom have been striving to reclaim land rights and get legal settlements.

6. New Zealand's economic landscape has adapted to align with international market prices and demands, as well as the domestic corporate and political environment. The primary economic activity has been agricultural exports; nevertheless, the nation has been expanding into industrial processes and tourism. The development of natural gas has been increasing.

13 Chapter THE PACIFIC AND ANTARCTICA

13.1. THE PACIFIC ISLANDS

The Pacific region encompasses numerous islands and archipelagos. New Guinea is the largest island and houses the majority of the realm's people. A number of Pacific islands have attained independence, although others continue to be governed by their colonial powers. The Pacific Theater of World War II served as a battleground between Japanese and American forces, significantly influencing the present situation of numerous islands. The United States has been a significant force in the post–World War II dominance and governance of numerous island groups. The Hawaiian Islands were admitted as the fiftieth state of the United States in 1959.

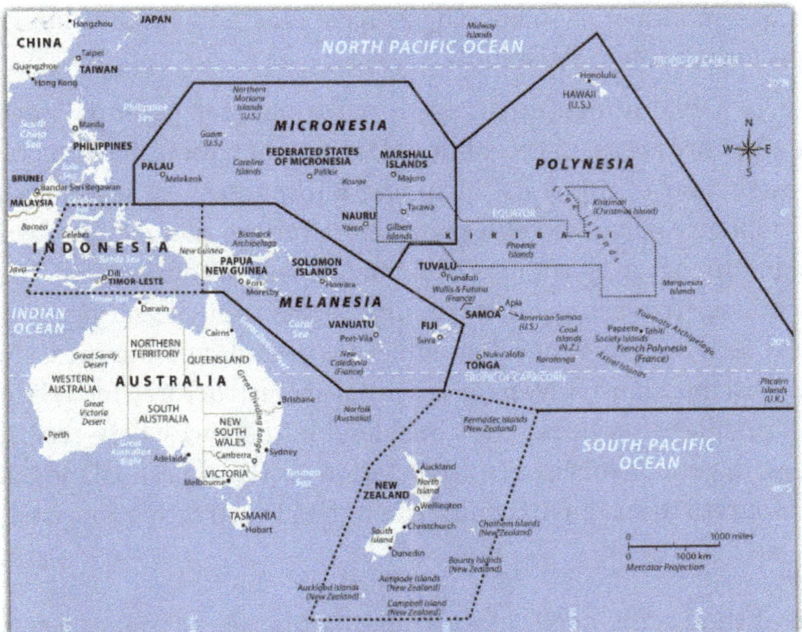

Figure 13.1 The Tropical Realm of the South Pacific with the Three Main Regions of Islands

The numerous islands can be categorized into three primary groupings according to physical geography, indigenous populations, and geographical location: Melanesia, Micronesia, and Polynesia. Indigenous cultural legacy persists robustly in the South Pacific; yet, Western culture has significantly infiltrated the lives of the populace. The process of globalization significantly impacts the economic conditions that shape the cultural dynamics of the Pacific. Islands or archipelagos that remain under external political authority are predominantly shaped by European or American cultural influences. Western influences in fast food, popular music, fashion, and social practices frequently prevail in television, radio, and movies. Intrusive Western cultural influences detract from the traditional indigenous culture and legacy of the inhabitants of these remote islands, who have been there for millennia.

Historically, the islands were economically autonomous. The primary economic activity were fishing and agriculture, while neighboring islands frequently engaged in trade and shared natural resources. Fishing has been a prevalent method of bolstering the economy. National boundaries have been modified to safeguard offshore fishing rights for each sovereign state. Numerous bodies of water have been overfished, thereby diminishing the islands' capacity to supply sustenance for their inhabitants or to generate national revenue. The growth of population and the advent of contemporary technologies have created a reliance on the world's core regions for economic assistance.

The Pacific is a highly marginal region with minimal potential for economic exploitation by core areas. In recent decades, certain nations have accrued wealth from the extraction of minerals, including phosphates, on several islands. The primary resources include a temperate temperature, picturesque beaches, and tropical island topography, all of which may appeal to tourists and visitors from various locations. Tourism is an expanding segment of the service industry and a significant source of revenue for numerous island communities. To enhance tourism, the islands must invest in essential infrastructure, including airports, hotels, and ancillary services. The considerable distances between islands and isolated regions render tourism transportation costly. Not all islands possess the financial resources to sustain expenditures aimed at attracting tourists.

Melanesia

The region of the Pacific north of Australia that borders Indonesia to the east is called *Melanesia*. The name originally referred to people with darker skin but does not adequately describe the region's current ethnic diversity. The main island groups include **Fiji**, **New Caledonia**, **Vanuatu**, the **Solomon Islands**, and **Papua**

New Guinea. All are independent countries except New Caledonia, which is under the French government. The island of New Guinea is shared between Papua New Guinea and Indonesia. Many islands on the eastern side of Indonesia share similar characteristics but are not generally included in the region of Melanesia.

Independent Countries of Melanesia

- Fiji
- Papua New Guinea
- Solomon Islands
- Vanuatu

Other Island Groups

- New Caledonia (*France*)

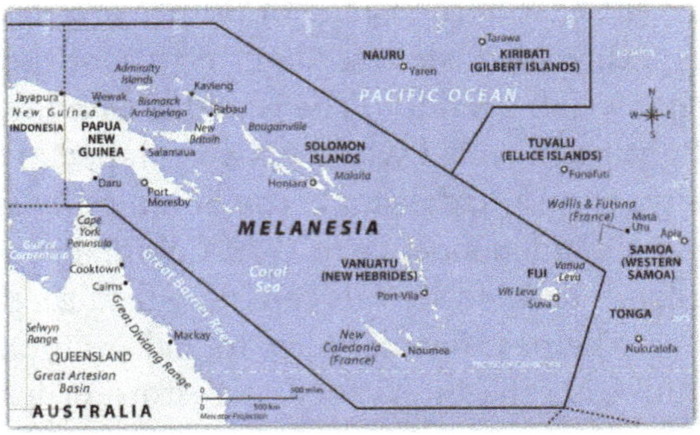

Figure 13.2 The Region of Melanesia

Papua New Guinea

Papua New Guinea is the largest country in the Pacific realm and therefore the largest in Melanesia. It is diverse in both physical terrain and human geography. The high mountains of the interior reach 14,793 feet. Snow has been known to fall in the higher elevations even though they are located near the equator. Many local groups inhabit the island, and more than seven hundred separate languages are spoken, more than in any other country in the world. Indigenous traditions create strong centripetal forces. Many islands of Melanesia are recently independent of their European controllers; Papua New Guinea received independence in 1975 and is working toward fitting into the global community.

Papua New Guinea is a diverse country that still has many mysteries to be revealed in its little-explored interior. The country's large physical area provides greater opportunities for the exploitation of natural resources for economic gain. The interior of the island has large areas that have not been exploited by large-scale development projects. In the past few decades, oil was discovered and makes up its largest export item. Gold, copper, silver, and other minerals are being extracted in extensive mining operations, often by outside multinational corporations. Subsistence agriculture is the main economic activity of most of the people. Coffee and cocoa are examples of agricultural exports.

A number of islands off Papua New Guinea's eastern coast—including Bougainville—have valuable mineral deposits. Bougainville and the islands under its jurisdiction are physically a part of the Solomon Island archipelago but are politically an autonomous region of Papua New Guinea. Volcanic vents deep under the sea continue to bring hot magma and minerals to the surface of the ocean floor, creating valuable exploitable resources. Papua New Guinea has laid claim to these islands and the underwater resources within their maritime boundaries. Rebel movements have pushed for the independence of the Autonomous Region of Bougainville but have been unsuccessful. The islands remain under the government of Papua New Guinea.

Solomon Islands

To the east of the island of Guinea are the Solomon Islands, a group of more than one thousand islands. About eighty of them hold most of the population of more than a half a million. The island of **Guadalcanal** was the site of some of the fiercest fighting in World War II between Japan and the United States. Honiara, the capital city, is on Guadalcanal. The Solomon Islands were a colony of Great Britain but gained independence in 1978. Colonialism, World War II, and ethnic conflict on the islands created serious centrifugal cultural forces, divisions, and political tensions over the past few decades. In 2003, military and police troops from other islands and Australia intervened to restore order after ethnic tension erupted into civil unrest.

Figure 13.3: Malaitan Chief on the Solomon Islands The heritage and history of these islands includes local cultures complete with isolated traditions and ethnic organization. Wikimedia Commons – public domain.

Shifting tectonic plates are the source of environmental problems. Active seismic activity has created earthquakes and tsunami conditions that have brought devastation to the region. An earthquake of 8.1 magnitude hit the Solomon Islands in 2007, bringing high waves and many aftershocks. The tsunami killed at least fifty-two people, and as many as one thousand homes were destroyed. The islands contain several active and dormant volcanoes. Tropical rain forests cover a number of the islands and are home to rare orchids and other organisms. There is concern that these resources might be harmed by deforestation and the exploitation of resources for economic gain.

Vanuatu

Figure 13.4: Saint Joseph's Bay on the Isles of Pines, New Caledonia These remote islands have moderate climates and beautiful coastal settings. Many have been unduly romanticized by works of fiction. These islands are isolated and can lack resources; life can be more difficult than it is often portrayed. Wikimedia Commons – public domain.

The country of Vanuatu was inhabited by a large number of South Pacific groups; as a result, many languages are spoken within a relatively small population. The French and the British both colonized the island archipelago. It was called the New Hebrides before independence in 1980, when the name was changed to Vanuatu. These small volcanic islands have an active volcano and have experienced earthquakes in recent years. One of Vanuatu's means of bringing in business has been to establish offshore banking and financial services, similar to what is found in the Caribbean. Many shipping firms register their ships there because of the advantages of lower taxes and flexible labor laws.

New Caledonia

New Caledonia is still a colony of France and was once a French prison colony. Under a current agreement, sovereignty is slowly being turned over to the local island government. Periodic reevaluations of the local government will be conducted to see if independence can be granted.

New Caledonia has historically relied on subsistence agriculture and fishing for its livelihood. About 25 percent of the world's known nickel resources are located here. Nickel resources will substantially affect the economy, bring in foreign investments, and raise the standard of living.

Fiji

Fiji is located in the eastern sector of Melanesia and has almost one million people. The country includes more than one hundred inhabitable islands, but two are home to most of the population. Colonialism heavily impacted the population's ethnic makeup. During British colonial rule, thousands of workers from South Asia were brought in by the British to work on the sugar plantations. After a century of British rule, Fiji became independent in 1970. The people of South Asian descent remained in Fiji and now make up more than one-third of the population. Ethnic conflicts erupted on the political scene between the Melanesian majority and the

South Asian minority. Political coups and coalition governments have attempted to work out political solutions with limited success. Fiji is quite well developed and has a substantial tourism industry that augments the other agricultural and mining activities.

Micronesia

North of the Solomon Islands and Papua New Guinea is the large region of Micronesia. The "micro" portion of the name refers to the fact that the islands are small in size—often only one square mile or so in physical area. The region has more than two thousand islands. Most of the islands are composed of coral and do not extend above sea level to any large extent. These low islands dominate the high islands. The high islands are usually of volcanic origin and reach elevations in the thousands of feet.

Guam

The largest island in Micronesia is **Guam**. It is only 210 square miles in area and reaches an elevation of 1,335 feet at its highest point. Coral reefs surround Guam's volcanic center. Guam is not an independent country but a US possession. The island was a strategic location during World War II, and the United States has major military installations located on the island.

Figure 13.5: Guam is a US possession. This photo shows the front of the University of Guam's School of Nursing. The university has more than three thousand students and is accredited by the Western Association of Schools and Colleges in the United States. Guampedia Foundation – UOG School of Nursing – CC BY 2.0.

Northern Mariana Islands

The **Northern Mariana Islands** are next to Guam and are current US possessions, along with **Wake Island** in the northeast. The US administers the United Nations

Trust Territory of the **Federated States of Micronesia**. Implemented in 1986 and renewed again in 2004, the islands entered into the Compact of Free Association with the United States and established an independent status.

Nauru

The independent island country of **Nauru** is only about eight square miles in physical area, but its large phosphate deposits created enormous wealth for its small population. Once the phosphates had been mined, however, there was little means to gain wealth on such a small island with a devastated landscape. Many on Nauru are trying to live off the investments from their mining wealth or have moved to find a livelihood elsewhere.

Independent Countries of Micronesia

- Federated States of Micronesia
- Kiribati (Western)
- Marshall Islands
- Nauru
- Palau

Other Island Groups

- Guam (US)
- Gilbert Islands (Kiribati)
- Northern Mariana Islands (US)
- Wake Island (US)

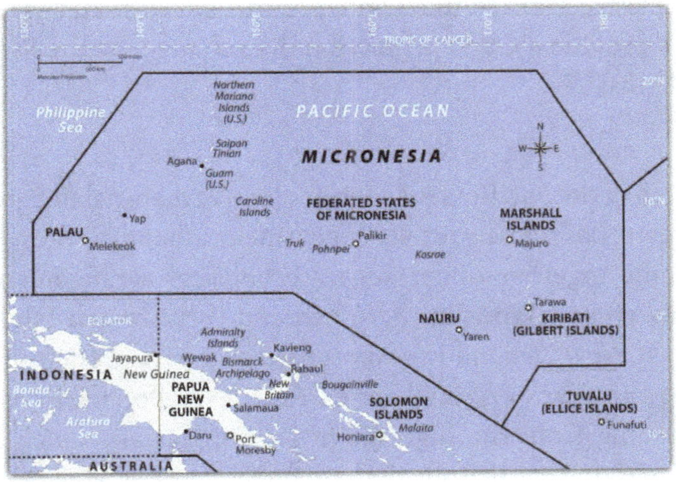

Figure 13.6: The majority of islands in Micronesia are low islands composed of coral.

Palau

Palau, located in western Micronesia, has a population of about twenty thousand people and an area of about 177 square miles. Its early inhabitants included people from Asia and from the Pacific realm. British explorers arrived early on the island, but Spain dominated it during the colonial era. After losing the Spanish-American War, Spain sold the island to Germany, which implemented mining operations on the island. After its defeat in World War I, Germany lost the island to Japan. Japan used it as a strategic outpost but was defeated in World War II and had to give up all its external possessions. After 1945, Palau was held by the United States and the UN. In 1994, the island opted for independence and retained an agreement of free association with the United States. The United States has held strategic military installations on Palau and other islands in Micronesia. Palau's economic and geopolitical dynamics are highly reflective of US activities in the region.

Marshall Islands

Marshall Islands, on the eastern side of Micronesia, experienced serious devastation from the conflict between Japan and the United States during World War II. The Marshall Islands became a testing ground for US nuclear weapons. Atomic bombs were tested on various atolls, rendering them uninhabitable. An atoll is a coral island that surrounds a lagoon. From 1946 to 1958, the United States conducted sixty-seven atmospheric nuclear tests in the Marshall Islands. The largest was known as the Bravo test, which included the detonation of a nuclear device over Bikini Atoll that was one thousand times more powerful than the atomic bomb dropped on Hiroshima during World War II. There are concerns about radioactive fallout that may still affect the people who inhabit nearby atolls. The Marshall Islands were granted independence in 1986 with an agreement with the United States to provide aid and protection in exchange for the use of US military bases on the islands.

Polynesia

The largest region of the Pacific is Polynesia, a land of many island groups with large distances between them. The root word *poly* means "many." Numerous groups of islands have come together under separate political arrangements. The region includes the **Hawaiian Islands** in the north and the **Pitcairn Islands** and **Easter Island** to the east. New Zealand is now studied as a part of the Austral realm, but the Maori living there are originally from Polynesia. Polynesia has a mixture of island types ranging from the high mountains of Hawaii, which are more than 13,800 feet, to low-lying coral atolls that are only a few feet above sea level. Islands that have enough elevation to condense moisture from the clouds receive adequate precipitation, but many islands with low elevations have a shortage of fresh water, making habitation or human development difficult.

Polynesian culture stems from island resources. Fishing, farming, and an understanding of the seas created a way of life that gave Polynesia its identity. Polynesians created innovative maps that provided a means of sailing across large expanses of open seas to connect with distant islands. Their lifestyle revolved around natural resources and the creative use of natural materials. Polynesian art, music, and language reflect a diversity of cultural trends derived from a common heritage. The warm climate and beautiful islands contrast with violent destructive storms and a lack of fresh water or resources, which can make life difficult.

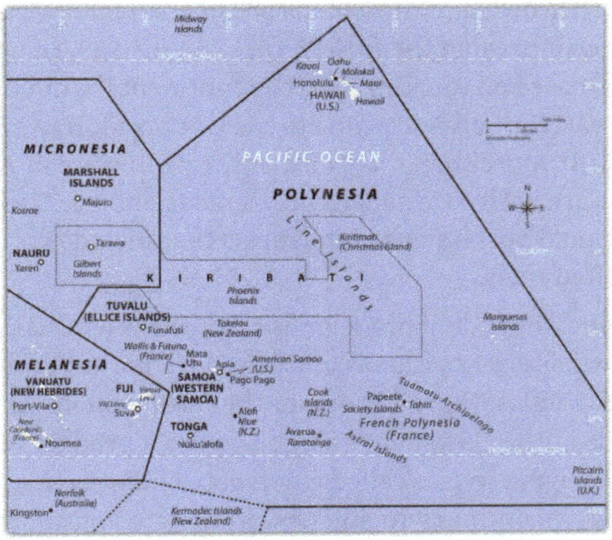

Figure 13.7: The region of Polynesia has island groups that are high islands with mountainous interiors.

Independent Countries of Polynesia

- Kiribati (eastern)
- Samoa
- Tonga
- Tuvalu

Main Island Possessions

- American Samoa (US)
- Cook Islands (NZ)
- Hawaiian Islands (US)
- Pitcairn Islands (UK)
- French Polynesia (FR)
    - Austral Islands

- The Marquesas
- Society Islands and Tahiti
- Tuamotu Islands

Hawaii

At the beginning of the twenty-first century, Polynesia only had four independent island groups: **Kiribati**, **Samoa**, **Tonga**, and **Tuvalu**. The rest of the many islands and island groups in Polynesia are claimed by or under the control or jurisdiction of other countries: mainly the United States, France, Great Britain, or New Zealand. Hawaii was a sovereign and independent kingdom from 1810 to 1893, when the monarchy was overthrown and the islands became a republic that was annexed as a US territory. Hawaii became the fiftieth US state in 1959. Hawaii's development pattern is modern, based on tourism from the continental United States and the US military base on Pearl Harbor. According to the US Census, Hawaii had a population of 1.3 million in 2010. More than one-third of the people are of Asian descent, and at least 10 percent are native Hawaiians or Pacific Islanders. The United States has a number of additional possessions in Polynesia that include various small islands, atolls, or uninhabited reefs.

The Hawaiian Islands include more islands than the few usually listed in tourist brochures. Approximately 137 islands and atolls are in the Hawaiian chain, which extends about 1,500 miles. Hawaii is one of the most remote island groups in the Pacific. The islands of the Hawaiian archipelago are a product of volcanic activity from an undersea magma source called a hotspot, which remains stationary as the tectonic plate over it continues to shift creating new volcanoes. Mt. Kilauea, an active volcano on Hawaii, the largest island in the Hawaiian chain, is considered by geologists to be one of the most active volcanoes in the world. The active volcano of Mauna Loa and two dormant volcanoes, Mauna Kea and Hualālai, are on the same island. Mauna Kea is Hawaii's tallest mountain at 13,796 feet above sea level, which is taller than Mt. Everest if measured from its base on the ocean floor.

Hawaii, like most islands of the Pacific realm, has a tropical type A climate, but snow can be found on the tops of its highest mountains during the winter months. The island of Kauai receives more than 460 inches of rain per year and is one of the wettest places on Earth. The rain shadow effect created by Mt. Wai'ale'ale is the reason for the high level of precipitation. All the rain falls on the windward side of the mountain, creating a rain shadow on the leeward side of the mountain, which is a semidesert.

Kiribati

Kiribati includes three sets of islands located in both Micronesia and Polynesia. The main component of Kiribati is the Gilbert Island chain in Micronesia, where the capital city and most of the population are located. The other two minor island chains are the **Phoenix Islands** and the **Line Islands** in Polynesia. Both island chains were US possessions before being annexed with the Gilbert Islands to become Kiribati. The Line Islands were used for testing of British hydrogen bombs starting in 1957. Three atmospheric nuclear tests were conducted by the British on Malden Island, and six were conducted on Christmas Island. There is concern about how radiation affected people present during the tests and thereafter. The Phoenix Islands have few inhabitants. In 2008, Kiribati declared the entire island group a protected environmental area, which made it the largest protected marine habitat in the world. Kiribati is the only country with land in all four hemispheres: north and south of the equator and on both sides of the 180° meridian.

Samoa

After the colonial era, Samoa was divided into **Western Samoa** and **Eastern Samoa**. The United States controlled the eastern islands, which are referred to as **American Samoa**. Before World War I, Germany gained control of the larger, more extensive western islands only to lose them to New Zealand after the war. Western Samoa was under the New Zealand government until 1962, when it gained independence. The name was officially changed from Western Samoa to Samoa in 1997.

The Samoan Islands are volcanic, and the most active volcano last erupted in 1906. In Samoa, three-quarters of the nearly two hundred thousand people live on the larger of the two main islands. Colonialism has had a major impact on the culture, especially in the case of religion. Christianity became widespread once it was introduced and is now the religion of about 99 percent of the population. American music and societal trends are also a major influence on the islands because of migration between Hawaii and the US mainland. Many Samoans have moved to the United States and established communities. Cultural traditions have been preserved and are often integrated into modern society. Samoa has some of the oldest history and traditions of Polynesia. For many years, the United States has held an extensive naval station in the bay of Pago Pago on American Samoa. During World War II, there were more US military personnel on the islands than Samoans. American Samoa became a key military post for the United States. American Samoa remains a US possession; however, Samoans are not US citizens unless one of their parents is a US citizen.

Tonga

South of Samoa is an archipelago that is home to the Kingdom of Tonga. Only about 36 of the 169 islands are inhabited by a total population of about one hundred twenty thousand people. Tonga is ruled by a monarchy that never lost its governance powers throughout the colonial era. Tonga is the only monarchy in the Pacific. The two main methods of gaining wealth are by remittances from citizens working abroad and tourism.

Tuvalu

The island nation of Tuvalu comprises four reef islands and five atolls for a total land area of about ten square miles. In 2008, it had a population of about twelve thousand people. These statistics indicate that Tuvalu is one of the four smallest countries in the world. Nauru is only about eight square miles in area. Only the Vatican and Monaco are smaller. The low elevation of the islands of Tuvalu make them susceptible to damage from rising sea levels. The highest point is only fifteen feet in elevation. Any increase in ocean levels as a result of climate change could threaten the existence of this country.

French Polynesia

The South Pacific is home to many islands and island groups that are not independent countries. The biggest and most significant group in the southern region is **French Polynesia.** France colonized a large number of islands in the South Pacific and has continued to hold them in its control or possession as external departments or colonies. In western Polynesia, the French maintain control over the islands of Wallis and Futuna. French Polynesia consists of four main island groups: the **Society Islands**, the **Austral Islands**, the **Tuamotu Islands**, and the **Marquesas.** There are around 130 islands in French Polynesia, and many are too small or lack resources to be inhabited.

Figure 13.8: The Moorea Ferry in Papeete Harbor, Tahiti The only ways to get to the islands are by aircraft or by ship. Transportation costs can be high for imported goods or for tourism development. Wikimedia Commons – CC BY 2.5.

Tahiti, located in the **Society Islands**, is the central hub of French Polynesia. Papeete is the capital and main city with a population of almost thirty thousand. Tahiti is a major tourist destination with a mild climate that stays at 75 °F to 85 °F year-round and receives adequate rainfall to sustain tropical forests. Most of the people live along the coastal areas; the interior is almost uninhabited. The Society Islands include the island of **Bora Bora**, which is considered by many to be a tropical paradise and one of the most exotic tourist destinations in the world.

Figure 13.9: Bora Bora in the Society Islands in French Polynesia with Mount Otemanu in the Background Bora Bora is a world-class tourist destination catering to the international traveler. Tensaibuta – Bora Bora – CC BY 2.0.

The volcanic **Marquesas Islands** to the northeast are the second-most remote islands in the world after the Hawaiian Islands. The weather pattern in the Pacific does not bring enormous amounts of precipitation to the Marquesas, a reality that restricts human expansion in the archipelago. The higher elevations in the mountains—the highest is 4,035 feet—draw some precipitation from the rain shadow effect, giving rise to lush rain forests on portions of the islands. With less than ten thousand people, the Marquesas do not have a large population to support and rely on financial support from outside to sustain them. French painter Paul Gauguin is buried there, and the islands are remembered as his home during the last years of his life.

The **Austral Islands** are the southernmost group of islands in French Polynesia and are home to only about 6,500 people. French Polynesia also includes the **Tuamotu Archipelago**, between the Society Islands and the Marquesas, which comprises about 75 atolls and an uncounted number of coral reefs that extend for about nine hundred miles. The islands have a population of fewer than twenty thousand people, and the main economic activity is the cultivation of black pearls and coconuts.

The French government used islands in the Tuamotu Archipelago as test sites for nuclear weapons. From 1966 to 1974, the French tested 41 atomic devices above ground in the atmosphere, and from 1974 to 1996, they tested 137 atomic devices below ground. Radiation concerns are the same here as they are on the Marshall Islands, where the United States tested atomic weapons. Scientific testing monitored by the World Health Organization has determined the humans living closest to the atolls are not presently in danger of radioactive materials either in the environment or in their food supply. The long-term effects of the underground tests continue to be monitored.

The Pitcairn Islands, Easter Island, and the Cook Islands

To the east of French Polynesia are the four Pitcairn Islands, controlled by Great Britain. The main island, Pitcairn, is the only inhabited island in this chain and is one of the least inhabited islands in the world; the total population is fewer than fifty people. Mutineers from the HMS *Bounty* escaped to Pitcairn in 1790 after taking various Tahitians with them.

Even farther east than Pitcairn, on the edge of Polynesia, is Easter Island. Now under the government of Chile, Easter Island was historically inhabited by Polynesians who built large stone heads that remain somewhat of a mystery. At the center of Polynesia are the fifteen small **Cook Islands**, which are controlled by New Zealand and are home to about twenty thousand people, many of whom claim Maori ethnicity.

Key Takeaways

- Melanesia includes the islands from Papua New Guinea to Fiji. Micronesia includes small islands located north of Melanesia. Polynesia includes island groups from the Hawaiian Islands to the Pitcairn Islands. Papua New Guinea is the largest country in the Pacific, approximately seven hundred languages are spoken by the many local groups that live there.
- Low islands in this region are usually composed of coral and low in elevation. High islands are usually volcanic in origin and mountainous with high elevations. Micronesia consists mainly of low islands, while Polynesia consists of many high islands, such as Hawaii.
- Tourism is the main economic activity in the Pacific, but minerals and fossil fuels provide some islands with additional wealth. Fishing and subsistence agriculture have been the traditional livelihoods. Offshore banking has also been established in the region.
- The United States, the United Kingdom, and France used various islands for nuclear testing. Radiation fallout continues to be an environmental concern. Typhoons, tsunamis, volcanic activity, earthquakes, and flooding

create devastation on the islands. Fresh water can be a valuable resource, as it is in short supply on many islands.

Discussion and Study Questions

1. What are the three main regions of islands in the Pacific? What island nations belong to each?
2. What are the traditional methods of making a living on the Pacific islands?
3. What is the difference between a low island and a high island? Give examples of each.
4. Why are so many islands in the Pacific governed by the United States?
5. Name three major environmental concerns of these islands.
6. What has been a growing sector of the economy for many of the Pacific islands?
7. Which island group is the largest protected environmental marine habitat in the world?
8. What islands did the United States, Great Britain, and France use for testing nuclear weapons?
9. On which island are more than seven hundred separate languages spoken? Why are so many languages spoken?
10. What main factors have determined the economic activities of the Pacific?

Geography Exercise

Identify the following key places on a map:

- American Samoa
- Austral Islands
- Bora Bora
- Cook Islands
- Easter Island
- Fiji
- French Polynesia
- Gilbert Islands
- Guadalcanal
- Guam
- Hawaiian Islands
- Kiribati
- Marquesas
- Marshall Islands
- Melanesia
- Micronesia
- Nauru
- Northern Mariana Islands
- Palau
- Papua New Guinea
- Pitcairn Islands
- Polynesia
- Samoa
- Society Islands

- Solomon Islands
- Tahiti
- Tonga
- Tuamotu Islands
- Tuvalu
- Vanuatu
- Wake Island
- Western Samoa

13.2. ANTARCTICA

Antarctica, the southernmost continent, is a land of striking contrasts and extreme conditions. With its vast, icy expanse covering about 14 million square kilometers, it is the coldest, windiest, and driest continent on Earth. Home to unique wildlife such as penguins, seals, and various seabirds, Antarctica serves as a vital area for scientific research, offering insights into climate change and ecosystems. The breathtaking landscapes, including towering icebergs and expansive glaciers, create a hauntingly beautiful atmosphere that captivates adventurers and researchers alike. Despite its harsh environment, Antarctica plays a crucial role in regulating the planet's climate and is a key area for understanding global environmental changes.

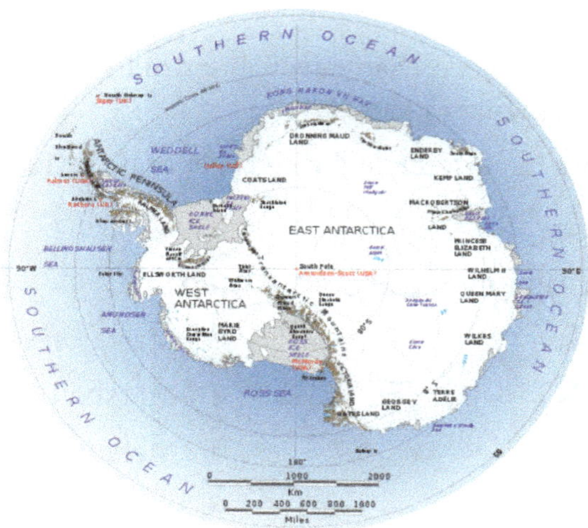

Figure 13.10: The Southern Ocean and Antarctica. Wikimedia Commons - public domain.

The Southern Ocean

The Southern Ocean, which surrounds the continent of Antarctica, is often misunderstood or not included on many maps of the Southern Hemisphere. The cold waters off the coast of Antarctica move from west to east in a clockwise rotation around the continent in a movement called the West Wind Drift, or the Antarctic Circumpolar Current. The Southern Ocean's northern boundary does not border a land mass but meets up with the Atlantic, Pacific, and Indian Oceans. Most geographers accept the northern boundary of the Southern Ocean to be located south of 60° latitude even though the actual limit has not yet been firmly agreed upon.

The Southern Ocean's northern boundary has more to do with the marine conditions of the realm. There is a transition called the Subtropical Convergence in which the cold, dense waters of the Southern Ocean meet up with the warmer waters of the Pacific, Atlantic, and Indian Oceans. The cold, dense water from the south sinks below the warm waters from the north to create a zone of upwelling and mixing that is conducive to high levels of productivity for organisms such as phytoplankton and krill. The zone of Subtropical Convergence can be visually observed by the grayish, cold southern waters meeting up with the bluish-green, warm northern waters. The krill, which thrive on phytoplankton, are an important link in the food chain for marine organism such as fish, penguins, seals, albatrosses, and whales in the Southern Ocean.

Physical Geography

The world has seven focal continents. Rated by physical area from the largest to the smallest, they are Asia, Africa, North America, South America, Antarctica, Australia, and Europe. Antarctica, which is larger than Australia and 1.3 times larger than Europe, is located entirely south of 60° latitude and is surrounded by the **Southern Ocean.** Antarctica has the highest average elevation of any continent; there are many mountain ranges. The two-thousand-mile-long **Transantarctic Mountain** range divides Antarctica into a small western region and a larger eastern region. At both ends of the Transantarctic Mountains are the two main ice shelves: the **Ross Ice Shelf** and the **Ronne Ice Shelf.** The **Ellsworth Mountains** are located in the western region and are home to **Mt. Vinson** (or the **Vinson Massif**), which is the highest peak on the continent, reaching an elevation of 16,050 feet. This is higher than any mountain in the contiguous, continental United States, Europe, or Australia.

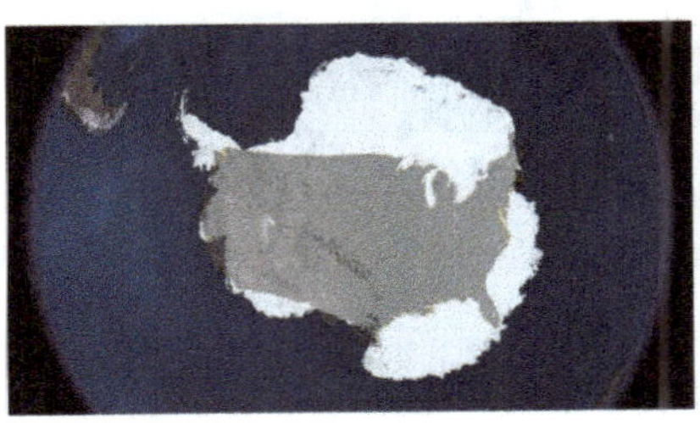

Rank of 7 Continents in Square Miles	
1. Asia	17,139,445
2. Africa	11,677,239
3. North America	9,361,791
4. South America	6,880,706
5. Antarctica (*with ice sheet*)	5,400,000
6. Europe (*with Russian region*)	3,997,929
7. Australia	2,967,124
The United States (*land area*)	**3,537,438**

Figure 13.11: Size Comparison of Antarctica and the United States Antarctica without the ice sheet would be considerably smaller; some estimate it would be only about one hundred thousand square miles in land area, or the equivalent of the physical area of the US state of Colorado. NASA – public domain.

The **Antarctic Peninsula** is actually an extension of the Andes Mountains of South America and is home to active volcanoes. The peninsula is the location of a volcano on Deception Island that erupted in the late 1960s and destroyed research stations in the area. There was an additional large eruption in 1970. The volcano continues to show activity, and sightings of lava flow continue to be reported. There may be more volcanic activity than what has been recorded. An underwater volcano in the Antarctic Peninsula was discovered in 2004. **Mt. Erebus** (12,448 feet), located on Ross Island on the other side of Antarctica from the Antarctic Peninsula, is the world's southernmost active volcano. Mt. Erebus has been active since 1972 and has a large lava lake in its inner crater.

About 98 percent of Antarctica is covered by an ice sheet that is, on average, up to a mile deep. In some areas, it is nearly three miles deep. In the winter season, the ice sheet's area might double as it extends out from the coastline. The Antarctic ice sheet holds about 70 percent of the earth's fresh water. If the ice sheet were to melt, the sea level could rise considerably and cover many of the earth's low-lying islands, peninsulas, and coastal regions with low elevations. Antarctica is considered a desert because it usually averages fewer than ten inches of precipitation per year. Coastal regions annually receive as much as four feet of snow, while the interior near the **South Pole** might only receive a few inches.

Figure 13.12: Mt. Erebus (12,448 feet), located on Ross Island, is Antarctica's most active volcano and has a lava lake in its inner crater. Lin padgham – Mt Erebus – CC BY 2.0.

There are areas in Antarctica that are not covered with ice but have a landscape of bare ground. This nonice portion of the continent protrudes above the ice sheet and only covers a combined physical area equivalent to about half the US state of Kentucky. The only plant life that exists here are the many different mosses and lichens that grow during short periods of the year. Below the giant ice sheet are dozens of subglacial lakes. **Lake Vostok**, the largest lake discovered in the Antarctic so far, was found two miles below the ice sheet and is the size of Lake Ontario. It is unknown what aquatic life might exist in these lakes. If all the ice and snow were removed from the continent, the total land area would be considerably smaller and would consist mainly of mountain ranges and islands. Some estimate that this land area altogether would only equate to about one hundred thousand square miles, roughly equivalent to the physical area of the US state of Colorado. This does not account for the fact that if all the ice were to melt, the sea level would rise and cover more land area. The land portion of the continent would also expand upward because of the loss of the weight of the ice, which has been compressing the continent.

Not only is Antarctica the driest continent with the least average annual precipitation and the highest continent in average elevation; it is also the coldest of the continents. The lowest temperature ever recorded on Earth was −128 °F in 1983 at a Russian research station in Antarctica. Temperatures reach a minimum of less than −110 °F in winter in the interior and greater than 55 °F near the coast in summer. No permanent human settlements exist in Antarctica other than research stations from a number of countries.

The Antarctic Treaty

The continent is not politically controlled by any one government. Early seafaring explorers sailed in these waters, and various countries laid claim to sections of the continent. The continent was first sighted by explorers in 1820, and the South Pole was first reached in 1911. Land claims to the continent were established by the home countries of early explorers. Forty-six countries are now included in the **Antarctic Treaty**, which was originally signed by twelve countries in 1959. The treaty, designed to protect the environment and encourage scientific research, prohibits military activities, mineral mining, and the disposal of waste products. All land claims were suspended when the Antarctic Treaty was initiated, but the claims are not without political ramifications. Antarctica is divided into pie-shaped sections, and each of the original claimant countries is allocated a portion, according to their claim. The countries with original claims are Norway, New Zealand, France, Chile, Australia, and Argentina. Other countries, including Brazil, Peru, Russia, South Africa, Spain, and the United States, have reserved their right to submit claims on the continent in the future if the issue of territorial claims becomes significant. A large sector of **West Antarctica** called Marie Byrd Land remains unclaimed.

Research stations account for the entire human population in Antarctica. Approximately one thousand people live in Antarctica year-round, and up to five thousand or more live there during the summer months. Many of the research stations rotate their personnel, and tours of duty last anywhere from a few months to a year or more. Various family groups have worked there as well as other service workers, including Russian Orthodox priests, who have rotated every year at one of the Russian research stations.

Figure 13.13 Emperor Penguins, Ross Sea, Antarctica. Wikimedia Commons – public domain.

Tourism brings the largest number of additional people to the continent. Tourists come for short-term visits to experience the conditions or see the many species of penguins or fauna that exist here. More than forty-five thousand tourists visit the Antarctic Treaty area yearly. Most arrive on commercial ships that specialize in tours of the region. Tours only last one or two weeks.

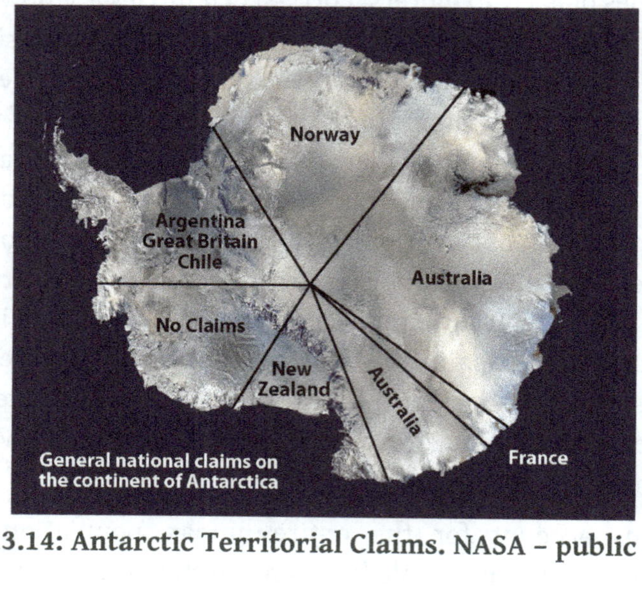

Figure 13.14: Antarctic Territorial Claims. NASA – public domain.

Research has revealed that mineral resources are to be found under the ice in Antarctica, and oil and natural gas are found in offshore deposits. Antarctica is a frontier for economic development that is not under the jurisdiction of any one government. The Antarctic Treaty has been the determinant of the level of human activity. The current treaty restricts any extractive activity. Fishing is also regulated within the treaty, but without enforcement procedures, there have been questions about its effectiveness. Whaling was once a major industry in this realm. Whaling stations were established on the Antarctic Peninsula and nearby South Georgia Island. However, the increased use and extraction of petroleum reduced the need for whale oil and the industry collapsed. Some whaling continues in the waters of the Southern Ocean, which has led to questions about how to manage these natural resources. In 1998, negotiations between interested countries met in Madrid, Spain, and created the **Protocol on Environmental Protection to the Antarctic Treaty** (known as the Madrid Protocol). The protocol designates Antarctica as a natural reserve that can only be used for peaceful purposes and for science. All mining or economic activity is banned.

Climate Change

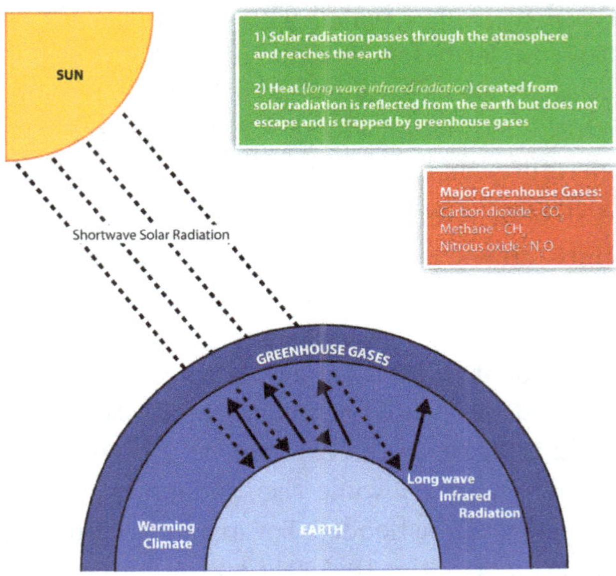

Figure 13.15: The Greenhouse Effect and Climate Change

Climactic conditions on Earth have varied widely during the planet's history. There have been long periods of heating or cooling. The last ice age, which ended about ten thousand years ago, created large ice sheets that covered much of the Northern Hemisphere. The earth then entered into the current interglacial period with warmer temperatures that melted the ice sheets; the polar regions have the last remaining ice on the planet. The earth has experienced large fluctuations in its temperature at various times in its past. Natural changes in the conditions that affect climate can include but are not limited to the dynamics of the sun, changes in the earth's orbit, and volcanic eruptions.

Human activity has impacted conditions both locally and globally. Since the Industrial Revolution, humans have been pumping enormous amounts of carbon dioxide into the atmosphere, which affects the planet's climate and temperature. The Industrial Revolution introduced the burning of coal as a fuel to boil water to operate steam engines. This allowed power to become more versatile and mobile. The introduction of the automobile increased the burning of petroleum, which released carbon dioxide into the atmosphere in the form of engine exhaust.

Large-scale deforestation and the burning of fossil fuels have increased the quantity of heat-trapping "greenhouse gases" in the atmosphere. Nitrous oxide, methane, carbon dioxide, and similar gases act like the glass panels of a greenhouse that allow short-wave radiation from the sun to enter but do not allow the long-wave radiation of heat to escape into space. Deforestation reduces the number of trees that use carbon dioxide and store carbon in plant fibers. The burning of wood or carbon-based energy sources such as oil or coal releases the carbon back into the atmosphere. Fossil fuels are created when dead plant and animal life have been under pressure and decay for long periods and have retained their carbon component. This all leads to a rise in the activity of the carbon cycle. Carbon is a key component to the regulation of the earth's temperature. Life on Earth is dependent on temperature conditions that are regulated by the atmosphere. This natural cycle has been augmented by human activity.

Changing global temperatures are one aspect of climate change that has received attention in recent years. Technically speaking, global warming is an average increase in the temperature of the atmosphere near the earth's surface. Few environmental effects could impact Antarctica as much as the phenomenon of changing temperatures. There would be major ramifications for the entire world if temperatures would increase to the extent that the Antarctic ice sheet would melt and dissolve away. Sea levels would rise, and many areas of the planet with large urban populations would be flooded. For instance, many of the largest urban centers in the world are port cities that rely on the import and export of goods and materials. These cities would be in danger of being under water if the sea level were to rise even a few feet, let alone up to two hundred feet (approximated to be equivalent to the volume of water in the polar ice caps). It should be noted that light-colored surfaces such as snow and ice reflect more radiation than do darker surfaces. If the Antarctic ice sheet were to melt, this would certainly exacerbate global climate change, as more radiation would be absorbed at the surface (land and water). Climate change might also affect agricultural production. Global changes in temperature would alter ecosystems and the habitats of organisms, changing the balance of nature in many biomes.

The scientific community continues to study the dynamics of climate change. In the last decade, some of the warmest annual temperatures in the past century have been documented. Data gathered from both the National Oceanic and Atmospheric Administration (NOAA) and the National Aeronautics and Space Administration (NASA) indicate that the earth's surface temperatures have increased by more than one degree in the past one hundred years.

Ozone Depletion

Climate change can have a major impact on atmospheric conditions. It should be noted that good ozone depletion in the stratosphere (the second layer of the earth's atmosphere) has different causes and conditions than temperature change in the troposphere (the layer just above the earth's surface). The two concepts have separate and distinct dynamics that are not directly related. Ozone (O_3) is a simple molecule consisting of three oxygen atoms. Common oxygen gas molecules have two oxygen atoms (O_2). Depending on where ozone is located in the atmosphere, it can be either a protective safeguard from ultraviolet (UV) radiation from the sun or an element in smog that causes health problems. Good ozone in the stratosphere provides a protective shield preventing harmful UV radiation from reaching the earth. UV rays from the sun are known to cause skin cancers, eye damage, and harm to organisms such as plankton. Bad ozone molecules in the troposphere mix with various chemicals to create smog, which reduces visibility and can cause respiratory health problems. Tropospheric smog might contribute to climate change but does not serve the same function as ozone molecules in the stratosphere. The depletion of good ozone in the stratosphere is a different chemical process than the creation of bad ozone in the troposphere.

Ozone in the stratosphere is vital to the protection of living organisms from damaging UV radiation from the sun. In the stratosphere, UV radiation is absorbed in a continuous cycle in which oxygen molecules are turned into ozone molecules and then back into oxygen molecules. Oxygen molecules (O_2) absorb UV radiation in the stratosphere and separate into two oxygen atoms ($O_2 \rightarrow 2\ O$). Each of these oxygen atoms (O) will attach to another oxygen molecule to create an ozone molecule ($O + O_2 \rightarrow O_3$). Each ozone molecule (O_3) will absorb UV radiation, which separates it back into an oxygen molecule and an atomic oxygen atom ($O_3 \rightarrow O_2 + O$). The separate oxygen (O) atom will attach to another separate oxygen atom (O) to become an oxygen molecule (O_2) again ($O + O \rightarrow O_2$). This cycle will continue to absorb UV radiation and keep UV radiation from reaching the earth's surface. It is a chemical reaction that gives off heat and increases the stratosphere's temperature.

Chemicals such as chlorine and bromine interact with protective ozone molecules in the stratosphere and break them down in a chain reaction that depletes the stratosphere of ozone molecules and stops the cycle that absorbs the UV radiation. One chlorine molecule can destroy one hundred thousand ozone molecules, and bromine atoms can destroy ozone molecules at a rate of many times that of chlorine. Chlorine and bromine enter the stratosphere through the discharge of chlorofluorocarbons (CFCs), hydrofluorocarbons (HCFCs), and other chemicals that deplete ozone molecules. These chemicals have been mainly used in industrial processes such as refrigeration and air conditioning and in solvents and insulation foam. In the stratosphere, radiation from the sun breaks HCFCs and CFCs apart, releasing chlorine atoms that destroy ozone molecules.

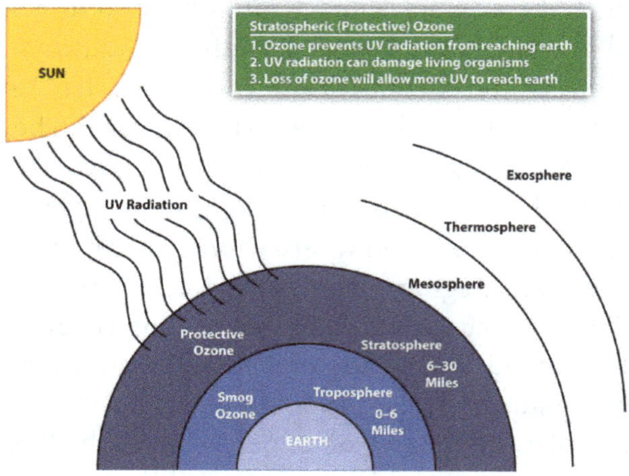

Figure 13.16: Ozone in the stratosphere protects the earth from harmful ultraviolet radiation from the sun.

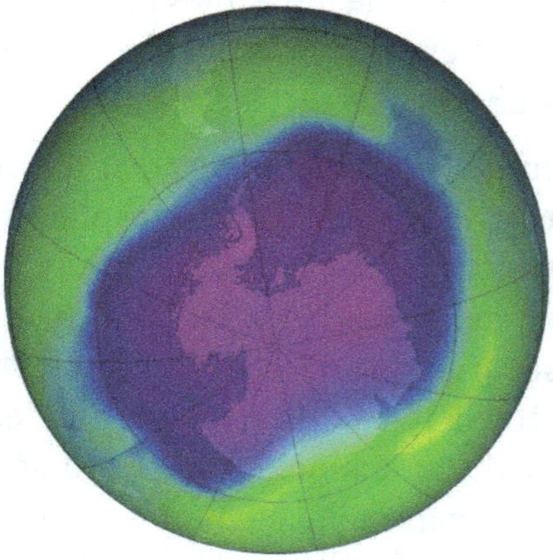

Figure 13.17: One of the largest ozone holes was observed in 2006 over Antarctica. NASA – public domain.

This process also occurs naturally when volcanoes erupt and emit sulfur aerosols into the atmosphere. The sulfur aerosols break down CFCs and halon; this results in the release of chlorine and bromine, which deplete ozone molecules. In recent years, nitrous oxide has become a major chemical that can reach the stratosphere to act as an agent in the ozone depletion process. Nitrous oxide can be released into the atmosphere from human activities such as the use of nitrogen fertilizers in agriculture or from vehicle exhaust from burning fossil fuels. Nitrous oxide is also released naturally from the soil or from ocean water.

Ozone depletion has been especially noticeable over Antarctica. A large area of ozone depletion in the upper atmosphere is often called an ozone hole. The ozone hole over Antarctica is not exactly a total depletion of ozone but a major reduction in the amount of ozone found in the stratosphere over the South Pole. Recent ozone levels in the stratosphere over Antarctica have decreased and are lower than they were in 1975. Polar stratospheric clouds, which sometimes develop over Antarctica during the extremely cold winter months, severely reduce ozone levels. The clouds trap chlorine and nitric acid in their ice crystals. As the circulation of westerly winds starts in the spring, an atmospheric vortex is created over the polar area. The ozone hole increases during the spring when sunlight increases—from September to early December. The sunlight speeds up the chemical reactions that deplete the ozone molecules. During this time, as much as half of the lower stratospheric ozone can be destroyed, creating an ozone hole.

Global Impacts and Organizations

Many governments around the world have established agencies to address environmental issues and have invested resources in continued research and scientific discovery. Many nongovernmental agencies and organizations such as Greenpeace International have also been established to address the development or management of Antarctica and to address environmental concerns. International organizations such as the United Nations Environment Programme (UNEP) have been formed to enhance cooperation between countries and concerned entities. In 1998, the Intergovernmental Panel on Climate Change (IPCC) was established by the UNEP and the World Meteorological Organization to address the work of the United Nations Framework Convention on Climate Change (UNFCCC). Through an international treaty, the agency focuses on the harmful effects of climate change. One outcome of the UNFCCC's work was the Kyoto Protocol developed in 1997 in Kyoto, Japan, which created a legal commitment by participants to reduce greenhouse gases and address climate change issues. As of 2010, 191 countries have signed on to the initiative. The United States has not yet ratified the protocol. Enforcing the Kyoto Protocol is a matter of debate. The 2009 United Nations Climate Change Conference (also referred to as the Copenhagen Summit) was held in Denmark to address some of the same issues that were discussed in the Kyoto Protocol.

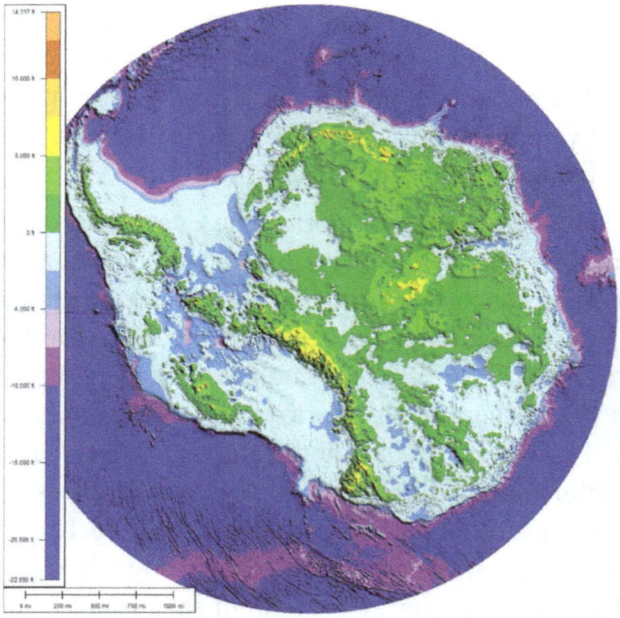

Figure 13.18: One Projection of the Subglacial Topography of Antarctica below the Ice Sheet This is what Antarctica might look like if all the ice were removed. The physical area remaining is estimated by some to be about one hundred thousand square miles, which is about the same physical area as the US state of Colorado. It is unclear what the actual physical area would be if the ice melted. Wikimedia Commons – CC BY 3.0.

The depletion of stratospheric ozone in polar regions or the thinning of the ozone layer in midlatitudes would have global repercussions for human activities and environmental conditions. Climate change is a global phenomenon, transcending national boundaries and governmental jurisdictions; it influences the entire globe and affects all individuals, regardless of their contribution to the issue. A increase in sea level, as expected due to global warming, will impact every coastal nation globally. It would not be confined to any certain sort of nation, whether developed or developing. Challenges posed by heightened UV radiation and climate change transcend political borders and economic circumstances; they necessitate global awareness and collaboration, as they cannot be addressed by a single nation independently—if they can be addressed at all.

Key Takeaways

- Antarctica has the highest average elevation of any continent and has many mountains. The continent includes active volcanoes, high mountains, and the Transantarctic Mountain range. The continent is surrounded by the Southern Ocean.
- The ice sheet that covers Antarctica is more than a mile thick and holds about 70 percent of the earth's fresh water. Antarctica is still considered a

desert because it receives so little precipitation on an average basis. Large bodies of water are also located below the ice sheet.

- The concept of global warming is one aspect of climate change that indicates an increase in greenhouse gasses such as carbon dioxide, which help regulate the temperature of the earth's atmosphere. The end result is warmer temperatures on the earth's surface. The burning of fossil fuels is a main source of carbon dioxide that enters the atmosphere. Climate change may result in the melting of the polar ice sheet over Antarctica, which could raise sea levels considerably.
- Ozone in the stratosphere protects the earth from harmful UV radiation from the sun. Various chemicals reduce the amount of protective ozone in the stratosphere, which allows more UV radiation to reach the earth. A seasonal cycle varies the amount of ozone in the stratosphere over the South Pole, causing an ozone hole when ozone is not abundant.

Discussion and Study Questions

1. Where are the main mountain ranges in Antarctica? What is significant about Mt. Erebus?
2. How can Antarctica be considered a desert region with all that ice? How thick is the ice sheet?
3. What percentage of the world's fresh water is located in Antarctica?
4. What was the Antarctic Treaty designed to accomplish? Who agreed to the treaty?
5. What seven countries have laid claims to the territory of Antarctica?
6. How does the burning of fossil fuels contribute to climate change?
7. How would continued global warming affect Antarctica and the rest of the planet?
8. How does ozone protect the planet? What damage would ozone depletion cause?
9. Why is there an ozone hole over the South Pole during certain months?
10. What does the Kyoto Protocol attempt to accomplish?

Geography Exercise

Identify the following key places on a map:

- Antarctic Peninsula
- East Antarctica
- Ellsworth Mountains
- Lake Vostok
- Mt. Erebus
- Mt. Vinson

- Ronne Ice Shelf
- Ross Ice Shelf
- Southern Ocean
- South Pole
- Transantarctic Mountains
- West Antarctica

13.3: CONCISE SUMMARY

The low coral islands and high volcanic islands of the South Pacific can be categorized into three regional classifications: Polynesia, Micronesia, and Melanesia. The Pacific islands represent the largest geographic area worldwide, with significant distances between various island groups.

The Pacific has considerable ethnic diversity, comprising hundreds of unique languages and cultural traditions. Papua New Guinea is the largest island and is home to nearly seven hundred languages. Many Pacific islands are possessions of France, the United Kingdom, New Zealand, or the United States.

The Pacific islands are susceptible to natural events such as typhoons, volcanic eruptions, earthquakes, tsunamis, and tropical storms. Some islands have radiation issues owing to their use as nuclear testing locations by the United States, France, and Great Britain. Further environmental issues involve deforestation and the depletion of freshwater resources.

Antarctica is a glaciated continent that exceeds both Europe and Australia in area. The Southern Ocean surrounds Antarctica, located entirely south of 60° latitude. The Antarctic ice sheet surpasses one mile in thickness but does not cover the entire land area. Prominent mountain ranges feature volcanic peaks that have erupted in recent years. The only permanent human structures in Antarctica are research stations.

The phenomena of global warming may impact the Antarctic continent if temperatures rise and the ice sheet experiences rapid melting. Increased atmospheric temperatures may result from heightened greenhouse gas levels caused by the combustion of carbon-based fossil fuels.

Ozone in the stratosphere protects the Earth from ultraviolet (UV) radiation. Stratospheric ozone depletion may occur due to the release of industrial pollutants. A significant ozone hole intermittently develops above Antarctica with a seasonal increase of ozone-depleting chemicals.

INDEX

A

Abortion 144

Adapted 458

Aerosols 766

Alcohol 500, 707

Alkaline 408

Animals 348, 409-410, 421, 424, 602, 703, 718

Aquifers 448, 451

Assembly 105, 470, 582

Atmosphere 26-27, 764-765, 767, 769

Automation 701

Automobile 168, 210, 325

B

Banknote 51

Biomes 736

Biosphere 603

Bromine 766

Buffer 639

C

Capricorn 8, 16, 439

Carbon 27-29, 764, 769

Cattle 281, 318-319

Channels 575

Chemicals 250, 301, 533, 765-766

Children 34, 53-54, 75-76, 141, 242, 315, 359, 468, 567, 582, 661

Chlorine 766

Church 78, 82, 142, 165, 182-183, 222, 284, 414

Cinema 598

Climate 2, 7-8, 17-18, 20, 23, 27-30, 58, 63, 85, 93, 99, 106, 157, 205, 239-240, 286, 302, 305, 335, 443, 466, 476, 491, 534, 572-573, 618-620, 630, 654, 670, 673, 724, 731, 738, 758, 763-765, 767-769

Clouds 29, 281, 553, 767

Coastline 734

Cobalt 398

Cocaine 301

Coffee 97, 117, 227, 246, 262, 264, 282, 302, 310, 316, 401, 404, 570

Collision 562

Combustion 28

Cooking 588, 602

Cooler 241

Copper 226, 337, 349

Cricket 557

Crossroads 528

Cruise 264-265

Cuisine 308

Culture 30, 35, 56, 82, 84, 86-88, 97, 103, 161, 164, 173, 195-196, 218, 308-309, 333, 374, 412, 473, 500, 526, 530, 538, 554, 567, 626, 630, 650, 652, 697, 726, 737, 744

Customer 596

Cyclone 269, 272

Cyprus 522

D

Darwin 724

Deaths 365, 371, 397, 405

Deepwater 408

Density 559, 563

Depletion 32, 226, 765, 767, 770

Depression 273

Deserts 225, 578, 722

Dharma 612

Diamonds 380, 387, 427

Dimensions 599

Dioxide 27, 29, 764, 769

Dishes 601

Diversity 107, 129, 182, 230, 268, 343, 441, 678

Doctor 579

Domains 2, 5

Donkey 577

E

Earthquake 263, 584, 656

Ecosystems 410

Ecuador 294, 304-308

Elections 110, 115, 261, 394, 489, 544

Electronic 399

Elements 14, 212, 474

Elephants 369, 603

Empire 70, 76, 87, 92, 100, 121, 128-129, 132-133, 139, 150, 152, 155, 220-221, 224, 277, 283-285, 304, 348-350, 459, 465, 493, 520, 523, 548, 555-556, 594

Epidemic 433

Erosion 26, 637

Explosion 471

Extension 225

External 486

F

Family 31, 34, 36, 53-54, 75-76, 80, 125, 232, 242, 309, 315, 359-360, 386, 402, 453, 468, 473, 499, 503, 510-511, 524, 560, 570, 587, 593-594, 634, 661, 690, 699

Famine 106

Farmers 609, 624

Farming 66, 330

Farmland 66, 187, 619, 666

Festivals 267

Firewood 23, 25, 509

Fishery 67

Flooding 270, 449, 554, 587-588, 623

Football 317, 399, 726-727

Forestry 212

Fossil 28, 448, 597, 764, 769

Free-trade 252, 508

G

Generation 105, 669

Genocide 152, 394

Geography, 1-3, 5, 11, 14-16, 30, 44, 55-56, 58, 107, 162, 182, 245, 279, 305, 340, 343, 356-357, 496, 527, 639, 654, 673, 678-679, 696, 718, 721, 741

Geological 736

Governance 125, 632

Grasses 66, 722

Grasslands 424, 736

Grazing 347, 476

Greenhouse 27, 764

Greenland 71, 83-84

Greenwich 9

Growth 31-32, 34, 52-54, 59, 75, 159-160, 171, 178, 208, 232, 264, 266, 309, 360, 367, 413, 430, 445, 499, 560-561, 568, 596, 600, 602, 631, 633, 637-638, 641, 659-661, 687, 691, 699

H

Habitats 26

Helicopter 419

Hemisphere 7-8, 330, 337

Heritage 74, 77, 82, 86-88, 99, 103-104, 109, 198, 236, 255, 289, 333, 368, 386, 414, 476, 504-505, 569, 684, 723, 737

Heroin 301, 543

Himalayan 561-562

Hotels 265

I

Indicator 559

Instrument 267

Invaders 219

Irrigation 204, 449

Islands 63, 85, 102, 164, 215-216, 219, 253-255, 264, 266-268, 274, 296, 393, 503, 509, 572, 618, 654, 659, 673-674, 676, 696, 702-703, 705, 707, 709, 714, 733-734, 737, 743-758, 770

K

Kingdoms 349-350, 355, 417, 430

L

Landfall 273-274

Landforms 63, 69

Landlocked 309, 532, 628, 653

Landmass 551, 562

Landscapes v, 1, 101, 198, 673

Landslides 26

Lawyers 501

Layers 4, 320

Leopards 603

Livestock 241, 476

Long-wave 27

M

Madagascar 424-425, 442

Mainland 98, 216-218, 224, 275, 424, 439, 632, 644-647, 653, 671, 673-674, 677-678, 695, 697, 718

Marine 759

Markets 46, 109, 233, 399, 594, 600, 643

Marriage 608

Meditation 611

Migration 46, 57-58, 165, 172, 181, 467, 634

Military 120-121, 172, 236, 251-252, 334, 353, 378, 380, 395, 416, 471, 491,

493-494, 499, 504-505, 515, 521, 542, 546, 580, 582-583, 626, 647, 658, 664, 667, 669, 681, 689, 692-694, 705, 709, 750, 753

Missiles 258, 541

Monogamous 402

Movement 28, 112, 428, 705

N

Newspaper 12

Nuclear 93, 132, 258, 488, 525, 565, 623, 656, 750, 756

Nutrients 320

O

Operations 10, 66, 252, 419, 730

Orthodox 78, 142, 165, 414, 494

P

Parliament 102-104

Pilgrimage 460, 607

Planet 27, 29-30, 32, 340, 763-764, 769

Police 237

Policy 431, 435, 633-634, 641, 646, 666

Pollution 67, 131, 229, 602, 663

Polygamous 402

Powerhouse 671

Predators 736

Protocol 763, 767

Pyramids 42, 55, 232, 360, 447

Q

Quaternary 168, 170

R

Radiation 27, 29, 756, 764-766, 769-770

Railroad 591

Railway 148, 167

Rainfall 20, 205, 254, 270, 281, 336, 424, 465, 491, 508, 596, 722, 734

Rainforest 688

Refugees 122, 365, 392, 394-395, 478, 483, 489, 497

Revolution 44, 61, 66, 72-73, 76, 80, 99-100, 166, 230, 257, 452-453, 525, 764

Rivers 64, 69, 280, 346, 445, 451, 513, 521, 553, 563, 578, 585, 587, 622, 662, 674

Robust 125

Rubber 317

S

Salary 56

Satellite 498

Silver 283, 305

Spectrum 519

Stadiums 438

Sulfur 766

Systems 3, 5, 218, 294, 354-355, 529, 561, 632

T

Tectonic 28, 30, 442, 562, 584, 618, 656, 674, 678, 708, 735

Television 173, 195-196, 474, 727

Tourism 103, 106, 168, 201, 227, 230, 260, 262, 264-265, 268-269, 275, 306-307, 311, 367-370, 410, 412, 433, 438, 474, 567-568, 572-573, 597, 684, 687, 690-691, 723, 729, 731, 737, 742, 744, 754

Tourists 264-265, 367-368, 744, 762

Traditions 40, 86, 198, 400, 402, 404, 412, 458, 501, 606-607, 753

Traffic 594, 624

Training 542

Traits 458

Tribal 577, 580, 684, 711

Tropics 8, 345

Typhoons 274, 709

U

Uranium 373

V

Valentine 526

Variations 23, 58

Volcanic 29, 215, 239, 408, 654, 675, 702, 707, 748, 756, 770

Volcanoes 147, 226, 390, 708-709, 752

W

Walmart 45

Warfare 76, 219, 397

Warlord 380

Waterfowl 534

Waterways 131

Wealth 47, 49-51, 71, 103, 116, 168, 236, 243, 245, 266, 277, 298, 305, 310, 316, 324, 336, 348, 377, 413, 427, 440, 497, 504, 537, 545, 631, 640, 643, 660, 705, 728-729, 739, 749

Weather 2, 18, 27, 269, 688

Wildlife 368-369, 409-410, 425, 427, 603

Worldwide 93, 173, 367

Wrestling 400, 652

Z

Zealand 717-720, 732-742, 753

Zimbabwe 358, 369, 426, 430-432

www.ingramcontent.com/pod-product-compliance
Lightning Source LLC
LaVergne TN
LVHW062214200926
848484LV00006B/1393